Principles of Water Law and Administration

Principles of Water Law and Administration

National and International

Dante A. Caponera

3rd Edition, revised and updated by
Marcella Nanni

Routledge
Taylor & Francis Group

LONDON AND NEW YORK

Cover Illustration:
The Nile: A Source of Life, watercolour by Alleyne Caponera

First published 2018 by Routledge

2 Park Square, Milton Park, Abingdon, Oxon, OX14 4RN
605 Third Avenue, New York, NY 10017

Routledge is an imprint of the Taylor & Francis Group, an informa business

First issued in paperback 2020

Copyright © 2019 Taylor & Francis

Typeset by Apex CoVantage, LLC

All rights reserved. No part of this book may be reprinted or reproduced or utilised in any form or by any electronic, mechanical, or other means, now known or hereafter invented, including photocopying and recording, or in any information storage or retrieval system, without permission in writing from the publishers.

Notice:
Product or corporate names may be trademarks or registered trademarks, and are used only for identification and explanation without intent to infringe.

Library of Congress Cataloging-in-Publication Data
Names: Caponera, Dante Augusto, 1921–2003, author. | Nanni, Marcella, author.
Title: Principles of water law and administration : national and international /
 Dante A. Caponera (1921–2003).
Description: 3rd edition / revised and updated by Marcella Nanni. |
 Boca Raton : CRC Press/Balkema, [2019] | Includes bibliographical references and index.
Identifiers: LCCN 2019005622 (print) | LCCN 2019007140 (ebook) |
 ISBN 9780429465703 (ebook) | ISBN 9781138610569 (hardcover : alk. paper)
Subjects: LCSH: Water—Law and legislation. | Water conservation—Law and legislation. |
 Water resources development—Law and legislation.
Classification: LCC K3496 (ebook) | LCC K3496 .C37 2019 (print) |
 DDC 346.04/691—dc23
LC record available at https://lccn.loc.gov/2019005622

Published by: CRC Press/Balkema
 Schipholweg 107c, 2316 XC Leiden, The Netherlands
 e-mail: Pub.NL@taylorandfrancis.com
 www.crcpress.com – www.taylorandfrancis.com

ISBN: 978-1-138-61056-9 (hbk)
ISBN: 978-0-367-72932-5 (pbk)
DOI: https://doi.org/10.1201/9780429465703

Contents

Acknowledgements	xix
Biographical notes	xxi
Preface	xxiii

1 Introduction — 1
- 1.1 Water and the society — 1
- 1.2 The need for a water policy, legislation and administration — 2
- 1.3 The interdisciplinarity of the subject — 4
- 1.4 The physical context — 5
    - 1.4.1 The hydrologic cycle — 5
    - 1.4.2 Definition and analysis of precipitation — 5
    - 1.4.3 Analysis of flow — 6
    - 1.4.4 Groundwater and its behaviour — 6
    - 1.4.5 The notion of aquifer — 7
    - 1.4.6 Weather modification — 8
    - 1.4.7 Climate change — 8
    - 1.4.8 Integrated water resources management (IWRM) — 9
- 1.5 The socio-economic context — 9
    - 1.5.1 Water demand — 10
    - 1.5.2 Water pricing — 10
    - 1.5.3 Externalities — 11
    - 1.5.4 Cost analysis — 11
- 1.6 The purpose of the book — 11
- References — 12

2 Earliest water regulations and management — 13
- 2.1 The importance of water regulations throughout history — 13
- 2.2 The difficulty of studying early water regulations — 14
- 2.3 The development of earliest water law principles — 15
- 2.4 Ancient Egyptian water regulations and management — 16
    - 2.4.1 Earliest dynasties (3400–2650 BC) — 16
    - 2.4.2 Later dynasties (2650–300 BC) — 17

2.5	Ancient Mesopotamian water regulations and management	17
	2.5.1 Historical development	17
	2.5.2 The Hammurabi Code	18
	2.5.3 Subsequent codifications	19
	2.5.4 Detailed water regulations	20
2.6	Ancient Hindu water regulations and management	21
	2.6.1 The Hindu legal system	21
	2.6.2 The water regulations in the Code of Manu	22
2.7	Ancient Chinese water regulations and management	23
	2.7.1 Introduction	23
	2.7.2 Water regulations	24
	2.7.3 Basic principles	25
2.8	Hebrew water regulations and management	25
	2.8.1 Introduction	25
	2.8.2 The importance of water	25
	2.8.3 Water law principles	26
	2.8.4 Conclusion	27
2.9	Pre-Columbian water regulations and management	28
	2.9.1 Coastal Peru (Inca)	28
	2.9.2 Meso-America (Maya-Aztec)	28
2.10	Other early systems of water regulations and management	29
2.11	General conclusion	29
References		30

3 Roman and intermediate period   33

3.1	Introduction	33
3.2	Roman water law principles	34
	3.2.1 The origins and Regal period (1000–500 BC)	34
	3.2.2 The Republican period (509–27 BC)	35
	3.2.2.1 The classification and ownership of water	36
	3.2.2.2 The right to use water	37
	3.2.2.3 Protection from harmful effects of water and control of waterworks and structures	38
	3.2.2.4 Water administration	38
	3.2.3 The Principate (27 BC–286 AD)	39
	3.2.3.1 The classification and ownership of water	41
	3.2.3.2 The right to use water	41
	3.2.3.3 Protection of existing water rights: the interdicta	43
	3.2.3.4 Protection from harmful effects of water and control of waterworks and structures	45
	3.2.3.5 Water administration	46
	3.2.4 The Absolute Monarchy or Late Empire (286–565 AD)	48
	3.2.4.1 The classification and ownership of water	49
	3.2.4.2 The right to use water	50

		3.2.4.3	Protection from harmful effects of water and control of waterworks and structures	51
		3.2.4.4	Water administration	51
3.3	Intermediate water law principles in Europe (565–1812)			52
	3.3.1	First period (fifth century AD–1158)		52
		3.3.1.1	Introduction	52
		3.3.1.2	Principles of water law under Roman-Barbaric rule	54
		3.3.1.3	The feudal system	54
		3.3.1.4	Water law principles under the feudal system	55
	3.3.2	Second period (1158–1812)		56
		3.3.2.1	Introduction	56
		3.3.2.2	The classification and ownership of water	57
		3.3.2.3	The right to use water	58
3.4	Intermediate water law principles in Ibero-America			59
	3.4.1	Early legal principles		59
	3.4.2	Water regulations		59
References				60

4 Definition and sources of water law — 61

4.1 Introduction — 61
4.2 The content and concept of water law — 61
4.3 The relationship between water law and other legal disciplines — 63
    4.3.1 Constitutional law — 63
    4.3.2 Administrative law — 63
    4.3.3 Civil law — 63
    4.3.4 Criminal or penal law — 63
    4.3.5 Agrarian law — 63
    4.3.6 Mining law — 64
    4.3.7 Natural resources and/or environmental law — 64
    4.3.8 Public health law — 64
    4.3.9 Other legal disciplines — 64
4.4 Sources of water law — 64
4.5 Legislation in general — 65
4.6 International and interstate agreements — 66
4.7 Customary law — 67
4.8 Case law and arbitral awards — 68
4.9 Doctrine, or scholarly opinion — 69
4.10 Common law — 70
4.11 Conclusion — 71
References — 71

5 Existing systems — 73

5.1 Introduction — 73

5.2	Original Roman water law principles and their influence in subsequent legislation		74
5.3	Customary water law: its importance		74
5.4	Water law principles in the Islamic system		76
	5.4.1	Introduction	76
	5.4.2	The origin and sources of Islamic water law	77
	5.4.3	Water ownership and the right of use	78
	5.4.4	Groundwater law	79
	5.4.5	The process of codification and the Ottoman Civil Code 'Mejelle'	80
		5.4.5.1 The Ottoman Civil Code 'Mejelle'	80
		5.4.5.2 The legal status of water	81
		5.4.5.3 The right to use water	81
		5.4.5.4 Maintenance of waterways	81
		5.4.5.5 The harim	82
	5.4.6	Islamic water administration	82
		5.4.6.1 Customary water administration	82
		5.4.6.2 Recent developments in the administration of water	84
		5.4.6.3 Government action in Moslem countries	84
5.5	Water law principles in civil law countries		85
	5.5.1	Introduction	85
	5.5.2	The legal status of water resources	87
	5.5.3	The right to use water	87
	5.5.4	Water quality and pollution control	88
	5.5.5	Water administration	89
	5.5.6	Conclusion	90
5.6	Water law principles in common law countries		91
	5.6.1	Introduction	91
	5.6.2	The legal status of water resources	92
	5.6.3	The right to use water	93
	5.6.4	Water quality and pollution control	94
	5.6.5	Recent developments of the common law system	95
	5.6.6	Water administration	96
5.7	Water law principles in the former Soviet system		97
	5.7.1	Fundamentals of Soviet water law	97
	5.7.2	The legal status of water	98
	5.7.3	The right to use water	98
	5.7.4	Order of priorities	99
	5.7.5	Harmful effects of water	99
	5.7.6	Water quality and pollution control	100
	5.7.7	Enforcement	100
	5.7.8	Centralized inventory and planning	101
	5.7.9	Water administration	101
	5.7.10	Evolution and trends	101

5.8	Water law principles in the Hindu subak system in Bali		103
	5.8.1	Legal-historical background	103
	5.8.2	Definition and origin	104
	5.8.3	Organization	105
	5.8.4	The legal status of water resources	105
	5.8.5	The right to use water	106
	5.8.6	Order of priorities	107
	5.8.7	Water quality and pollution control	107
	5.8.8	The water distribution system	107
	5.8.9	Financial aspects	108
	5.8.10	Water law implementation	108
	5.8.11	The settlement of disputes	108
	5.8.12	The statutory subak	108
	5.8.13	Conclusion	109
References			109

6 Development by region — 111

6.1	Africa		111
	6.1.1	Introduction	111
	6.1.2	Customary law	112
	6.1.3	Countries following principles of the civil law system	113
	6.1.4	Countries following principles of the common law system	115
	6.1.5	Countries following principles of other systems	116
	6.1.6	Countries influenced by principles of the Islamic water law system	118
6.2	Asia and the Pacific		118
	6.2.1	Introduction	118
	6.2.2	Countries following principles of the civil law system	119
	6.2.3	Countries following principles of the common law system	120
		6.2.3.1 Australia	120
		6.2.3.2 Bangladesh	121
		6.2.3.3 India	122
		6.2.3.4 Sri Lanka (Ceylon)	123
		6.2.3.5 Other countries	123
	6.2.4	Countries following principles of other systems	124
		6.2.4.1 People's Republic of China	124
		6.2.4.2 Japan	125
		6.2.4.3 The Philippines	125
		6.2.4.4 Other countries	126
	6.2.5	Water administration	127
6.3	Central and South America		129
	6.3.1	Introduction	129
	6.3.2	Central and South American water law principles	130
		6.3.2.1 Latin American countries	130
		6.3.2.2 Other countries of Central and South America	132

		6.3.3	The legal status of water resources	132
		6.3.4	The right to use water	133
		6.3.5	Order of priorities	133
		6.3.6	Legislation on water use, quality and pollution control	134
		6.3.7	Water administration	135
	6.4	Europe		137
		6.4.1	Introduction	137
		6.4.2	The legal status and the right to use water	138
		6.4.3	Water quality and pollution control	139
		6.4.4	Institutional framework	141
			6.4.4.1 Federal states in Europe	142
			6.4.4.2 Unitary states in Europe	144
			6.4.4.3 States undergoing a process of decentralization	146
			6.4.4.4 Basin level	147
		6.4.5	The process of transposition of the European legal framework	149
			6.4.5.1 Background	149
			6.4.5.2 Developments	152
	6.5	United States of America		154
		6.5.1	Introduction	154
		6.5.2	Federal water law principles	155
		6.5.3	State water law principles	158
			6.5.3.1 Riparian water law	159
			6.5.3.2 The appropriation doctrine	160
			6.5.3.3 Groundwater management	162
			6.5.3.4 The conjunctive use of surface and underground water	164
	References			165

7 Possible contents of and reasons for water law — 167

7.1	Introduction		167
7.2	The contribution of the lawyer		168
7.3	General considerations		168
7.4	Water policy		170
7.5	Collection and use of data and information		171
7.6	Water resources planning		172
7.7	Ownership or other juridical status of water		173
	7.7.1	Public waters	173
	7.7.2	Private waters	174
	7.7.3	*Res nullius*	174
	7.7.4	Common waters: community or tribal waters	174
	7.7.5	Conclusion	174
7.8	The right to use water		175
	7.8.1	Basic concepts	175

	7.8.2	Legal régimes governing the right to use water freely or by virtue	
		of a simple declaration/registration	176
		7.8.2.1 The free use of water	176
		7.8.2.2 Declaration/registration	176
	7.8.3	The permit system	176
		7.8.3.1 Applicability	177
		7.8.3.2 Different types of permit	177
		7.8.3.3 The procedure for granting permits	178
	7.8.4	Characteristics of permits and concessions	179
		7.8.4.1 The difference between permits and concessions	179
		7.8.4.2 Common characteristics of permits and concessions	179
	7.8.5	Recognition or reallocation of pre-existing water rights	180
		7.8.5.1 Customary and riparian rights	181
		7.8.5.2 Rights under the appropriation doctrine	181
		7.8.5.3 Powers of the administration	181
		7.8.5.4 The reallocation of water	182
	7.8.6	Water markets	182
7.9	Limitations to the right to use		183
7.10	Priorities		184
7.11	Beneficial uses		185
	7.11.1	Water and other natural resources	185
	7.11.2	Domestic and municipal uses	185
	7.11.3	Agricultural uses	186
	7.11.4	Industrial uses	186
	7.11.5	Hydropower production	187
	7.11.6	Setting of minimum flow requirements	187
	7.11.7	Other public uses	187
	7.11.8	Conclusion	188
7.12	The right to water		188
7.13	Harmful effects of water		190
7.14	Water quality and pollution control		191
	7.14.1	Waste and misuse of water	191
	7.14.2	Recycling, reuse of water and recharging of aquifers	192
	7.14.3	Wastewater and effluent control	192
	7.14.4	Health preservation	192
	7.14.5	Pollution control	192
7.15	Environment protection: the 'greening' of water law		195
7.16	Underground waters		197
	7.16.1	General considerations	197
	7.16.2	Exploration or prospecting permits	198
	7.16.3	Drillers' licences or permits	199
	7.16.4	Groundwater found incidentally	199
	7.16.5	Metering	199

	7.16.6	Groundwater pollution control	199
	7.16.7	Reservation of good quality groundwater for drinking purposes	200
	7.16.8	Zoning mechanisms	201
7.17		Control and protection of waterworks and structures	201
7.18		Protected zones/areas	202
7.19		Legislation on financial aspects	203
7.20		Implementation of water legislation	204
	7.20.1	General remarks	204
	7.20.2	Judicial and administrative control over water ownership, use and distribution	205
	7.20.3	Protection of individual rights and water rights	205
	7.20.4	Administrative procedures for claims against the water administration	205
	7.20.5	Water tribunals or courts	206
	7.20.6	Penalties and sanctions	206
7.21		The interconnection between water law and other legal enactments relevant to water law	207
7.22		Customary water law and institutions	207
7.23		Water users' associations	208
7.24		National water resources administration	208
References			208

8 Water resources planning and water law — 211

8.1	The rationale of water resources planning	211
8.2	Objectives of a water resources plan	213
8.3	Types and characteristics of plans	214
8.4	The relationship between water resources planning and economic and social sectors	216
8.5	Methods for planning processes	217
8.6	Relevant administrative and institutional issues	217
	8.6.1 Administrative and institutional issues	218
	8.6.2 Other factors contributing to a better administration and planning of water resources	219
8.7	Water resources planning under the European Water Framework Directive	220
References		221

9 National water resources administration — 223

9.1	Introduction	223
9.2	Different types of water resources administration	223
	9.2.1 Institutions according to their powers	223
	9.2.2 Institutions according to their functions	224
	9.2.3 Institutions according to uses	225
	9.2.4 Institutions according to their territorial level of jurisdiction	225

		9.2.5	Institutions according to their legal régime	227
		9.2.6	Special water development agencies	227
		9.2.7	Water users' associations	228
	9.3	Major issues of water resources administration		229
		9.3.1	The need for coordination	229
		9.3.2	The question of centralization, decentralization and deconcentration of the water administration	230
		9.3.3	The water rights administration	231
		9.3.4	The need for a water resources 'regulatory' institution	232
			9.3.4.1 Definitions	232
			9.3.4.2 Major objectives and functions of a regulatory institution	232
		9.3.5	The role of water law in institution building	234
	9.4	A possible institutional solution		234
		9.4.1	Institutions at the national level	234
			9.4.1.1 A national water resources council	235
			9.4.1.2 A national water committee or commission	235
			9.4.1.3 A central water administration	235
		9.4.2	Institutions at the regional, basin, sub-basin and local levels	236
			9.4.2.1 At the regional level	236
			9.4.2.2 At the basin or sub-basin level	236
			9.4.2.3 At the aquifer level	237
			9.4.2.4 At the local level	237
			9.4.2.5 At the international level	237
		9.4.3	Conclusion	238
	References			239

10 International water resources law in general 241

	10.1	Introduction		241
	10.2	The concept of 'international water resources' and other definitions		242
		10.2.1	A historical review	242
		10.2.2	The drainage basin concept	244
		10.2.3	The expression 'international water resources'	245
	10.3	The sources of international water resources law		246
		10.3.1	Introduction	246
		10.3.2	International conventions	246
			10.3.2.1 General conventions	247
			10.3.2.2 Particular conventions	249
		10.3.3	International customary water law	250
		10.3.4	The codification of international water resources law	251
		10.3.5	The law-making activity of the European Union	251
		10.3.6	General principles of international water resources law	253
		10.3.7	Resolutions of intergovernmental organizations	254
		10.3.8	Judicial decisions	255

		10.3.8.1	Decisions of international courts	255
		10.3.8.2	Arbitral awards	257
		10.3.8.3	Decisions of national tribunals	258
	10.3.9	\multicolumn{2}{l	}{Contributions of publicists and international non-governmental organizations}	259
		10.3.9.1	The work of the Institute of International Law	259
		10.3.9.2	The work of the International Law Association	260
		10.3.9.3	The work of the Inter-American Bar Association	261
		10.3.9.4	The work of the Asian-African Legal Consultative Committee	262
		10.3.9.5	The work of the Pan American Union	262
		10.3.9.6	The work of the Council of Europe	262
		10.3.9.7	The work of the International Association for Water Law	263
References				263

11 International water resources law: major issues — 265

11.1	Boundary demarcation	265
	11.1.1 Introduction	265
	11.1.2 The boundary on a successive river	265
	11.1.3 The boundary on a contiguous river	266
	11.1.3.1 The boundary at the banks (river res nullius)	266
	11.1.3.2 The boundary at the banks (river res communis)	266
	11.1.3.3 The boundary at one of the banks	266
	11.1.3.4 The boundary at the median line	267
	11.1.3.5 The boundary at the thalweg	267
	11.1.4 Natural modifications of the boundary on a contiguous river	268
	11.1.5 The boundary on a bridge over a contiguous river	268
11.2	Navigation	269
	11.2.1 Origins	269
	11.2.2 The internationalization of navigation	269
	11.2.3 The Congress of Vienna (1815)	270
	11.2.4 The Treaty of Paris (1856): the régime of the Danube	272
	11.2.5 The navigation régime after 1856	273
	11.2.6 The Act of Berlin (1885)	273
	11.2.7 The régime after World War I	274
	11.2.8 The régime established at Barcelona (1921)	274
	11.2.8.1 The administration of international waterways	275
	11.2.8.2 Evaluation	276
	11.2.9 Developments after Barcelona	276
	11.2.10 The régime after World War II	277

11.3	Non-navigational uses of water		278
	11.3.1	Introduction	278
	11.3.2	The theory of absolute territorial sovereignty	278
	11.3.3	The theory of absolute territorial integrity	279
	11.3.4	The theory of limited territorial sovereignty and integrity	279
	11.3.5	The shared natural resources concept	280
	11.3.6	Equitable and reasonable utilization and participation	280
	11.3.7	Obligation not to cause significant harm	281
	11.3.8	Floating	282
	11.3.9	Production of energy and industrial uses	282
	11.3.10	Procedural rules	284
	11.3.11	Conclusions	286
11.4	Harmful effects of water		287
	11.4.1	Definition	287
	11.4.2	Evolution	287
	11.4.3	The emerging rule	289
11.5	Quality control of water		290
	11.5.1	Definition	290
	11.5.2	Evolution	291
	11.5.3	The emerging rule	296
11.6	Armed conflict		297
	11.6.1	Definition	297
	11.6.2	Precedents	298
	11.6.3	The emerging rule	299
11.7	Environmental aspects		299
	11.7.1	Definitions	299
	11.7.2	Evolution	300
	11.7.3	The emerging rule	305
11.8	The right to water in international law		306
	11.8.1	Definitions	306
	11.8.2	Evolution	306
	11.8.3	The emerging rule	308
	References		309
12	**Developments in the law of transboundary aquifers**		**311**
12.1	Introduction		311
12.2	Sources and evolution of international groundwater law		312
12.3	The experience of federal countries		320
12.4	The codification of the law of transboundary aquifers		320
12.5	Institutional issues		323
12.6	The emerging rules		325
12.7	Conclusion		325
	References		326

13 International water resources administration — **329**

- 13.1 Introduction — 329
- 13.2 Institutional developments — 329
  - 13.2.1 Institutional developments in Europe — 329
    - 13.2.1.1 The Rhine Commissions — 331
    - 13.2.1.2 The Danube Commissions — 332
    - 13.2.1.3 Other commissions — 332
  - 13.2.2 Institutional developments in the Americas — 333
    - 13.2.2.1 The International Joint Commission between the USA and Canada — 333
    - 13.2.2.2 The International Boundary and Water Commission between USA and Mexico — 334
    - 13.2.2.3 The Plata River Basin — 336
    - 13.2.2.4 Other commissions — 336
  - 13.2.3 Institutional developments in Africa — 337
    - 13.2.3.1 The Nile Commission — 337
    - 13.2.3.2 Post-1960's basin institutions — 338
    - 13.2.3.3 The Liptako-Gourma Authority — 340
    - 13.2.3.4 Institutional framework for cooperation in Southern Africa — 340
    - 13.2.3.5 The contribution of regional economic integration organizations — 341
  - 13.2.4 Institutional developments in Asia — 342
    - 13.2.4.1 The Mekong River Commission — 342
    - 13.2.4.2 The Indus Commission — 343
    - 13.2.4.3 Joint commissions between Nepal and India — 344
    - 13.2.4.4 The India-Bangladesh Joint Commission — 344
    - 13.2.4.5 Institutional arrangements for the Aral Sea basin — 344
    - 13.2.4.6 The Helmand River Commission — 345
    - 13.2.4.7 Other commissions — 345
- 13.3 Evaluation of existing arrangements — 346
- 13.4 Objectives and purposes — 348
  - 13.4.1 Technical responsibilities — 348
  - 13.4.2 Economic and financial responsibilities — 349
  - 13.4.3 Legal and administrative responsibilities — 350
  - 13.4.4 Possible options — 350
- 13.5 Duration, constitution and decision-making procedures — 350
  - 13.5.1 Duration — 350
  - 13.5.2 Constitution — 350
  - 13.5.3 Procedures for decision making — 351
  - 13.5.4 Legal status — 351
- 13.6 Territorial competence — 351
- 13.7 Functions and powers — 353

13.8	Form	353
13.9	Major institutional requirements for rational international water resources administration	354
13.10	Economic and financial requirements	355
13.11	Prevention and settlement of disputes	355
13.12	Conclusion	356
References		356

Index 357

Acknowledgements

Acknowledgements to the second and the present editions

As the redactor of the second and the present editions of the book, Principles of Water Law and Administration, National and International, I wish to acknowledge the support and editorial advice so graciously offered to me by Mrs Alleyne Caponera for the enhancement of the text.

Marcella Nanni
Rome, 2018

Acknowledgements to the first edition

Before going to press, it is my pleasant duty to express a few words of acknowledgement and gratitude to Prof. ir W.A. Segeren, Director of the International Institute for Hydraulic and Environmental Engineering (IHE) of Delft, where I have been teaching for many years, and particularly to Prof. Mr. J. Wessel, Director of the Centre for Comparative Studies on River Basin Administration (RBA Centre), who provided facilities for me to concentrate on the finishing of the text and without whose personal interest and encouragement this publication might not have seen the light.

The IHE was established in 1957 to offer international post-graduate education principally to developing countries in civil and environmental engineering. The RBA was created in June 1989 at the Delft University of Technology as a research centre in the field of both theoretical and practical studies on river basin administration. Both of these institutions also deal with legal and administrative aspects of water resources management, hence their interest in the publication of this book.

Dante A. Caponera
Rome, 1992

Biographical notes

The Author: Dante A. Caponera (1921–2003)

Dante A. Caponera began his career in the field of water law and administration in the 1950's, undertaking missions for the Food and Agriculture Organization of the United Nations and other UN agencies. He became Chief of the FAO Legislation Branch, which, thanks to his efforts, is now a provider of legal assistance to developing countries, not only in water law but also in all of the fields in which FAO is involved. He was the initiator of the FAO Legislative Studies series.

As a consultant, Dr. Caponera assisted governments and international basin institutions in the drafting of water legislation and international water agreements, leading to water codes embodying principles such as those of river basin management and users' participation and improved river basin management institutions. He used a methodological approach which he developed and which is now known world-wide. A firm believer in the importance of law in the management of water resources, he also promoted training in the field of water law and administration as a vehicle for dialogue and cooperation leading to improved legal and institutional frameworks.

A prolific writer, Dr. Caponera also published a collection of selected writings in 2003.

The Reviser: Marcella Nanni

Marcella Nanni is a recognized international expert in water law and administration and related disciplines. She has provided consultancies to a number of governments, river basin institutions and regional bodies for projects financed by international organizations such as FAO, the European Union, the World Bank, the Asian Development Bank and by bilateral donors. This advice has covered the drafting of water policies, legislation and international agreements, the development of proposals for the approximation of legislation of EU accession countries to the EU water *acquis*, the restructuring of water resources management institutions, water law implementation requirements and the formulation and conduct of capacity-building programmes.

Dr. Nanni is a member of the board of directors of the International Association for Water Law/Asociación Internacional de Derecho de Aguas (AIDA) and is the editor of the association's newsletter, 'Aquaforum.' She collaborated with Dr. Caponera for the research on the first edition of this book and produced the second revised edition of the same.

Preface

Over time, water law and associated institutional mechanisms have evolved in response to specific challenges. The earliest water regulations catered to the need to construct, operate and maintain irrigation networks, facilitate water distribution and organize the defence against floods. With socio-economic development, population growth, urbanization and technological progress, the need arose for more elaborate legal and institutional frameworks. Initially, during the modern era, water legislation was mainly use-oriented; that is, it dealt with specific water uses or with the harmful effects of water. Subsequently, growing industrialization generated the need to introduce rules to prevent and fight against water pollution. More recently, the concern for the state of the environment, coupled with the recognition of the interdependence of all its elements and of the relationship between a healthy environment and the life and well-being of people, have brought about the 'greening' of water law, i.e., the integration into it of environmental considerations with a view to finding a balance between water demands and environmental sustainability.

Because of the challenges imposed by the overexploitation and increasing deterioration of water resources, climate change, environmental and health issues and technological progress, water laws and institutions have reached an unprecedented level of complexity. However, since more challenges lay ahead, they should not be considered as static tools that, once in place, function effectively under all circumstances. Rather, they follow an evolutionary process as required by changing needs. Accordingly, a number of developments have taken place since the second edition of this book in 2007, and further developments are to be expected in the years to come.

Already in its first edition, the book emphasized that most countries, including those following the civil law tradition according to which the landowner owns the water located on or under his land, have shifted to public ownership of all water resources, whether surface or underground. Thus, the state as the owner or trustee of these resources has the authority to allocate or reallocate them by issuing permits or concessions. The first edition further explained that the power of the state to grant permits has expanded to all the activities that may cause water pollution. In parallel with this, it called attention to the need to consider water not in isolation from other natural resources and the broader environmental context.

The concept of integrated water resources management (IWRM), which promotes the coordinated development and management of water, land and other natural resources within hydrological and hydrogeological units, i.e., by river basin or aquifer, has now gained the support of countries worldwide, as is highlighted in this edition of the book. Legislation recently enacted or to be enacted prescribes that water resources are to be managed by river basin and calls for the development of river basin plans setting clear objectives in terms of water quantity and quality, and having binding effects on governmental decisions. Particularly in European countries, integration tends to be sought between river basin plans and flood management plans. Water legislation further provides for the establishment of institutional mechanisms for basin planning and management and for the participation of stakeholders in the planning process.

Environmental considerations are increasingly being absorbed into water legislation. Among other things, this translates into the requirement to assess the impact of water resources development on the environment, either through provisions embedded in the relevant texts, or, by reference, to environmental protection legislation. A further 'green' element of contemporary water legislation is the ecosystem approach, which considers the relationship between water and related resources and promotes their conservation and sustainable use in view of the benefits that society may derive from them. Thus, the setting of minimum flow requirements to preserve aquatic ecosystems and protect living resources is gradually becoming a feature of water legislation, alongside a high rank being assigned to the environment in the order of priorities among competing water uses. Some countries have gone as far as recognizing juridical personality and autonomous rights to rivers, and even to nature. These new elements, which were not present in the second edition of the book, are illustrated in this edition, which shows that water law is expected to become 'greener and greener'.

Finally, water legislation normally accords top priority to the satisfaction of drinking and domestic water needs. However, recent enactments go further, by providing for the reservation of water in sufficient amounts and of a suitable quality for the satisfaction of these needs. A human right to water is now recognized in a number of national constitutions and in an increasing number of water statutes, particularly in developing countries. This new development is also considered in this edition of the book.

This book also shows that those which once were challenges for only industrialized countries are now spreading to the developing world, so that the water legislation recently enacted in a number of developing countries presents features common to that of industrialized countries. Thus, this legislation supports IWRM within a river basin context, seeks to provide the flexibility needed to adapt to the effects of climate change and, to varying degrees, addresses environmental issues.

International water law has also undergone steady development since the second edition of this book, which placed emphasis on the progressive consolidation, as a result of consistent state practice, of the principle of equitable and reasonable utilization, of the obligation not to cause significant harm and of the general obligation of states to cooperate. Together with procedural rules concerning the exchange of information and the notification of planned measures with possible adverse effects, these basic tenets were enshrined in the UN Watercourses Convention, which was adopted in 1997 and entered into force in 2014. They may be considered as the pillars of international customary water law and are now being incorporated into a growing number of international water agreements worldwide.

Both the convention and existing agreements consider groundwater only in so far as it is connected to surface water. Only few agreements deal specifically with groundwater, although their number might increase in the near future. The most recent of these agreements was concluded between Jordan and Saudi Arabia in 2015 and concerns the Al-Sag/Al-Disi Layer. In 2008, the International Law Commission (ILC) of the United Nations adopted the Draft Articles on the Law of Transboundary Aquifers, of which the UN General Assembly took note in the same year, at its sixty-third session. Although not legally binding as such, the Draft Articles represent an authoritative statement of the law of shared groundwater resources, since they reflect prevailing international customary water law as applicable to transboundary aquifers. Therefore, when seeking to reach agreement on the management of their transboundary aquifers, countries are now inclined to take them into consideration, as evidenced by the reference made to them in the preambles to the Guarani Aquifer Agreement (2010) and to the Memorandum of Understanding for the Iullemeden, Taoudeni/Tanezrouft Aquifer System (2014). These are new developments with respect to the second edition of this book, but the law of transboundary aquifers is still going through a process of evolution and the question as to whether the Draft Articles should be turned into a convention remains open.

Like domestic water law, international water law is undergoing a 'greening' process which translates into the obligation to prevent, reduce and control pollution and, in general, to prevent transboundary harm, and into the obligation to conduct environmental impact assessments when planned developments are likely to cause significant transboundary harm. These obligations may now be considered as part of international customary water law, as evidenced by consistent state practice and case law. Another emerging principle which is being absorbed into an increasing number of treaties refers to the duty to protect and preserve ecosystems, including through the release of minimum ecological flows from dams. Given the growing concern for the environment, there is little doubt that, in the future, environmental law elements will be integrated into international water law to a greater extent than they were before.

Until recently international water law was not concerned with the right to water. This right was to be derived from international legal instruments dealing with human rights. Although mention of it was made on a number of occasions, such as at the 1977 UN Water Conference of Mar del Plata and other international discussion fora, it is with General Comment No. 15 of 2002, interpreting Articles 11 and 12 of the 1966 International Covenant on Economic, Social and Cultural Rights, that the right to water was brought to the attention of the international community and the debate was ignited as to its existence as a self-standing human right. The right to water was explicitly recognized by the UN General Assembly and by the Human Rights Council in 2010, but many countries abstained from voting the relevant resolutions. This, together with the very few contemporary water agreements acknowledging it as a self-standing human right, leads to the conclusion that the debate is still open. This new development was not considered in the second edition of this book.

It is worth noting that although international water law does not require the establishment of institutional mechanisms for the management of transboundary water resources, countries now recognize that these mechanisms can play a crucial role for the definition of what is equitable and reasonable, the prevention of adverse transboundary impacts and the prevention of water disputes. Thus, a vast majority of contemporary water treaties require the setting-up of international institutional mechanisms.

To conclude, water law is in constant evolution. As law in general, it is influenced by physical, social, economic, climatic, political, religious and other factors. Since circumstances change over time, it is subject to continuous adjustments; therefore it needs to be conceived as a dynamic and flexible tool capable of addressing new challenges when they arise.

Marcella Nanni
Rome, December 2018

Preface to the second edition

This second edition of 'Principles of Water Law and Administration, National and International,' aims at highlighting the developments which have taken place in the field of water law and administration since the book was first published in 1992, as well as present trends. As it was conceived in 1992, the book remains unique, in that it deals with a wide range of legal and institutional issues and discusses them without losing sight of the context in which these issues arise. Ample coverage is provided in the first chapters to the sources of water law and to the historical evolution of the existing legal systems, as Dante A. Caponera, the author and an eminent scholar, firmly believed that a legal system may not be fully understood if due attention is not paid to its roots. In turn, he felt that it is impossible to develop a new law without taking into account the context in which it will have to be implemented and that, besides the legal system, non-legal aspects, including geographic, physical, climatic and socio-economic elements, should be carefully considered.

These considerations also apply to the development of legal and institutional frameworks for the management of international river basins and aquifers, but in this case the situation may be more complex because the basin and/or aquifer states concerned may present differences as to their geography, hydrology (or hydrogeology), climate and socio-economic conditions. Traditions and religion also play an important role, as they may influence international relations considerably.

It was my privilege to know Dante A. Caponera, to cooperate with him in a number of undertakings relevant to water law and administration and to participate in the development of some of the ideas which were incorporated in the first edition of the book. Therefore, I feel a few words deserve to be written about the spirit which was – and is - behind this book.

Dante A. Caponera was of the opinion that there is a need for people acquainted with principles of water law and administration and able to advise governments and international organizations and institutions on how to handle the legal issues arising in connection with water resources management. Mainly for this reason he decided to write a book on this complex subject, through which he could share his vast experience with others, whether they be lawyers or non-lawyers.

Water law is a multidisciplinary subject, but it is difficult for the lawyer, who sees water as a liquid flowing from the tap, to fit it into a non-legal – hydrologic, engineering, socio-economic – context. On the other hand, the engineer is likely to address water management issues through physical solutions – dams, hydropower plants, canals – without taking into consideration the individual rights and legitimate interests at stake. These considerations were a major stimulus in the process of preparation of the book and resulted in a text that was, and still is, palatable to all, including water resources policy makers, planners and water administrators. After a few years in circulation, the book became a 'bible' for many government officials at various levels, to be resorted to in order to arrive at informed decisions, and has continued to play this role. Indeed, no other text is more comprehensive than 'Principles of Water Law and Administration.'

Given its clear language and the fact that it refrains from using a strictly 'legalese' jargon when dwelling on the various topics, the book may also be considered as a manual of easy access to anyone enrolling in a course on water law and administration. At present, courses of this kind and seminars on the subject are becoming more and more frequent, as the demand for an enhanced knowledge of the legal issues involved in the management of water resources is growing on a par with the threats to which these resources are increasingly exposed.

This second edition maintains the same logic and the spirit of the first one, although acknowledging the developments that have occurred since 1992. In particular, it covers progress in the codification of international water law, including the work of the International Law Commission on a possible new groundwater convention, and legal developments in the European Union, which are also relevant for countries applying for membership. Furthermore, it analyzes the new laws and institutions of the countries of the former Soviet Union, keeping in mind that the situation is still fluid and subject to change. In addition, it strives to update information relating to national water laws, institutional developments and legal and institutional arrangements for international river and lake basin management and transboundary aquifers in Africa, Asia and Latin America.

Dante A. Caponera passed away in 2003. This edition of the book intends to be a tribute to him and to the experience that he accumulated during his vast and successful career, which he left behind for the benefit of future generations of water lawyers and non-lawyers with a stake in water law and administration.

Marcella Nanni
Rome, January 2007

Chapter 1

Introduction

1.1 Water and the society

The importance of water in all aspects of human activities is well known; one basic condition for human, animal and plant survival is the availability of water. It is through the combination of water with one or more basic natural resources that other 'secondary' resources are made available. Water, combined with land, provides plants and forests, which, in turn, are indispensable to sustain human and animal life. Water is also an important element for social stability, and the economic development of any community, country or civilization depends largely on its availability.

Water resources are not evenly distributed, so that while in some areas of the world there may be excess water, in other areas there may be a shortage. On the other hand, the amount of water available in a state, area or basin is invariable, while water demands increase continually.

Water demands for drinking purposes augment at a faster pace than population growth, since higher standards of living require increased amounts of water for food production and other water-consuming activities, such as the watering of lawns and gardens, and for hydro-power generation and recreational purposes (golf courses, swimming pools, etc.).

In all countries, agricultural and industrial development requires the construction of hydraulic works, such as irrigation systems, reservoirs and tanks. Industries, particularly those based on chemical processes such as oil refineries, the manufacturers of synthetic materials and paper mills, utilize considerable quantities of water, in spite of water-saving technologies. Air-conditioning and air-cooling plants also use considerable amounts of water. Thus, the availability of water is both a prerequisite of and the limiting factor to the economic development of a country.

Human activities involving the use of water have direct or indirect effects on aquatic ecosystems and on the environment, which may lead to a loss of biodiversity. Changes in the environment caused by overexploitation of natural resources such as land, water and forests in turn contribute to further destruction of land and the spread of deserts. The indiscriminate dumping of urban and industrial wastes turns rivers into sewage canals, with the result that at a certain point the water is lost for further use. Landfills and waste disposal into the subsoil lead to the same result and to environmental degradation.

Irrigation practices not sustained by adequate drainage may cause siltation, soil erosion and the loss of previously good lands. Excess chemicals in the form of weed killers or fertilizers cause water contamination with disastrous consequences for downstream domestic, agricultural, fishing and industrial uses and the contamination of groundwater. Inadequate sewerage or drainage systems, or, even worse, the absence of such systems, have made water bodies carriers of waterborne diseases. The reduction of the discharge of rivers into the sea caused by excessive domestic, agricultural and industrial uses may lead to the increase of infiltration of brackish water into deltaic groundwater areas.

On the other hand, excessive floods, which are often caused by lack of watershed protection measures and by the improper use of land and forests, can destroy or render less useable what was once high-production potential land. Radioactive contamination and toxic wastes spoil atmospheric water and cause air pollution and acid rain, endangering human, animal and plant life. Finally, the overpumping of groundwater may cause its exhaustion and the intrusion of seawater in coastal areas.

Most countries have reached the end of the era during which water was considered an unlimited resource and are at the beginning of a new era in which it must be used more thriftily and protected from pollution. It may be said that no country, region or basin in the world can be fully satisfied with the quantity or quality of water at its disposal to meet present and foreseeable future water demands.

If not adequately planned and managed, both quantitatively and qualitatively, water use may cause detrimental side effects to the water itself and to other natural resources. Since these negative effects can be avoided through the enactment of adequate water legislation and the establishment of an appropriate water administration, it is safe to say that success in the development, protection and conservation of water resources in a country depends to a large extent on the effectiveness of its water laws and institutions. Moreover, where supplies are scant or almost fully utilized, pressures of new demands require greater efficiency in use and legal mechanisms for the reallocation of the available water from polluting and less productive uses to new and more desirable ones.

1.2 The need for a water policy, legislation and administration

As a consequence of these realities, the need is being increasingly felt for more careful consideration of all the problems related to the use of water resources, among which, in first rank, are the legal and administrative aspects involved. The need for formulating policies in support of water resources development programmes and activities is also felt.

The overall objective of a water policy is to achieve the maximization of benefits deriving from available water resources and to promote their rational and sustainable management.

A sound and well-balanced water resources policy should be viewed, as the case may be, at the national, basin, regional, local or project levels. It should be designed in line with the existing situation and requirements in any particular state, region or basin, and be concerned with finding ways and means to satisfy existing and future water demands on the basis of water availability, existing uses, water quality, estimates of population growth and technical and financial possibilities. This calls for proper water resources management planning and the allocation of financial resources where needed.

Just as a national water resources policy must be viewed within the context of an overall development plan, as it may constitute either a catalyst or a barrier to plan implementation, the institutional setup for water resources management may act either as a stimulus or as a constraint upon the national development process. Likewise, a well conceived water legislation may constitute a means to implement water policy decisions and facilitate the rational utilization of water resources, while an inadequate water legislation can act as a hindrance to this utilization.

In many cases, water legislation has come down from the days when the resource was considered to be inexhaustible. Time has overtaken the laws which gave users a free hand on waters, and advances in knowledge and technology have outdated many early types of control. The search for new sources has led to the extensive use of groundwater, to transbasin water transfers, to water storage and distribution schemes of formerly undreamed size, to the recharging of aquifers, the recycling of water and the use of treated wastewater. Many countries have no laws which provide for the management of these new sources or for controlling these projects.

To meet new water needs, innovative water laws must be designed, not only to facilitate and achieve efficient allocation or reallocation of resources and environmental protection, but also to aim towards the attainment of social, economic and other national and international goals. Water law reform may accompany land reform, in that the redistribution of land may call for a redistribution of the water rights appurtenant to that land. Settlement schemes to open up new land to irrigation or to turn nomadic or pastoral people to farming may require special provisions in water laws or special organizations to manage or distribute water. Programmes aiming at the welfare of indigenous populations may need special provisions to enable them to compete with proposals for industrial development or commercial irrigation projects. Future holders of water rights may have different degrees of sophistication that require different treatment.

Modern planning, development and management of water resources must be based on water law principles which are implemented through legal procedures that authorize and facilitate these processes. It is important to recall that in earlier days 'water laws' referred mainly to legal rules governing the relationships among water uses, such as the riparian doctrine or the prior appropriation doctrine. Nowadays, the interests and objectives of governments must also be taken into consideration. As a result, the system must regulate the relationship between water users and the state and clearly identify and define the powers of the state concerning all uses of water, both public and private.

In some countries there may be a need for abolishing riparian rights and for adopting a system of state control over water. In others, the existing controls should be modernized in order to adapt the existing water rights to new priorities. Constitutional questions may need to be handled in connection with the termination of private water rights or the taking over of private waters by the state. Government institutions and agencies may have to undergo changes, this giving rise to political and legal problems when interjurisdictional conflicts for the sectorial management of water resources arise.

Due to the wide range of water utilizations, harmful effects and quality aspects, many government ministries, autonomous institutions, private corporations and individuals are involved, concerned, interested or users of water. On the other hand, legal enactments often purport to govern specific water uses, harmful effects or misuses, and each of these laws may be administered by a different ministry or department without apparent or enforced coordination. The results of this situation are the overlapping of responsibilities, inadequate or poor planning and coordination, a sectorial approach to water projects, detrimental effects of one project on another, waste of natural, financial and human resources, insecurity in the rights to use waters and uncertainty as to the successful implementation of projects.

Often water resources projects, though they may have been technically and economically well conceived, have been hampered, delayed or doomed to failure as a consequence of inadequate water legislation or other legal constraint. Investors and international agencies underwriting large projects now seek from the law the security once provided by a seemingly inexhaustible stock of water. In fact, any water development or conservation project needs capital, the investment of which is only feasible if the legal rights are well defined, water is allocated in volume, time and quality to satisfy the demand and the recovery of the capital originally invested is secured.

In the case of international drainage basins, the lack of adequate conventions and institutional arrangements renders the development of water resources projects problematic and, sometimes, is a cause of international water disputes.

For these reasons, a rational water resources policy must take into consideration the legal and institutional aspects of water resources management, both with respect to their implication in the technical, economic and social aspects and at every level of water resources management, including data collection and processing, planning, the implementation of policies and decisions and the monitoring of such implementation.

The issues connected with water administration and legislation are being increasingly dealt with by the United Nations system of agencies for the benefit of member countries. The Food and Agriculture Organization of the United Nations (FAO) was the first to initiate water law studies in 1951; the United Nations and its Economic Commissions for Asia and the Pacific (ESCAP, formerly ECAFE), for Europe (ECE) and for Latin America and the Caribbean (ECLAC) all undertake studies and activities in this field. The same applies to the United Nations Educational, Scientific and Cultural Organization (UNESCO), which, *inter alia*, has assisted the International Law Commission (ILC) in the preparation of the Draft Articles on the Law of Transboundary Aquifers. Through these and other organizations and financing institutions, such as the European Union, the World Bank and the Asian Development Bank, technical assistance is provided to countries wishing to modernize their water administration and legislation and for formulating national water policies.

1.3 The interdisciplinarity of the subject

It is increasingly recognized by managers, technicians and economists that water resources legislation and institutions play a key role in the planning, operation and maintenance of projects connected with water resources development, conservation and protection. The inadequacies in the laws are considered as constraints in the development of water resources, and because these constraints are usually encountered, not by legally trained people but by water managers (water engineers, hydrologists, hydrogeologists) and economists, the exact content, spirit, implications and extent of water legislation are either not well known or overlooked altogether.

On the other hand, lawyers generally lack the knowledge of basic hydrologic, technical and economic data which are indispensable for dealing with water management. Therefore, they are not always equipped to provide the legal and institutional solutions sought by water technicians and economists responsible for water management.

The science of water resources policy, administration and law is a relatively new one, necessitating an interdisciplinary approach. Starting from the technical aspects of water (hydrologic cycle, different types of uses, single or multi-purpose projects, etc.) it purports to provide water resources planners and managers with those legal and institutional tools necessary to overcome the constraints encountered by the technicians and economists.

In addition to technical aspects, socio-economic elements should be considered by the lawyers having responsibilities in the field of water resources management. These aspects are explained in the following two sections.

1.4 The physical context

Whenever dealing with water law, one should take into consideration the natural context of water and the way in which it occurs. Because the purpose of water law is to regulate the use, conservation and protection of water, it is necessary to know what the subject of regulation is.

1.4.1 The hydrologic cycle

The hydrologic cycle is, in broad terms, the cyclic movement of water in the globe, from the sea to the atmosphere, from the atmosphere to the earth and subsequently back to the sea.

With the heat of the sun, water evaporates from the oceans and other bodies of water and rises into the atmosphere. Atmospheric water condenses in clouds which are moved by the wind and then descends to the earth in the form of rain, snow, hail or dew, part of which evaporates immediately, some is absorbed by plants (evapo-transpiration), part of it infiltrates into the ground to form underground aquifers, and some of it flows on the surface (water flow) forming watercourses, rivers, lagoons, wetlands, etc., which in turn flow into the seas. From the seas, the hydrologic cycle starts again.

Both surface and underground water flowing into the sea are part of the hydrologic cycle. Some underground waters, however, are trapped in the subsoil and do not participate in the hydrologic cycle.

In certain areas, periods of drought or flood may occur, during which one might think that the cycle has ended. This is, however, only the result of local or regional situations; the hydrologic cycle continues and is a dynamic phenomenon.

The imaginary line which connects all the points of higher elevation on the surface of land within which all the water flows into a common terminus, either to the sea or to an internal body of water, forms a river basin, a drainage basin or a watercourse system.

Presently, in terms of law and administration, man's control over water is limited to that part of the cycle when water falls to the earth. Activities relating to non-renewable groundwater, the recharge of aquifers and water desalinization are not adequately controlled, although groundwater is increasingly being scrutinized by lawmakers and scholars. Likewise, the behaviour of atmospheric water, which can be influenced, thanks to modern techniques such as cloud seeding to produce weather modifications, is rarely controlled by law.

1.4.2 Definition and analysis of precipitation

Precipitation is a general term describing all forms of water falling from the atmosphere to the earth. The factors contributing to the production of precipitation are humidity, the rise of humid air, cooling and condensation.

The amount of water available, whether it be surface, underground or atmospheric, is not always possible to know, because different institutions deal with these aspects. Meteorologists are only interested in atmospheric water, hydrologists and their institutions deal with surface water, while groundwater is handled by hydrogeologists and their institutions. In most countries, therefore, three separate bodies of specialists deal with the same resource. For one it is only a mineral, for another a matter of weather forecasting and yet another is interested in the surface flow.

From the viewpoint of water administration, when contemplating land use planning, soil erosion control, forest conservation and watershed management, it is indispensable to know the amount of flow at any required level: regional, basin, national, international. This is in order to allocate water among various users.

Furthermore, the question of the relationship of water with other natural resources arises, in connection with land management, soil erosion, salinization, floods, coastal areas and ocean management, for instance. In addition, water has a direct influence on fauna and flora. If possible, it is necessary to determine an ecological balance for the use of water.

Therefore, when preparing a water law, one must coordinate and interconnect water law and administration with the legislation concerning other natural resources, because problems may arise when uses of different natural resources are licenced separately, by different authorities, without coordination being sought.

1.4.3 Analysis of flow

All hydrologic phenomena are aleatory: their future behaviour cannot be accurately calculated on the basis of mathematical laws. However, they can be predicted on the basis of statistical analysis that takes into account historical data. One can also analyze these phenomena by using mathematical probabilities.

Flow is the precipitation that remains on the earth before returning to the sea. It may remain on the surface of the ground, infiltrating the soil or gathering in lakes or ponds, or it may encounter a river and flow naturally towards a sea or inland water body. This flow will influence the type of vegetation, soil and gradient of the land.

Rivers are classified according to their velocity of flow and continuity and may be perennial, intermittent or recurrent. For water allocation purposes, it is important to calculate the monthly or annual average of the river flow.

1.4.4 Groundwater and its behaviour

General awareness of the importance of groundwater has notably increased in recent times, due to the fact that surface water resources have become insufficient to meet the numerous needs stemming from population and economic growth. The attention of governments and international organizations has been drawn to groundwater, the utilization of which is fundamental in order to satisfy basic needs, particularly in densely populated areas and in the presence of conditioning factors such as the lack of surface water in arid and hyper-arid regions.

Facilitated by modern technology, groundwater exploitation has greatly expanded, causing a number of problems. In fact, if not adequately controlled, abstraction activities may cause, *inter alia*, the depletion of aquifers, the deterioration of groundwater quality, salt water intrusion in coastal areas and land subsidence. Activities other than groundwater abstraction, such as mining activities, the discharge and disposal of solid and liquid wastes, the use of fertilizers and pesticides in agriculture and other land uses can be responsible for the deterioration of groundwater quality.

For a better understanding of the legal problems which may arise in connection with these activities and of the consequent need for subjecting them to public control, it is necessary to shed light on non-legal aspects of groundwater, along with some definitions.

The expression 'groundwater' includes all waters located below the land surface. The question as to whether or not a spring is a 'groundwater' remains open and left for the decision of hydrologists and hydrogeologists. Groundwater is meant to include two large categories. To the first belong those groundwaters which are directly connected to the hydrologic cycle, i.e., those which derive from, and are replenished by, rainfall, snow, hail and surface water. If underground waters of this kind are withdrawn faster than they are naturally recharged, the groundwater level will be lowered, and the groundwater will be gradually exhausted. In this case, the legal regime should ensure the application of the principle of 'safe yield.' Such expression designates the quantity of water that may be extracted from an aquifer annually without producing detrimental consequences to the yield or to the quality of the water, and therefore to the community at large.

To the second category belong those groundwaters which are not part of the hydrologic cycle and consequently are not replenished. They are to be found in aquifers receiving negligible contemporary recharge and are located mainly in arid regions.[1] These groundwaters are not connected with surface water and do not reach the surface by natural processes. In other words, they are in storage and are usually a relic of past climatic régimes. If abstracted they are depleted and eventually will be exhausted.[2] Thus, to a certain extent[3] they have to be treated as minerals, i.e., water which, after use, is no longer available. A legal régime different from that of renewable groundwater has to be applied.

Hydrogeologists have used many names to designate the various types of groundwater, such as, phreatic to designate those generally connected to the hydrologic cycle; and fossil, magmatic, juvenile, etc., to designate those which are apparently not renewable.

These definitions are quite relevant from a legal viewpoint, as they have repercussions on the different legal régimes applicable to each category of groundwater. For example, is a groundwater ownership right confined to the groundwaters to be found by the landowner immediately beneath his land, or does it include fossil, magmatic and juvenile groundwater? Likewise, depending on whether the groundwater under consideration is renewable or non-renewable, a different legal régime as regards its use may apply.

Although they follow different routes, surface and underground waters originate from the same source. Thus, whenever feasible they must be managed and exploited jointly. Surface water is used in preference for agricultural and industrial purposes, while groundwater is generally used only for domestic supply.

1.4.5 The notion of aquifer

Another important definition is that of 'aquifer.' An aquifer is a geologic formation receiving, retaining and storing groundwater. All aquifers are characterized by a flow component and a storage reserve. Since the storage of an aquifer is often remarkably large, the time employed by the water to flow from the area of recharge to that of discharge may be extremely long, also considering the fact that groundwater moves slowly. During this time frame, the flow

---

1 Such as the Nubian Sandstone Aquifer System and the North-Western Sahara Aquifer System. According to scientists, 'non recharging' aquifers are those which receive less than 200 mm annual rainfall; 'contemporary' refers to a 100 year time span.
2 Matthews, O.P. (1984) *Water Resources Geography and Law*. Washington, DC, Association of American Geographers, 6.
3 Non-renewable groundwater differs from mineral resources because (a) in arid regions it may be the only source of water supply to satisfy vital human needs and (b) unlike minerals, it may become polluted and therefore unfit for use. See Nanni, M. & Foster, S. (2005) Groundwater Resources – Shaping Legislation in Harmony with Real Issues and Sound Concepts. *Water Policy* 7:5.

(and the discharge) may be influenced as a result of human activities. If an aquifer is hydraulically connected to one or more other aquifers, we are in the presence of an aquifer system.

The notion of aquifer is also important as regards water pollution, because what is polluted is not only the water, as non-specialists may think, but the aquifer itself, and this is a pollution which can be eliminated with difficulty and only over a long period of time. Due to the complexity of groundwater flow and the time lag between the occurrence of pollution and when the effects are felt, it is technically difficult to prove liability for pollution. Finally, it is essential to identify the linkage between aquifer recharge, which is practiced to increase the volume of water that may eventually be exploited, and land use, because urban and agricultural development may radically modify recharge rates.

Aquifers are underground water reservoirs and must be managed as hydrologic units; their indiscriminate exploitation must not be allowed.

1.4.6 Weather modification

Since the 1950's, studies have been carried out in the United States concerning rainfall artificially increased through cloud seeding. This process only confirms the fact that the dynamic physical factors that intervene in the natural rain process are far from simple. The idea of modifying the weather is not new; man has often tried to do this because rain is directly linked to the social and economic development of the society.

The matter of weather modification raises many legal issues: there may be conflicts among users, both at the national and international level if it is not adequately controlled by law.

1.4.7 Climate change

Climate change entails long-term continuous change to the global earth climate or to regional climates. Such change may be due to processes inherent to our planet, to external influences, or to human activities typical of the modern era, i.e., of the last 150 years. In the latter case, we are in the presence of anthropogenic climate change, brought about by greenhouse gas emissions originating in human activities.

The United Nations Framework Convention on Climate Change (1992) employs this expression to designate climate change due to human activities (anthropogenic change) only, while it uses the expression 'climate variability' to designate climate change due to natural causes.

Nowadays nobody would deny the fact that human actions are susceptible to bring about climate change. The extent and effects of this change are the subject of heated debate, even among experts in this area. Depending on regions, the effects of climate change may range from the increase of extreme natural phenomena such as floods and droughts, to seasonal variations in rain patterns, and may lead to an acceleration of the hydrologic cycle.

Human actions relating to water, such as water abstraction, pumping, water diversions, drainage, irrigation, water supply, wastewater treatment and discharge have certainly contributed to climate change, whether directly or indirectly, either because of a higher energy demand for different purposes of water use, or because of a higher incidence of economic activities producing greenhouse gases, which allow greater productivity in the agricultural, industrial and tourism sectors.

On the other hand, water may serve the purpose of mitigating the effects of climate change. For instance, hydropower generation may provide clean energy, provided that the context in which it takes place is favourable and that certain conditions are met. By the same token, navigation may constitute an 'ecological' mode of transportation of merchandise whenever modern fleets are employed. By regulating or promoting certain practices, water law and administration may influence the anthropogenic causes of climate change. Water law may have an influence on water demand and water quality, promote efficiency in water use and water recycling. By balancing efficiency and equity, it may provide the flexibility needed to adapt to climate change, as well as security in the tenure of water rights.

Water law and administration may promote modern water resources management principles, such as integrated water resources management, stakeholder participation, the 'polluter-pays' and 'user-pays' principles, education and capacity building. As far as the relations among states are concerned, climate change challenges may only be addressed through enhanced international cooperation. International water law plays therefore a particularly important role.

1.4.8 Integrated water resources management (IWRM)

According to the Global Water Partnership (GWP), a non-governmental organization, integrated water resources management (IWRM) is 'a process which promotes the coordinated development and management of water, land and related resources in order to maximize economic and social welfare in an equitable manner without compromising the sustainability of vital ecosystems.' This notion, which was the subject of extensive discussions after the International Conference on Water and the Environment in Dublin, 1992, entails that all water resources, whether surface, underground or atmospheric, all water uses and water quality aspects are to be considered and managed together in an integrated manner, taking into consideration the relationship between these resources, other natural resources and related ecosystems. By implication, the ideal unit for IWRM is the river basin.

This holistic approach requires the rejection of the traditional fragmentation of laws and sector-by-sector, top-down, resource management approaches, and the adoption of comprehensive water legislation and supporting institutional frameworks.

1.5 The socio-economic context

The complexity of the management of water resources through law is shown by its relationship with almost all the sectors of human life. Water is used for domestic purposes, water supply in municipalities and rural areas, and for agricultural purposes. When formulating legislation, decisions are to be made as to whether small gardens should be treated in the same way as large irrigation areas, or if the watering of a few animals should be treated differently from the watering of a herd. Hydropower production, generation and distribution is *per se* a water use, which differs from the use of water for mining, agro-allied industries, cooling and washing. In some developing countries, water mills constitute an industrial use. Water is utilized for transportation, navigation and timber floating, uses which compete with one another. Water is utilized for medicinal and thermal purposes, recreational purposes, public bathing, swimming pools and fishing purposes. We now have uses for geothermal energy production. Legislation enacted in recent decades recognizes that water also serves an ecological purpose. Therefore, it requires the maintenance of a minimum flow in water bodies, or the constitution of a water 'reserve,' such as is the case in South Africa.

If not adequately controlled by law, the uses of water may cause pollution and consequently health hazards and waterborne diseases. The question arises as to whether the control of pollution should be the responsibility of the water administration or of the health authorities. Water pollution may also be brought about by certain land uses and other activities which fall under the control of different authorities. Thus, the issue of coordination with the water administration arises.

Generally speaking, there are many laws and regulations which deal with the sectorial aspects of water resources management. There are, as well, various administrations dealing with the same public functions. The above-mentioned examples show the complexity of water resources management; any decision on specific aspects has repercussions on the whole social context, which is why governments are often reluctant to take measures concerning them.

1.5.1 Water demand

The demand for water varies in proportion to population density and the price that one has to pay for it. When water prices are low, the quantity of water utilized will be correspondingly high. On the contrary, as the price of water increases, there will be, up to a certain point, a decrease in the quantity requested. The relationship between price and quantity is called 'price elasticity.'

Water demand is different from water requirement; the latter is fixed and immutable and does not depend upon the price or the quantity of water available. For example, the water requirement for plants relates directly to the amount of water that plants need to survive.

Water demand is affected by a number of socio-economic implications, such as the legal system, the institutional framework, customs and traditions, religious beliefs and economic and financial considerations prevailing in any one country. Climate is another factor affecting water demand: dry areas will require more water than humid ones. Finally, water demand is influenced by water quality; thus, water containing a high percentage of salt will not be of great domestic or agricultural use.

Water demand for irrigation purposes is affected by soil characteristics: the texture, density, structure, permeability, fertility, salinity, drainage effectiveness and topography of the soil.

Water is directly linked to the economy. When prognosticating, decisions must be made on the amount of capital and equipment to be allocated for the creation of new facilities.

Three factors may intervene to increase the amount of available water to meet future water demand. Firstly, there may be an improvement in the use of water through the elimination of water losses; secondly, the authorities may reallocate water; finally, new water projects may be constructed and new water sources discovered. New water projects should always be a factor for more equitable water distribution and for economic growth.

1.5.2 Water pricing

Water resources, like any other natural resource, have a price.

Water pricing is a political issue which requires the collaboration and consultation of the government departments involved in water resources management. It is obvious that the hypothetical water price is the one resulting from free competition on the market. Such hypothetical price reflects production costs and optimal water allocation. However, the real price of water is not explicitly determinable. While the cost of production of water may be high, the price the consumer pays for it is often very low. Thus, the price paid for drinking water is not directly connected to the production cost.

It is generally felt that, in the case of towns, the users are in a position to pay for water and electricity. In areas where irrigation is practiced, this is not always true. In some countries, irrigators can pay for water, but for newly irrigated areas in developing countries one cannot expect immediate payment for water. Thus, those who are in a position to pay for a particular use will do so, subsidizing the water price for other uses.

1.5.3 Externalities

When the activity of one economic unit directly affects the utility or the welfare of other economic units, economists will say that there is an 'externality.' To eliminate or 'internalize' this externality produced by the use or pollution of water, it is necessary to intervene in the price charged for water. Intrinsicly, this means that all the users must recognize the influence that they have on other users. For instance, if user A, being located upstream, creates problems to user B in the use of water or in the pollution of the river, the idea is that user A has to pay for or diminish the water use (or pollution) corresponding to the prejudice caused to user B.

Similar situations may arise in the use of groundwater. According to the economists, any activity creating an externality has to be considered a 'diseconomy.'

Various ways are available to respond to these situations: payment of a corresponding fee, rationing, or, in the case of industrial installations, the abstraction of waters downstream and the return of them upstream, so that the waters must be treated before their use.

This can be achieved through sound economic analyses and legislation which promotes flexibility and minimizes the damages that may occur.

1.5.4 Cost analysis

If one tries to allocate water in accordance with its maximum cost benefit, i.e., at its highest price, then water would have to be assigned to those users who would be able to pay the most. This system would only be adequate if all users had the same monetary capacity. It is obvious that domestic water allocation cannot be linked to users' incomes.

In the neo-classic approach, prices are connected to the forces of the market. In such context, offer and demand for goods do not depend upon political and social factors, but determine the price of goods directly. Thus, the market is an objective and natural mechanism where economic intervention is prohibited.

Water prices are based on many criteria, including social, religious, cultural, political, economic and legal elements. Therefore, the neo-classic economic analysis cannot be applied here. The law must take into consideration these factors.

1.6 The purpose of the book

The purpose of this book is to provide a tool for dealing with the legal and institutional aspects of water resources management to those who are called upon to carry out functions within the context of water resources administrations and to face the legal issues raised by water management.

These issues are generally encountered not by lawyers, but by water administrators, who are, of course, aware of the problems of the allocation of water for various uses, but usually do not know how to solve the issues, as these are of a legal nature. On the other hand, lawyers generally do not know exactly what water is. For them, it is something liquid flowing from the tap. Unfortunately, lawyers trained in the field of water resources are a rare commodity.

This book purports to be multidisciplinary in nature and considers general issues of water law and administration. In addition, it intends to indicate problems that may arise and guidelines for considering possible alternative solutions. Sometimes, only a brief description of or guidelines concerning particular topics will be offered, since an exhaustive treatment of the subject would require more than one volume.

References

Biswas, A.K. (2004) Integrated Water Resources Management: A Reassessment. *Water International* 29 (2): 8.

Burke, J.J. & Moench, M.H. (2000) *Groundwater and Society: Resources, Tensions and Opportunities*. Themes in Groundwater Management for the Twenty-First Century. New York, United Nations Publications ST/ESA/265.

Caponera, D.A. (1976) *Outline for the Preparation of a National Water Law Inventory*. Background Paper No. 7. Rome, FAO.

Duckstein, L. (1975) The Role of New Technologies for Improved Water Management and Related Effects on Water Law Systems. *Proceedings of the Conference on Global Water Law Systems*. Valencia.

Grays, L. & Nobe, C.K. (1975) Water Resources Economics, Externalities and Institutions. *Proceedings of the Conference on Global Water Law Systems*. Valencia.

Maas, A. et al. (1962) *Design of Water Resources Systems*. Cambridge, Mass., Harvard University Press

Matthews, O.P. (1984) *Water Resources Geography and Law*. Washington, D.C., Association of American Geographers.

Sewell, W.R.D. (1974) Water Resources Planning and Policy Making Challenges and Responses, Chapter 13. *Priorities in Water Management*, F.M. Laversedge, Western Geographical Series, Vol. 8, Dept. Of Geography, University of Victoria, Victoria, B.C.

World Resources Institute (1986) *World Resources 1986*. Washington, D.C., Basic Books.

Chapter 2

Earliest water regulations and management

2.1 The importance of water regulations throughout history

A quick perusal of history demonstrates the intimate connection between the economic and social development and the stability of a group of people, and the availability and reliability of water supply. This has rightly led many authors to define the first developed social groupings as hydraulic civilizations. Often, these civilizations are referred to by the name of the river valley around which they developed. Thus the Egyptian civilization is the civilization of the Nile, the Assyro-Babylonese or Mesopotamian is the Tigris and Euphrates civilization, the Hindu is the civilization of the Indus, the Chinese is the civilization of the Huang-Ho. Civilizations which developed during other and more recent historical periods and which had great social impact also grew around important water points; such is the case of the pre-Columbian, Peruvian and Meso-American civilizations around the coastal valleys of central and South America, the Khmer civilization around the Mekong River, the Elam and Suziana civilizations around the Dez River (in south-west Iran) and the Helmand civilization along the Helmand River between present Afghanistan and Iran. Arabic civilizations, which originated in desertic areas, also developed and expanded from oases to river valleys, i.e., from well-watered points to better-watered ones.

All major human migrations and the birth of towns and communities have been closely correlated with the search for and the settlement around naturally irrigated areas and valleys adequately supplied with water. Early transportation was facilitated by waterways with consequent benefits derived from trade and interexchanges.

Civilizations developed wherever river valleys were subject to recurring floods which, in turn, brought natural flood irrigation to areas into which the rivers carried, together with water, fertile alluvial soil suitable for agricultural development.

As soon as human groups settled around a water point or a river valley, the need arose for minimum water control in order to satisfy water demands and to ensure an equitable water distribution between different uses and users. It is from this need that the earliest water law systems developed. Their growth, persistence and character varied and were dependent upon many factors, such as local geo-physical and climatic conditions, socio-economic and managerial situations, and the religious-philosophical beliefs of the populations concerned.[1]

---

1 For a description of earliest hydraulic civilizations: Pan American Union (1955) *Irrigation Civilizations, a Comparative Study*. Washington D.C.; Steward, J. (1949) Cultural Casualty and Law: A Trial Formulation of

In regions where water was abundant, water control was largely directed towards defence against harmful effects of water such as flood warning and control and fight against water invasion, land reclamation, embankment and dyke construction and maintenance. In areas where water was scarce, this control developed towards the conservation of water supplies and adequate distribution of the little water available; water regulations were more detailed and restrictive. It must be pointed out that, in early water regulations, the religious character of water, either as a gift, a reward or as a punishment by nature, God or the gods, was always present, possibly with the exception of China.

Furthermore, the amount of labour required for successful water resources development and conservation activities was an important factor which determined and influenced the socio-economic organization and growth of the hydraulic civilizations.

It may also be said that the development and growth of early hydraulic civilizations were closely related to the degree of effectiveness of the administrative-managerial, religious and legal controls imposed on water use; on the other hand, diminished social concern over the management of water was one of the main causes of the subsequent decay, and sometimes disappearance, of early hydraulic civilizations.[2] A typical example is the disappearance of the Mesopotamian civilization as a consequence of diminished administrative control over the canals, which became burdened with silt. Likewise, throughout history, the intervention of external invaders with the destruction of waterworks and disruption of existing water management also made certain hydraulic civilizations disappear. Examples include the destruction by Tamerlane of the retention works on the Helmand River in the fourteenth century AD in the lower valley of present Afghanistan, which caused the Helmand valley community to disappear and the desertification of the area; the destruction of the canalizations in the Neghev, of the Roman aqueducts by the barbarians and of the Singhalese waterworks.

2.2 The difficulty of studying early water regulations

The sources for the study of ancient law in general are many; for the analysis of ancient water law in particular, three sources seem to be relevant. The first are the observations by past and present authors on the life and laws of ancient people; the second consists of the ancient texts and compilations commonly referred to as 'codes' and which are a continuance of the historical development of primitive law; the third is represented by legal or other relevant documents which have come down to us, especially those embodying legal transactions and administrative or managerial instructions.

However, this study is difficult, as it involves the interpretation and analysis of scattered passages of ancient texts. It is even more difficult to abstract basic principles from them. A thorough knowledge of the works in their entirety would be necessary, not only from translated, deciphered or transliterated texts, but also from direct sources. This task presupposes

---

Development of Early Civilizations. *American Anthropologist* 51, 1; Steward, J. (1953) Evolution and Progress. *Anthropology Today*. Chicago, University of Chicago Press; Wittfogel, K. (1957) *Oriental Despotism*. Yale University Press; Drower, M.S. (1956) Water Supply, Irrigation and Agriculture. *A History of Technology*, Vol. I. Oxford Clarendon Press.

2 For a description of this situation see: Caponera, D.A. (1978) Water Laws in Hydraulic Civilizations. *Festschrift für Karl Wittfogel*, Society and History. The Hague, Paris, New York, Mouton Publishers.

the knowledge of ancient languages, and those who know these languages are seldom lawyers. To these difficulties may be added the need to regard the legal and institutional aspects of water ownership, utilization and administration within a historical framework, which includes the political, religious and geo-hydrographical background of the period, in relation to the regions and civilizations within which they originated, developed and eventually disappeared. Therefore, a thorough and scientific legal analysis of ancient texts to ascertain or abstract irrefutable legal principles on water laws of early times would require the cooperation not only of lawyers but also of anthropologists, linguists, philosophers, theologians, historians, political scientists, sociologists, economists and water technicians, qualities very seldom found in one person.

On the other hand, the importance of considering the values of primitive water regulations and institutions within the overall framework of the society in which they developed cannot be overlooked. A general knowledge of the background of the historic and socio-economic situation of primitive hydraulic civilizations is therefore a prerequisite for undertaking these studies. Finally, in the eyes of a modern writer the administrative and legal concepts of ancient civilizations may appear somewhat different and have diverse meanings if considered in the light of our current institutions within which our scientific, technological and legal preparation have developed.

2.3 The development of earliest water law principles[3]

The very nature of water presented two precisely opposite challenges to the evolving early hydraulic civilizations: how to convey it to the land for irrigation where it was needed, and how to control it where it threatened men or was likely to cause damage. In the first instance, hydraulic civilizations developed where natural and recurring floods brought water and alluvial soil to the lands; at a later stage man-made aqueducts and irrigation canals provided for such need. As a defence against harmful effects of water, dykes, dams and artificial hydraulic structures were built. As regards the harmful effects of recurring floods, populations were obliged to pool their efforts and organize themselves. These two physical aspects of water brought about the first quest and development of water law principles.

The incalculable amount of labour necessary to ensure water control obliged a whole community to work for a common end, and, as a consequence of the union of efforts in water activities, a decisive step took place toward the elevation of a community to the level of a state. In fact, defence and divine worship constituted the combined action for this progression. Hydraulic civilizations necessitated an authority which planned the works, supervised their execution and brought them by coercion to successful completion. Such coercion must have taken place by means of water regulations under an appropriate water administrative authority. Government control over agriculture and commerce was subsequent to the construction of canals and navigation structures.

---

3 For the development of primitive law systems: Diamond, A.S. (1935) *Primitive Law*. London, Longmans, Green and Co.; Armijo, Nolde & Wolff (1938) *Recueil d'etudes sur les sources du droit*, 3 vol. Publ. en L'honneur de F. Geni; Maine, H.S. (1906) *Ancient Law* and *Early Law and Custom*. With regard to primitive water regulations and administration, Pan American Union, *op. cit.*; Von Ihering, R. (1897) *The Evolution of the Aryan*, translated from German by A. Drucker. London, Swan Sonnenschein and Co. Ltd.

At the earliest stage of development of law, the chief valuable was land, including water, cattle and slaves (for which water was needed), ships or boats (utilizing water) and the rest of movable items (food, household furniture, equipment), etc. However, unlike movables, land accompanied by the availability of water was, and is, the essential source of life in all phases of human society. Land is fixed in extent, and all members of the community must have access to it in appropriate quantities. This is how the rights-duties relationship of land and waters developed. Public interest of the community in land and water gave rise to their becoming public property. Individuals within hydraulic civilizations practiced a certain mode of conduct towards the use of communal land and water for drawing, for transit, hunting, fishing, etc., and this behaviour came to be considered the right of any member of the community. Such public rights varied according to the character of the land and water and the socio-economic structure of the community; in most places development gave rise to the notion that all lands and waters belonged to the community and that the rights of individuals or groups were either dependent or partial. In those communities where there was no land property, what vague notion existed with respect to land ownership probably went in the direction of communal property under various forms: tribal, property of groups intermediate between the tribe and the family, cases where land and water were vested in the chief, or a combination of all of these.

2.4 Ancient Egyptian water regulations and management[4]

2.4.1 Earliest dynasties (3400–2650 BC)

Very little is known of the water regulations of ancient Egypt, although Egyptian records relate that they existed. The most important codifications referred to are those of Horemheb at Karnak. None of these regulations seem to have survived.

However, reliable records describe in detail the hydraulic structures built and the water administration through different periods, from which it may be assumed that a whole body of water regulations must have existed.

In the pyramid texts, the god Osiris is identified with the Nile waters. According to tradition, Menes, the legendary first of the Pharaohs, dammed the Nile to control floods. As early as the period of the first two dynasties (about 3400–2980 BC) there was a 'Constructor of the King' (*Medeh Nisout*) providing for public works. From that time the Nile was measured and recorded on the Palerm Stone.[5] During the third dynasty (2980–2900 BC) the water administration under the Constructor of the King developed into two departments, at the head of which were a Master of Canals and a Master of Lakes. While no changes occurred during the fourth dynasty (2900–2750 BC), under the fifth dynasty (2750–2625) a 'Master of Water Castles' was added to the water administration, and the state organization developed even further.

The civil status of the population, the cadastre of lands, the registrar of all deeds, the water administration and the public works department were coordinated by the Central

---

4 On primitive Egyptian laws and institutions: Pirenne, J. (1952) *Archives d'histoire de droit oriental*, Vol. 1–4; Pirenne, J., *Histoire des institutions anciennes d'Egypte*, Vol. 1–3; Revillut, E. (1903) *Précis de droit Egyptien comparé aux autres droits de l'antiquité*, Vol. I. Paris, Giard et Brière editeurs.
5 Borchardt, L. (1906) Nielmesser und Niels marken. *Preuss. Akad. Wiss. phil. hist. Abh. nicht zur Akadgehör*, Gelehrter, No. 1, Berlin.

Chancery. Branch offices of these services were scattered throughout the country, and officers recorded everything on papyrus conserved in the state archives. Through this organization and from the records, the state could assess and collect taxes through its revenue services. Under the fifth and sixth dynasties a larger degree of decentralization of services developed, together with a diminished concern for waterworks construction and maintenance. It is during this last period that the civilization decayed; for a while the deltaic towns became small autonomous states governed by an oligarchy.

2.4.2 Later dynasties (2650–300 BC)

During the twelfth dynasty, the centralization of water administration was renewed. The district governors of the Old Kingdom had as their chief title 'Digger of Canals' and were responsible for the upkeep of canals and dykes, for patrolling and inspecting the banks when water reached its height and for organizing aid when disaster threatened. In the case of floods a state of emergency was declared, and obligatory corvees were organized to fight against water invasion. A larger network for measuring the height of the Nile (in Nilometers) developed, the records of which were used to compare with past records and to forecast possible floods downstream. The deterioration of dykes and canals was regarded as an offence punishable by death, and this principle continued to be maintained up to and after the Roman period.

In ancient Egypt land and water belonged to the pharaoh who, as a living god on earth, granted its temporary use practically to whom he liked. Every community had to provide the pharaoh with the produce of the soil through its *Xerp* or public officer in charge of a district. Waterworks were carried out by groups of 1,000, 100, 10 or 5 men organized under a farm leader, in working groups to whom a plot of land was granted for cultivation. The income deriving from cultivated lands was controlled and assessed for taxation by public treasury officers who were themselves secretly controlled by the pharaoh. Although the scribes recorded everything, it is unfortunate that, so far, no written water regulations have come to light, probably, as some authors have suggested, because the pharaoh needed no law other than his own.[6]

2.5 Ancient Mesopotamian water regulations and management[7]

2.5.1 Historical development

The ancient Mesopotamian civilization was basically a fluvial civilization. The rivers of Mesopotamia, the Tigris and the Euphrates, provided water for irrigation and at the same time were means of transport. The distribution of water was one of the main concerns of the

---

6 Wilson, J.A. (1951) *The Burden of Egypt*, Chicago, Univ. of Chicago Press, 41–50.
7 On primitive Mesopotamian laws and institutions: Diamond, A.S., *op. cit.*; Woolley, C.L. (1929) *The Sumerians*, Oxford Clarendon Press; Von Ihering, *op. cit.*, 149–161; Driver, G.R. & J.C. Miles (1935) *The Assyrian Laws*. Oxford Clarendon Press; Driver, G.R. & J.C. Miles (1952) *The Babylonian Laws*. Oxford Clarendon Press; King, L.W. (1898–1900) *The Letters and Inscriptions of Hammurabi*, Vol. 2–4, London, Luzac; Adams, M. R., *Development Stages in Ancient Mesopotamia*, in Pan American Union, *op. cit.*; Cruveilhier, P. (1938) *Commentaire du Code d'Hammourabi*, Paris, Librairie E. Leroux.

governors during the time of the first Babylonian dynasty, in which many important legal and administrative documents were elaborated as sources of the Acadian-Sumerian law.

A series of records in the form of inscriptions are available concerning water regulations in the Babylonian, Assyrian, Hittite and related civilizations, and new ones are being discovered even today.

Among the legal documents available, the Hammurabi Code deserves special attention. Also, a wide range of contracts, sentences, administrative correspondence and private mail are to be taken into consideration as written sources of law. The series *ana ittisu*, meant for teaching the Acadian scribes the meaning of Sumerian terms, constitutes an authentic *de verborum significatione*, the value of which is incalculable.

The Sumerians worshipped the god of inundation, and the Sumerian King Gudea (2492 BC) is reported to have 'constructed a new canal and maintained in an excellent state of repair the irrigation system of smaller canals.' The Babylonian god Nun personified the idea that water is the source of life, of all blessings and the element of creation. During this period a codified law was developed under Shulgi, the second king of the third dynasty, laying down the basis of the Hammurabi Code. This code and the royal letters to local governors giving instructions for the upkeep of waterworks and canals constituted the basis for most of the water legislation in Mesopotamia, even when issued subsequently under new forms of codification.

2.5.2 The Hammurabi Code[8]

The prologue of the Hammurabi Code contains 26 reasons for praising the king, of which half except one refer to his activity as water ordainer.

In the introduction of this code, King Hammurabi describes himself as 'the gracious arbiter who has allotted . . . watering places to Lagash and Girsu . . .' and, 'the giver of the water abundance to . . . drink.'

Water is considered as possessing a divine character, and as means of punishment for various offences. Articles 53 to 56 are those dealing specifically with water control. Thus, Article 53 states that if a man has been slack in maintaining the bank of his fields and has not strengthened them and a breach has occurred whereby the waters carry away the harvest on the land, the man in whose land such breach has occurred shall replace the corn which he has caused to be lost. If he is not able to replace the corn, he and his goods shall be sold and the tenants of the water-land whose sesame the waters have carried away shall divide the proceeds from the sale (Article 54). A similar punishment is provided for in Article 55 in the case in which 'a man has opened his trench for irrigation and has been slack and has let the waters carry away the soil on his neighbour's land, he shall pay corn corresponding to his neighbour's loss.' Article 56 provides that if a man has released the waters and in so doing has caused the waters to carry away the works on his neighbour's field, he shall pay 10 *gur* of corn for every *bur* of land.

Finally, the code ends with a supplication to various divinities to punish anyone who would not comply with the regulations set forth. King Hammurabi exclaims: 'May Ea, the

---

8 The quotations of the Hammurabi Code are taken from Driver, S.R. & Miles, J.C. *The Babylonian Laws*; *op. cit.*, Vol. II, 11, 31, 33, 103–105.

great prince ... stop his river at the source, and cause bread-corn, the life of the people, fail to grow on his soil ...' and, '... may Ada, the lord of ever-flowing wealth, the controller of heaven and earth, my helper, deprive him of the rains from heaven and the flood water from the source, may he bring his land to ruin by famine and hunger, may the thunder in rage against his city turn his land into a heap left by the flood ...' The code ends up stating: 'I made it possible for my people to recline on well irrigated fields.'

2.5.3 Subsequent codifications

After the Hammurabi Code, some of the codifications of later civilizations in the area also contain similar provisions on water control.

Article 3 of the New-Babylonian Laws enacted about one thousand years later state that a man who has dug his cistern for irrigation but has not strengthened it, and by not doing so causes a flood in his neighbour's land, shall pay in flour the damage caused by him. This article may be compared with Articles 53–56 of the Hammurabi Code.

The Assyrian fragments containing legal provisions in our possession are not a complete code; they date from the period of the Middle Assyrian Empire (cir. 1500 BC). One tablet deals with land law, Clauses 7 to 15 specifically with wrongs committed against a person with respect to his land.

The Hittite laws,[9] dating back to the thirteenth century BC, also contain important articles relating to water. Special provisions exist relating to those who irrigated their land (Article 48), and while Articles 146–147 provide for an indemnity to be paid by the owner of the well to those injured while digging the well, Article 162 contemplates the payment of fines by those diverting waters unlawfully.

In the same region, the inscription of a *kudurru*[10] of Nebuchednezzar I (fifth century BC) contains a malediction for those having let mud fill the canals. This provision implies lesser concern for the upkeep of canals and other waterworks, resulting in the decay of the ancient Mesopotamian hydraulic civilization.

The remains of the impressive hydraulic works carried out in the Mesopotamian region indicate that specialized water engineers and a strong water adminstration must have existed. This is the case of the Nahrwan Canal (400 feet wide and 200 miles long), deriving water from the Tigris to supply Kalach (ancient Nimrod) built between 880–850 BC under King Assournazirpal; the Sennacherib aqueduct, which conveyed waters to Ninive from Kurdistan in the seventh century BC. Hammurabi's letters to local governors show that each district was responsible for the upkeep of its own waterworks; thus, to the governor of Larsa he writes: 'Summon the people who hold fields on this side of the Damanu Canal that they may scour it ... within this present month and let them finish it.' In another letter he complains that the bed of a canal has been imperfectly cleared so that boats cannot enter the city of Erech; the governor must have the necessary work done within three days ....[11]

---

9 Hrozny, F. (1922) *Code Hittite provenant de l'Asie Mineure*, 1ère partie. Paris, Genthner; and in German by Zimmerman & Friedrich (1922) *Hethitische Gesetze in der Alte Orient* 23, 2; 1925, 24, 3.
10 The *kudurru* were stone sculptures used as boundary markers in ancient Mesopotamia.
11 King, *op. cit.*, No. 71 and 5.

In the earliest periods, the direction of water administrations to construct, maintain and distribute waters for irrigation seems to have been under the authority of priests, thus implying a theocratic state. Subsequently a dynastic authority developed which gradually took the place of the priesthood.

The Mesopotamian state was born and grew as a consequence of organizing water distribution: '... the State was in the water and came forth out of it, no less than the earth itself according to the cosmology of the people.'[12]

More recent inscriptions found in the area refer to water rights owned and administered by bankers, and to sanctions for the owners of water rights for selling, renting or otherwise alienating them. It would appear that private water rights and localized water administrations came about in later periods and substituted previous communal rights over water and consequently the centralized water bureaucratic organization of earlier periods.

2.5.4 Detailed water regulations

In Mesopotamian thought the concept of private dominion of water did not exist. This determined restrictions to the use of land by private owners, as the water to be found thereunder was to be supplied to the communities.

The rules governing the use of water and maintenance works, both ordinary and extraordinary, were established by the administrative authority. As regards maintenance works, these were the responsibility of contiguous landowners, who were granted, as compensation, a monopoly on fishing in their part of channel.

The water administration was structured hierarchically. At the top of the hierarchy was the king, whose main responsibility was to take charge of the water administration by applying customary rules or, when these did not apply, by turning to analogy. Decentralized authorities were the circumscription chief or first civil authority and the local council, of which the mayor and the 'rector of the fields' were members. In Mari,[13] probably due to the strategic importance of the zone, the water authority was vested in the military chief.

As regards the harmful effects of water, these were attributed to the rain god (Adad) when caused by a major force. In the case of lease of land, floods were deemed not to alter the terms of the contract at all. The tenant was to bear the loss integrally when this occurred after the rent had been paid, while in the case the rent had not yet been paid or in the case of partnership, the owner of the cultivated field was to suffer the damage proportionally. The owner or tenant of the flooded field who had applied for a loan was granted a one-year moratorium during which he was exempted from paying interest. The same principle applied for droughts.

Harmful effects of water originating from human activities gave rise to an obligation to compensate any damages caused. So, the person who by watering his field caused, due to his negligence, the flooding of the land of his neighbour was under an obligation to pay for the damage caused. If he had acted deceitfully, he was condemned to pay a fine proportionate to the extension of the flooded surface, which was sometimes much higher in amount than the value of the crops.

---

12 Von Ihering, *op. cit.*, 161.
13 Ancient city in Syria, on the western bank of the Euphrates River.

Babylon law favoured the walling-in of fields. However, if the construction caused the neighbouring field to be inundated, its author was made responsible for compensating the damage, and if he did not take any action theron, his field was sold and the proceeds from the sale assigned to the owner of the neighbouring property.

Water-related violations were severely punished. Among the sixteen different forms of robbery contemplated by the Hammurabi Code, the theft of a *horia*, a bucket intended for collecting water from a well, was considered as a major crime.

On the other hand, it is worth noting that water was used as a means of punishment. In fact, alleged witches and women condemned for adultery were executed by immersion. The same fate was reserved to the tavern-keeper's wife who adulterated the wine she sold.

2.6 Ancient Hindu water regulations and management[14]

2.6.1 The Hindu legal system

The hydraulic civilization of the Mohan-Jo-Daro (about 5000 BC)[15] and the Indus Valley civilization (about 3000 BC) have many features similar to those of the Nile and Tigris-Euphrates civilizations: recurring annual floods, a strong bureaucratic water administration, a large number of waterworks and divine character of the waters.

A glance at some of the elements of the Hindu doctrine will help us understand the legal system established by this civilization, as it was deeply influenced thereby. Hinduism, or Brahmanism, is considered by its followers neither as a philosophy nor as a religion, but as a living tradition expressing one form of the Universal Truth. The Hindu doctrine is that of the Supreme Knowledge, of which all that is perceptible, materially or otherwise, is a component.

Man is considered both as a whole made of parts, and as a member of the community. Similarly, communities are seen as necessary components of Humanity, which, in the same way as minerals, plants and animals, is an element of the World, itself considered as a microcosm in relation to the Universe, or Cosmos, regarded as the substantial and formal expression of Truth. The Hindu, whatever his race, caste or personal degree of perception of the Supreme Knowledge, is well aware of his effective participation therein, together with his environment, as a necessary element of Universal Harmony.

This concept of Universal Harmony implies a necessary correspondence between the cosmic and human orders. As the Cosmos is represented as *Manu* or the prototype of Man (*manava*), such correspondence is necessarily present within the structure of every human community, all social groups having organs and functions similar to those of the human body. Harmony is achieved only when man's action, or *Karma*, corresponds to the functions

---

14 On primitive Hindu law and institutions: Bandayans, *Sacred Laws of the Aryans*, N.Y., 1987, trans. by G. Buhler; Visnu, *The Institutes of Vishnu*, Dr. Julius Jolly, New York, 1900; Mayne, J.D., *Hindu Law and Usage*, 8th Ed., 1914; Muller, M.F. (1886) *The Sacred Books of the East*, trans. by G. Buhler, Oxford Clarendon Press; Diamond, *op. cit*. On non-legal aspects of Hindu civilization, Piggot, S. (1950) *Prehistoric India*. Penguin Books. For a more recent analysis of some principles of Hindu law relating to water management see: Wohlwend, B.J. (1975) Hindu Water Law and Administration in Bali. *Proceedings of the Conference on Global Water Law Systems*, Valencia, K-1.

15 Also known as the Sindhus valley civilization, stretching, at its height, from present day Kashmir to Kutch.

assigned to him by his very nature. The same applies to communities, to Humanity, and to the total Order of the Universe.

To the integral harmony resulting from conformity with the above-described hierarchy corresponds, free of any moral connotations, the *Dharma*, or eminent principle of Order, or 'Law' around which the World accomplishes its revolution. In this sense, it constitutes the *Sanatana Dharma*, or Primordial Tradition and is regarded both as the fundamental Principle of the Hindu doctrine, or *Brahma*, the Divine Will and Supreme Master of the Universe, and as the substance of the doctrine as a whole.

As Principle, the *Dharma* governs all levels of the cosmic and human order, and embodies the necessary conformity of all elements of the Universe with their very nature and purpose. In this sense, it is possible to envisage a *Dharma* specific to each being (*swadharma*), to each collectivity or to the whole humanity, for the full duration of its cycle of existence. In this case, the *Dharma* appears as the specific Law or Norm of each particular cycle, as formulated from its origin by the *Manu* which governs it, i.e., by the Cosmic Intelligence reflecting in this world the divine Will expressed as the Universal Order.

As substance of the doctrine, the *Dharma* constitutes the content of the Sacred Books as a whole; in a restricted sense, however, it also expresses the legislative aspects of the doctrine as applicable to the social order.

The traditional Hindu scriptures constitute the *Veda*. The Sacred Books are however of two kinds: the *Shruti*, which emanate directly from the divine revelation, and the *Smriti* containing the collection of the traditional Hindu commentaries.

The *Shruti* contains the four *Vedas*, or Fundamental Books, and the six *Darshanas*, or 'points of view' (branches) of the doctrine. The last *Darshana*, entitled *Vedanta* (end and ultimate purpose of the *Veda*), contains the precepts of the *Sanatana Dharma* (*Lex Perennis*), as synthetized in the *Brahma-sutras*. The penultimate *Darshana*, known as *Mimansa* or *Karma/Mimansa*, refers to the field of action (*Karma*), and represents one of the sources from which the principles of Hindu jurisprudence were developed.

The *Smitri* contains, *inter alia*, various treaties on the six fundamental Sciences and four Practical Sciences or Arts, the Shivaite and the Vishnuite commentaries or books, and the *Dharma-Shastra*, or codes of law, which represent the Law of Manu as applicable to the social order.

2.6.2 The water regulations in the Code of Manu

The most important documents relating to water law are those contained in the *Manava-Dharma-Shastra* or Code of Manu, the ancient origin of which has been subsequently recorded in later texts.

For the Hindu, the Laws of Manu represent the expression of the divine will or principle of universal order, as the necessary norm of correspondence between the essential nature of man (*Dharma*) and his deeds (*Karma*).

According to *Rib-Veda* belief, the Goddess Sarasvati gave birth to the rivers; she was the peacock-riding wife of the four-headed Brahma, creator of all things.

The Code of Manu, after Brahamism arose, was set within a body of rules prescribing the various duties of the king, who was also law-giver and judge. The mythical Manu was besought by the sages to tell the sacred laws. He informed them how he was born from Brahma, how the world was created, and how he received the Laws from Brahma and communicated their content to the Ten Sages. He asked Bhrigu, one of the Ten, to repeat it to the other nine, after which the content of the code follows, recited by Bhrigu.

Chapter III, Sect. 151 of the code[16] states, 'Let him not entertain at a *sradda* (dinner) ... he who diverts water courses and he who delights in obstructing them ...' In Chapter IV, Article 226 reads, '... a rich man must always without avoiding it and with faith, do charitable works as constructing a reservoir or a well or building a public fountain ...,' and Article 229: 'He who gives water obtains satisfaction.' This provision is similar to the *Waqf* or religious endowment developed in Islamic law.

In these regulations is found the concept of public waters: '... water' along with other things 'they declare to be indivisible.' (Chapter IX, Sect. 219). In the same order of idea we find that kings used to collect rights of way for crossing rivers (Chapter VIII, Sect. 404) and that the law imposed upon the king special obligations with respect to public water such as to 'organize vigilance and guards, both stationary and patrolling, and spies ...' on waters and on 'houses where water is distributed' (Chapter IX, Sect. 264–266).

The emphasis of economic and social concern placed on water in this code may be seen by checking the rigorous punitive system by which the common utilization of public waters was ensured. We find thus a moral sanction to consider 'as a matter for exclusion from the society of the good people the sale of a consecrated reservoir' (Chapter VIII, Sects. 61 and 69); then, the obligation to punish with death 'he who breaks the dam of a reservoir and causes loss of water by drowning him in the water or have his head cut off.' The offender 'may repair the damage but he shall have to pay the highest fine' (Chapter IX, Sect. 279). But he 'who shall take away the water, must be made to pay the first (or lowest) fine' (Chapter IX, Sect. 281). Furthermore, 'the fine of a gold *Masha* and the obligation to restore everything to its original state' shall be imposed upon anyone who 'destroys a public fountain' (Chapter VIII, Sect. 309).

Finally, the unlawful appropriation of the water of a well or cistern shall be punished by a 'lunar penance,' that is, a minor one (Chapter XI, Sect. 164).

Waters were also considered an element of purification, as well as a means to ascertain the culpability of a person for certain crimes. Special ceremonials had to be performed in this latter case.[17]

The control over water utilization and distribution was under the responsibility of a powerful water administration headed by a water superintendent; this high official was vested with full and undisputed powers on all questions relating to water.

2.7 Ancient Chinese water regulations and management[18]

2.7.1 Introduction

The water regulations of the Chinese hydraulic civilization evolved together with the particular legal, administrative, philosophic and political framework of China.

---

16 The quotations of the Code of Manu are taken from Muller, M.F., The Laws of Manu, in *The Sacred Books of the East; op. cit.*, Vol. XXV, 103, 379.
17 Rules regarding the Ordeal by Water, in the *Institutes of Vishnu*, Muller, M.F., *op. cit.*, Vol. VII, 59.
18 For a detailed study on the historical development of Chinese water laws see: Caponera, D.A. (1960) Water Law Principles in the Chinese Legal System. *Indian Journal of International Law*. Delhi.

Ancient Chinese legal thought was based on the belief in a close interconnection between the human social order and the natural cosmic order. This concept was particularly relevant with respect to water ownership, distribution, utilization and administration. Harmony and unity, prevailing throughout creation, were believed to be reflected in human behaviour towards oneself, the neighbours, the family, the group, the society and the emperor. As a consequence, the foundation of social order was not necessarily the law (*fa*), but the *li*, a term variously translated as rules of propriety, rituals or customs, prescribing to everyone a behaviour in harmony with the natural order. These are the views of the Confucian theory, out of which Chinese legal thought developed.

Around 200 BC, these doctrines, based on moral and ethical influences as opposed to law enforcement, were for a time countered by the so-called school of Legalists which held that the law (written and known) to which everyone was subject, governed the relationships between human beings.

It was the combination of Confucianist and Legalist theories or, as it has been said, the Confucianization of the law (*fa*) that gave birth to the Chinese legal system.

2.7.2 Water regulations

The development of water regulations closely followed the evolution of this legal history and philosophy.

The first reliable records on Chinese water law are to be found in the *Li-Chi*, or treatise on ceremonial rules, in which interesting and suggestive passages refer to the behaviour of the ruler in the administration and regulation of water resources, in harmony with the behaviour of nature and recurring seasons. As an example we may quote, '. . . In spring, all life starts and rains of heaven fall on earth, and therefore, let the waters run and irrigate the fields . . . ;' in the summer months, 'build dams and dykes and store the waters for later consumption . . . ;' in the winter months, life ceases and therefore hardship arrives '. . . let inspection of works and collection of water rates and taxes be undertaken . . . punish offenders,' etc.

From these regulations it appears that no private ownership of water existed, and the government administration was responsible for the construction, repair and maintenance of hydraulic works, including bridges, navigation and fishing. Local officials derived their authority from the will of the emperor and could issue orders and regulations in compliance with the behaviour of natural order. Labour was provided by the people concerned under a system of corvees as part of their yearly obligations towards the state.

Under Shih-Huang-Ti, the first emperor of the Ch'in Dynasty (249–207 BC), the influence of the Legalists was at its peak; the centralized administration enacted a uniform and publicized system of codification.

The subsequent Han Dynasty (200 BC-618 AD) codified the previous Ch'in system of laws, but again introduced the Confucian *li* (rules of propriety, custom) in the *fa* (law-punishment), or, as it has been said, the humanization of law occurred. An ordinance on waters was enacted in 111 BC, and a new principle of 'water equalization' appeared in this text. Land and water resources were under the control of a director of agriculture, and a court architect was responsible for water control. 'The mountains and the seas, including lakes, streams, rivers, swamps . . . are storehouses of heaven on earth . . . and . . . their control ought to belong to the office of Shu-Fu . . .' (secretary of the treasury).

2.7.3 Basic principles

It is interesting to note that the basic principles of water ownership, use and administration as stated in detail in the original Li-Chi and pre-Han periods, reproducing the earliest concepts of water law, were incorporated into the water ordinances of subsequent major codifications of the T'ang, (618–1644 AD), the Manchu or Ch'ing (1644–1911 AD) dynasties and, with obvious modifications for improved and enlarged water administration, up to modern times.

Under Chinese water law principles the concept of private water ownership never appeared; the individual's duties in water matters would eventually lead to and enhance public welfare. Under the 'water equalization principle,' the upper riparians could utilize the water but without monopolizing the resource. The water administration was always in charge of all water activities, including navigation, flood control, floating, waterworks construction and upkeep, bridge control and water policing, while individual irrigation and fishing could be practiced, provided they fit into a collective pattern. The obligations with respect to opening and closing ditches and canals, cleaning and upkeep of waterways, and 'turns of duty conscription,' or corvees, had to follow the natural seasonal variations and requirements according to the natural order. Punishment was provided for in the case of offences under the water codes, ordinances and regulations.

2.8 Hebrew water regulations and management[19]

2.8.1 Introduction

The Hebrews, once comprising a loose federation of tribes of mixed origin, have left us an invaluable code of early law, partly influenced later (400 BC) by contact with Babylon. The basic principles of this code are contained in the Bible, which today serves both Jews and Christians. But in the matter of water law the Christians have adapted themselves and accepted first the rules enacted by the Romans and subsequently those of the countries where they live. The Jews evolved their own legal doctrines and opinions from the Bible. These are contained in the Talmud, a consolidated and unified digest of Jewish law, written between the fifth and third centuries BC. With respect to water regulation the Talmud includes a few references which are the development of the basic principles contained in the Torah, or revelation of Biblical texts.

2.8.2 The importance of water

The attribution to waters of an economic value, of stability, exchange and prosperity stems from the context of the following passages of the Bible: 'And ye shall serve the Lord thy God and He shall bless . . . thy water.'[20] The fact that the habits, customs and social organization of men have been influenced more by their association with water than with land is indicated by these passages: 'For the Lord thy God bringeth the into a good land, a land

---

19 On Hebrew law: Diamond, A.S., *op. cit.*; Hirsh, A. (1957) *International Rivers in the Middle East*, Thesis, Columbia University, containing many references on municipal water legislation. Most of the information provided in this section is abstracted from Hirsh's thesis.
20 *Book of Exodus* XXIII, 25.

of brooks of water, of fountains and depths that spring out of valleys and hills . . . ;[21] 'I did know thee in the wilderness, in the land of great drought . . .'[22]

Water is a divine gift for the subsistence of human, animal and plant life: 'He sendeth the springs into the valleys, which run among the hills. They give drink to every beast of the field: the wild asses quench their thirst. By them shall the fowls of the heavens have their habitation, which sing among the branches. He watered the hills from his chambers . . . He causeth the grass to grow for the cattle and the herb for the service of men . . .'[23]

2.8.3 Water law principles

As a consequence of the benefits that water provides to all living creatures, water resources seem to be excluded from private ownership. Thus, under Talmudic law: 'Rivers and streams forming springs, these belong to every man.'[24] The existence of wells belonging to the public domain and the right for every traveller to use them is also recognized.[25]

Servitudes and rights of way limited the use of water on privately owned lands, and the principle of compensation for the use of water also existed: 'And the children of Israel said unto Him, 'We will go by the highway, and if I and my cattle drink of thy water, then I will pay for it only, without doing anything else, go through on my feet'.'[26]

Restrictions on land ownership rights also appearing in Talmudic law prohibited riparian landowners from planting anything in a strip four cubits wide along an irrigation canal, and along the two banks of a navigable waterway for the purpose of both protecting the waterworks and facilitating water transport and navigation; in this manner, 'The haulers of barges can work along shore leaning inlandward as they pull their barges, unhampered by trees or vegetation;' public interest overcame private ones as '. . . the barge haulers may freely cut down any plant growing within the four cubit passageway, without warning the owner.' It is possible that such rules for protected areas also applied to wells or other water points.

Water utilized for domestic and irrigation purposes was subject to a certain order of preference. In the case of several irrigators receiving water from a common well, the one closest to the well conduit filled his cistern first, and the other irrigators did so in descending order. In the case of irrigation water coming from a stream, including the right of upstream riparians to divert the flow, Talmudic law does not seem to provide a definite principle. One authority gives priority to downstream users, another to upstream riparians.[27] Maimonides, a famous Jewish author, rules with the latter but adds that 'the stronger one's right is superior'

---

21 *Book of Deuteronomy* VIII, 7.
22 *Book of Hosea* XIII, 5.
23 *Book of Psalms* 104, 10–14.
24 *Talmud Bavli Shabbat*, 121b; *Beitza*, 39a; *Eiruvin*, 46a and 48a; *Tosephta Baba Qama*, 6, 15; cf. Maimonides, *Mishne Torah*; *Hilkhot Gezeila Vaaveida* I, 13; *Hilkhot Yom Tov*, 5, 14; *Orakh Hayim*, 397, 15; also *Karo, Shulhan Arukh, Orakh Hayim*, 397, 15; all quoted in Hirsh, *op. cit.*
25 *Talmud Bavli, Beitza*, 39a, and Rashi's commentaries there; *Talmud Bavli Nedarim*, 48a; *Tosephta Baba Qama*, 6, 15, and commentary *Magen Avraham*; all quoted in Hirsh, *op. cit.*, 155.
26 *Book of Numbers* XX, 19.
27 Hirsh, *op. cit.*, 1. 153; *Talmud Bavli, Gittin*, 60b.

in any dispute between riparians.[28] In another instance, a distinction is made between the law in Palestine, where the upper riparian has priority, and the laws in Babylon, which consider as the criterion the ease with which the respective owner may use the water.[29]

As for the use of a spring owned and utilized by a town for domestic purposes such as drinking, watering of animals, laundering, etc., the order of priorities is established as follows: 'A spring owned by the people of the city, their lives and the lives of others: their lives take precedence over those of others; their beasts and the beasts of others: their beasts take precedence over the beasts of others; their laundering and the laundering of others: their laundering takes precedence over the laundering of others; the lives of others and laundering: the lives of others takes precedence over their laundering.'[30]

For the maintenance of a well or of an irrigation ditch, all riparians shall assist those closer to the source of water; thus 'The most downstream owner repairs with all other riparians but repairs his own portion by himself.' In the case of drainage operations it is the most downstream riparian who benefits from the help of all; the most upstream riparian, being the sole beneficiary of the most upstream segment of the drainage conduit is alone responsible for its maintenance.[31]

2.8.4 Conclusion

It is interesting to note that many legal principles set forth in Talmudic law are similar to those later developed under Moslem water law, particularly with regard to the order of priorities in the uses of water, to the maintenance of waterworks and their protected area. This is understandable, as these legal obligations reflect the needs which arose in a similar climatic environment.

The early Jews, in their legal relationship with other Jews and with non-Jewish cultivators, were presumably subject to local water laws and customs, which on the other hand did not differ much from the principles contained in the Talmudic law. However, incapacities were sometimes imposed upon them by local rulers. Under the Sassanids they were barred from holding the office of waterworks warden;[32] according to one rabbi, however, this officer was a heavenly appointed official.[33] On the other hand the importance of such function is evidenced by the fact that Maimonides would appoint only Jews to this capacity in any restored Jewish theocracy.[34]

Talmudic law unified and consolidated Jewish doctrines and eliminated organized 'heresies.' It constituted primary law, encompassing the whole life, written 'as a fence around the Torah.' Although Jewish agriculture diminished as time went by, Talmudic doctrines influenced subsequent Moslem legislation and, after the eleventh century, also European law. It served as a guide to Jews and some non-Jews up to the industrial revolution.

---

28 Moshe B. Maimon (1135–1204), *Compendium of Jewish Law*, 5, 10.
29 Hirsh, *op. cit.*, 153; *Talmud Yerushalmi*, Gittin, 5, 9.
30 *Ibid.*, 154; *Talmud Bavli, Hedarim*, 80b; *Talmud Yerushalmi, Hedarim*, 11, 1.
31 *Talmud Bavli, Baba Metzia*, 108a; cf. Maimonides, *Mishne Torah, Hilkhot Shekheinim*, 3, 9, quoted by Hirsh, *op. cit.*, 154.
32 *Talmud Bavli, Ta'avit*, 20a, quoted by Hirsh, *op. cit.*, 154.
33 *Talmud Bavli, Baba Bathra*, 91b; Hirsh, *op. cit.*
34 *Mishne Torah, Hilkhot Melakhim*, 1, 4.

2.9 Pre-Columbian water regulations and management[35]

Although no written records of water regulation have come down to us from the civilizations existing in the Americas before the arrival of Columbus, the network of irrigation canals and other waterwork structures show that flourishing hydraulic civilizations existed.

2.9.1 Coastal Peru (Inca)[36]

On the southern coast of Peru, agricultural practices started around 1500 BC at Huaca Prieta, Cerro Prieto, and between 1200–800 BC flood irrigation seems to have been practiced under the leadership of priests at Cupisnique, Guanape, Nepena, Casma and other places. From 800 BC to 200 AD agriculture, based on canal irrigation, fully developed. The expansion of irrigation systems created the need for centralized water administration, always under the leadership of priests. From 200 to 1000 AD trans-valley irrigation systems and intensive agriculture took place. As a consequence of enormous population growth, the Mochica civilization expanded into a multi-valley state. From 100 to 1532 AD the organization of a new kind of society under military rule took place at Mochica, Maranga, Galinazo and Paracas.

The Inca conquest of the coast left the old administrative system managed through the local hereditary nobility largely undisturbed. Land and water were under state control; taxation was in the form of labour on state agricultural lands, public works, army and service to the emperor and nobility. Water engineers and other specialists were government civil servants.

2.9.2 Meso-America (Maya-Aztec)

In pre-Columbian Meso-America, extensive irrigation took place in Tecomaltepec, the practices of which have survived up to modern times. Water was diverted through a dam and a large network of canals from the Calderon River. Maintenance was continuous. Only those who had contributed to the construction of waterworks could benefit from irrigation water. Special arrangements were made for water allocation among the users whose lands were crossed by canals and those using the Calderon River directly.

A centralized water administration existed in Tecomaltepec, responsible for allocating irrigation water, upkeep of waterworks, and the organization of water distribution on a rotation basis. The system of sanctions for water offences ranged from the temporary suspension of water supply for a specified period to complete prohibition from using water. More than

---

35 On pre-Columbian civilizations and law in general: Central Andes (Peru): Wendell, B. (1946) The Archeology of the Central Andes. *Handbook of South American Indios*, Vol. 2. Washington DC, Bureau of American Ethnology; Same author (1948) The Peruvian Co-tradition. *American Antiquity*; Bird, J. (1948) Pre-ceramic Cultures of Chicana and Viru. *American Antiquity*; Willey Gordon (1951) *Peruvian Settlement and Socio-Economic Patterns, The Civilizations of Ancient America*. Chicago, Univ. of Chicago Press; Meso-America (Mexico): Palerm, A. (1954) *La distribucion del regadio en la area central de Meso-America*. Washington DC, Ciencias Sociales; Gomera Francisco Lopez (1870) *Conquista de Mexico*. Mexico DF, Impr. de J Escalante y Ca.; Armillas Pedro (1951) *Technologia, formaciones socio-economicas y religion en Meso-america, Civilizations of Ancient America*. Chicago, University of Chicago Press.
36 On coastal Peru: Collier, D. (1955) *Development of Civilization on the Coast of Peru*, in Pan American Union, *op. cit.*, Washington, DC.

382 villages where irrigation was practiced have been detected.[37] Irrigation works and practices were used in the early civilizations of the Tepanec Empire, the Toltec and Teotihuacan. Disputes among users and water transporters are also recorded.

2.10 Other early systems of water regulations and management

Other hydraulic civilizations have yet to be studied and investigated. Archaeologists and technicians often come across remains of impressive ancient hydraulic structures (canalizations, dykes, reservoirs, tanks, etc.,) which denote the past existence of hydraulic civilizations. In these cases also, the role of water administrations and laws ought to be kept in mind as they might provide useful grounds for a better understanding of the socio-political-religious and economic institutions of those civilizations.

As examples, we might quote the Helmand River valley civilization, a river flowing between what are now Afghanistan and Iran; the Khmer hydraulic civilization on the Mekong River in the territory of present Cambodia; the huge hydraulic works in Ceylon under Kanyan and Thesawalamai law particularly in the lower reaches of the Mahaweli Ganga; the Elam and Suziana civilizations around the Dez River in Iran; the Bratis river valley civilization in Java, dating from the inscriptions and edicts of Kings Dharmawangea Airlangga (1037 BC) and Tulodong (784 BC) referring to the construction of dams for navigation and irrigation; and the Naveteans of Neghev. In all cases, impressive hydraulic structures of the past demonstrate the existence of an organized water administration, and, most probably, of water regulations.

Historical and archaeological records and remains are replete with inscriptions and quotations of divinely inspired rulers who left records of their technical, institutional and legal achievements in the field of water.

The city states of ancient Greece, although living in isolation because of regional attitudes, treated in common irrigation as practised by different city states' organizations. They had common interstate water tribunals to adjudicate water matters for irrigation, and a common oath admonishing them 'not to cut, in time of war or peace, the irrigation water.'

2.11 General conclusion

In a brief section such as the present one, many topics of great relevance to the development of water regulations in early hydraulic civilizations have had to be omitted. One such topic would have been a comparative study of common patterns of parallel development; another, the persistence of early basic concepts up to modern codifications, either deriving from or expounded in the areas concerned. Yet another on the influence of early water law principles in the modern systems of water codification, and finally, other important aspects of water resources control in early stages such as those relating to various aspects of water utilizations.

For the study of this subject, only the available, scattered written records have been taken into consideration. Other records either not yet deciphered or not yet found might

---

37 Palerm, A, *The Agricultural Basis of Urban Civilizations in Meso-America*, Pan American Union, *op. cit.*

provide further important elements for the study of this particular aspect of early water control. Through the work of archaeologists and often of water engineers endeavouring to harness water resources for various purposes, remains of ancient waterworks, canalization and dam structures, etc., continue to be discovered.

As a general conclusion it may be said that with the probable exception of China, the earliest water control systems have been closely associated with religious beliefs; water constituted a gift of God or possessed a divine nature, and served as an element of purification or as a reward for a state of grace or as an instrument of punishment. It may also be said that, as soon as an organized hydraulic civilization evolved, some form of water regulation with legal obligatons was conceived, on which the bureaucratic water machinery could either justify its actions or impose its decisions. It would seem, therefore, that parallel to the development of hydraulic civilizations, water regulations and administrative water control occurred.

Often, the decay or disappearance of early hydraulic civilizations is reported as a consequence of decreased concern in the upkeep or maintenance of waterworks and canalization structures, which caused their filling up with mud, siltation or salinization. These consequences could have been avoided if the water administration had continued, as in periods prior to the decay, to maintain its concern for the safeguarding of the hydraulic structures and natural waters. In turn, this obviously means that for one reason or another, the people and the users of a water community did not feel obliged to follow the instructions received for undertaking the necessary works. As a consequence, the softening of or the lack of respect for existing water regulations caused the decadence and eventual disappearance of the civilization. The exception to this pattern is when factors foreign to the existing water community disrupted the existing structures, administration and regulations, as in the case of invasions by less water-concerned populations and the violent destruction of canals and dams.

The past history of water control and management should point out to present and future generations that an adequate management of water resources is a prerequisite for civilized progress and human survival.

References

Caponera, D.A. (1960) Water Law Principles in the Chinese Legal System. *Indian Journal of International Law*. Delhi.
Caponera, D.A. (1969) Water Laws in Early Hydraulic Civilizations. *Festschrift für Prof. Karl Wittfogel*. Washingon University.
Cruvelhier, P. (1938) *Commentaire du Code d'Hammourabi*. Paris, E. Leroux.
Diamond, A.S. (1935) *Primitive Law*, Longmans. London, N.Y., Toronto, Green and Co.
Driver & Miles (1955) *The Babylonian Laws*. Oxford, Clarendon Press; *The Assyrian Laws*. Oxford, Clarendon Press.
Drower, M.S. (1954) Water Supply, Irrigation and Agriculture, in *A History of Technology*. Oxford, Clarendon Press.
Hirsh, A. (1957) International Rivers in the Middle East, Thesis, Columbia University.
Maine, J.D. (1906) Hindu Law and Usage, 8th Ed., translated by G. Buhler. *The Sacred Books of the East*; *Ancient Law*.
Montañes, M.G. (1975) Water in the Classical Babylonian Law. *Proceedings of the Conference on Global Water Law Systems*. Valencia.
Muller, M. (1900) *The Sacred Books of the East*. Oxford, Clarendon Press.

Pan American Union (1955) *Irrigation Civilizations, a Comparative Study*, contributions by Beals, R.L., Palerm, A., Steward, I.H., Wittfogel, K, Robert, A.M. & others. Washingon, D.C.

Revillout (1903) *Précis de droit Egyptien comparé aux autres droits de l'antiquité*. Paris, V. Girard et Briereed.

Von Ihering, R. (1897) *The Revolution of the Aryan*, translated by A. Drucker, M.P. Swan. London, Sonnenschein & Co. Ltd.

Wittfogel, K.A. (1964) *Oriental Despotism, A Comparative Study of Total Power*. New Haven and London, Yale University Press.

Woolley, C.L. (1929) *The Sumerians*. Oxford, Clarendon Press.

Chapter 3
Roman and intermediate period

3.1 Introduction

The Roman system of laws has profoundly influenced the legal systems following it, both in Europe and in other parts of the world, right up to modern times, and for this reason the study of Roman water laws and institutions is particularly relevant.

In order to understand Roman water laws and the general principles to which these laws conformed, some knowledge of Roman legal history and legislative process is necessary.

As in any other system of law, Roman water law formed part of the whole Roman legal system, which evolved during a time period of almost 1,500 years. Many changes in the social, economic and political conditions occurred during that time, which in turn influenced the legal-institutional framework.

When considering Roman political and constitutional developments, Roman legal history may be divided into the following four main periods:

(i) the *early period* (about 500 years), which starts from the origins, first millennium BC, includes the legendary foundation of Rome in 753 BC, and ends with the Regal period and the beginning of the Republic in 509 BC;
(ii) the *Republican period* (about 500 years), which includes the period of the Republic (509 BC) up to the introduction of the Principate in 27 BC; the last century of this period constitutes, together with the period of the Principate, the 'Classical Period' of Roman law;
(iii) the period of the *Principate* (about 300 years), which starts when Octavian received the title of Augustus in 27 BC and ends in 286 AD with the accession of Emperor Diocletian;
(iv) the period of the *Absolute Monarchy* (about 250 years), which goes from the division of the Empire between the two Augusti (Diocletian and Maximian, 286–305 AD), includes the period of Western and Eastern Empires, and ends with the fall of the Western Empire (476 AD) and the Justinian Codification in 565 AD.

Within each one of these periods the legal régime of water is analyzed here under the following headings: (i) classification and ownership of water; (ii) right to use water; (iii) protection from harmful effects of water and control on waterworks and structures; and (iv) water administration.

3.2 Roman water law principles[1]

3.2.1 The origins and Regal period (1000–500 BC)

Ancient writers report that the earliest Roman people were divided into three tribes, the Ramnes, the Tities and the Luceres, and that these tribes were made up of elements drawn from the Latins, the Sabines and the Etruscans. These populations had a talent for military affairs and a genius for making law, both being instrumental to good organization.

During the subsequent Regal period, which goes from the legendary foundation of Rome, 753 BC, to 509 BC when the Republic is said to have been established, the Roman state was an agricultural community settled in Latium, dedicated to cultivation and stockbreeding. Except perhaps for a small holding allotted to each family as an inalienable homestead (*heredium*), there seems to have been no individual land ownership as we understand it. Areas under the name of *ager privatus* (private land) were apparently allocated to large groups or clans (*gentes*), consisting of a number of families descending from a common ancestor. *Ager publicus*, or public land, was divided into three classes: the first and largest set aside for the king; the second assigned to common pasture on which the citizens had the right to graze their cattle under some system of registration (*scriptura*); the third was an area of arable land apportioned among the clans, which periodically parcelled it out in plots to individual families for cultivation.

The territories conquered ranked as public land (*ager publicus*) and were divided in a similar way, although tracts were often left for occupation by individuals who could cultivate them as they liked upon payment of a rent to the state. In rare instances some areas were cut up into plots which citizens could not only possess but own.[2]

During this period, the sources of law were:

(i) custom (*ius non scriptum, mos maiorum*), i.e., non-written, habitudinary law, in which a distinction between law (*ius*) and religious law (*fas*) took place, in a parallel manner, with the existence of morality or rules of conscience (*boni mores*);
(ii) the so-called *Leges Regiae* or *Ius Papirianum*, of which no record remains. Pomponius, a jurist contemporary with the King Gaius, is quoted in the Digest – a later codification of Roman law – as saying that during the Regal period everything was regulated by the direct control of the kings through statutes eventually collected in a book of Sextus Papirius (contemporary with the King Superbus) called *Ius Civile Papirianum* (or *Leges Regiae*); fragments of these *Leges Regiae* show that they were religious regulations not created by any popular assembly.

No reliable information on water regulation has come to us from this period.

---

1 Caponera, D.A. (1975) Roman Water Law. *Proceedings of the Conference on Global Water Law Systems.* Valencia.
2 Turner, J.W. Cecil (1953) *Introduction to the Study of Roman Private Law.* Cambridge, Bornes and Bowes Publ. Ltd. 12; Girard, P.F. (1903) *Textes de droit Romain* (3rd Edition). Paris, Rousseau.

3.2.2 The Republican period (509–27 BC)

Although there is a good deal of uncertainty about Roman history before 390 BC when Rome was sacked by the Gauls, it is customary to regard the years following the Law of the Twelve Tables (cir. 499 BC) as the beginning of the historical era of Roman law.

The political institutions of the Republic consisted of three main groups: the Senate, the Comitia of the people and the magistracies.

The Senate and the Comitia (*Comitia Curiata, Centuriata, Tributa, Concilium Plebis*), i.e., assemblies, had judicial, legislative and electoral competence; the Senate also had competence in home and foreign affairs.

The magistracies were executive organs (public officials), and included the dictatorship, an extraordinary magistracy; the consulship (highest administrators); the censorship (census, public works, finance); the aedileship (maintenance of public services); the tribuneship (people's representatives); the quaestorship (treasury and finance), and the praetorship (justice). Some of the functions attached to particular magistracies varied with time. As a result of internal social development and external Roman conquests, the plebeians gradually attained equality with the patricians and could qualify for the magistracies.

As Roman economy evolved from agriculture to commerce, the city's constitutional machinery and its system of law evolved and expanded.

The sources of law were the following:

(i) *Mos*: long-established custom;
(ii) *Leges*: legislation *de iure* by the Comitia, forming the *ius civile*;
(iii) *Senatus consulta*: legislation *de facto* by the Senate;
(iv) *Edicta*: regulations enounced by magistrates in virtue of their *imperium*, forming the *ius honorarium*, or *ius novum*; and
(v) *Responsa prudentium* as an indirect source of law: interpretation by those skilled in the law, pontiffs until about 300 BC and jurists thereafter.

As far as water legislation is concerned, we may quote:

(i) the Law of the Twelve Tables, drawn up between 451 and 448 BC, of which only fragments have survived, thanks to the work of ancient historians (Livy, Dionisius) and of jurists (Cicero, Pomponius). This was the very first written law code drawn up by the Romans, a collection of public, private, criminal, religious and procedural customary rules which governed the relationships among the early inhabitants of Rome. It had a deep influence on the further development of Roman law until the rule of Emperor Justinian (sixth century AD). There exist two references to water in the Law of the Twelve Tables, showing that even then it was a matter of concern; one fragment thereof deals with what were probably the earliest rights which water users could claim with respect to private water (*rivus, aqua*).[3] Another fragment regulates damages caused by the flush

---

3 *De aqua, rivo, itinere actuque*: Table V, 3 in Diliberto, F.S. (1898) *Frammenti delle XII Tavole*. Palermo, Lo Casto, 48; also in Cogliolo, P. (1885) *Manuale delle fonti del diritto Romano secondo i risultati della piu recente critica filologica e giuridica*. Torino, Unione tipografico-editrice torinese, 5; Gravinae, J.V. (1708) *Origines Juris Civilis*. Lipsia, Gleditsch, 145.

of torrential rains on downstream lands.⁴ 'If the rainwaters cause damage, either they have to be directed elsewhere or compensation shall be paid.'
(ii) *Lex (Baebia) agraria* (cir. 112 BC), containing useful information on the legal condition of land. The importance of this law stems from the fact that during the Republican period the legal status of land directly influenced that of waters.⁵
(iii) *Lex Mamilia Roscia* (*Iulia agraria*) (cir. 58 BC), which stated that whoever prevented the regular flow of waters within the border ditches (*fossae limitaes*) of municipal land be fined, these fines to benefit the municipality.⁶
(iv) *Lex Coloniae Genetivae Iuliae* (cir. 43 BC). This law stated that all waters existing within newly allotted public land for settlers be subject to the same uses and burdens they were subject to under the former landowners (cir. 79 BC). The same law further empowered municipal authorities (*curia* and *duoviri*) to allocate surplus public waters (*aquae caducae*) to private utilizations (cir. 97 BC to 100 BC). Both laws shed light on the condition of municipal waters.⁷
(v) *Lex Sulpicia rivalicia* (cir. 50 BC), which stated that all the inhabitants of the city be allowed to make use of water pipes (*sifi*) for the distribution of domestic water; it probably had in view private diversions from public aqueducts.⁸

3.2.2.1 The classification and ownership of water

During the Republican period the legal status of water followed that of land. Under classical law,⁹ springs and artesian wells were expressly classified as appurtenances of land (*portis agri*). Accordingly, where land was public (*ager publicus*) all water running, springing, lying or gathering thereon was deemed public;¹⁰ all water which fell within private land, that is, within so many plots of *ager publicus* as had been allocated, was deemed private.

In this connection, apportioned public land (*ager publicus*) came under several main headings: *agri quaestorii*, public land which could either be sold by the quaestors, or often be given free to members of a colony (*ager divisus et ad signatus*) or to a single citizen (*ager viritanus*) for their ownership. *Agri limitati* were Italic lands under private ownership. Finally, those public lands which were subject to mere occupation (*agri occupatorii*) became owned by the occupiers under the legislation of the end of the Republic and the beginning of the Principate.

Ager publicus included all mountainous land and such strips of land marking the borders between existing colonies (*subseciva*) or, within a colony, between allotted plots of land

---

4 Table VIII, 9 in Diliberto, *op. cit.*, 96; also in Cogliolo, *op. cit.*, 6
5 The fragments of this law are reported by Mommsen, T. ed., *Corpus Inscriptionum Latinarum*, I, 79; also in Bruns, C.G. (1887) *Fontes iuris romani antique*. Friburgi in Brisgavia, 52. The law confirmed the former *Lex Thoria Agraria* (cir. 120 BC); it prohibited further allotment of *ager publicus* and confirmed former allottees' possession, subject to payment of a rent (*vectigal*) to the state; it suppressed the vectigal and turned the former possession of *ager publicus* into private land ownership (*ager optimo in re privatus*), subject to the ordinary tax (*tributum ex censu* – tax stemming from the registration of land in the Census); Cogliolo, *op. cit.*, 363, footnote c.
6 Costa, E. (1919) *Le Acque nel Diritto Romano*. Bologna, Zanichelli.
7 *Ibid.*, 16, 28.
8 *Ibid.*, 39.
9 Ulpian in the *Digest*.
10 Frontinus, *De condicione agrorum*: 20, 10.

(space *inter centurias exceptus*), often corresponding to a perennial river (*flumina*) and, sometimes, to streams (*rivi*). As a consequence, all rivers and some streams,[11] the springs feeding urban aqueducts, which were chosen by technicians at the foot of the mountains (*sub radicibus montium et in sexis silicibus*), mountain lakes and such rainwater as was collected by natural mountain pools or artificial tanks, were public.[12]

Springs, wells and pools located within private property were private.[13] Torrents (*flumina torrentia*) were also considered private, as land surveyors did not regard their irregular beds as reliable borders for delimiting lands (*subseciva*) or space (*inter centurias exceptus*).

During this period all water considered as public thing (*res publica*) entailed the vesting of the ownership thereof in the Roman people or in the autonomous Italic burgs (*civitates*: *municipia* and *coloniae*).[14] Private waters were considered appurtenances of private land ownership.[15]

The legal régime of private land ownership varied according to whether the land was under the status of Italic or of provincial soil. Italic soil could be the object of Roman private ownership (*dominium ex jure Quiritium*), which entailed freedom from land tax and the right of use and misuse (*utendi et abutendi*). Provincial soil (*praedia stipendiaria et tributaria*) could only be the object of a peculiar kind of private ownership (*dominium*) or later *dominium ex jure gentium* which entailed the subjection to land tax and to possession and use (*habere, possidere, frui, licere*).

Toward the end of the Republican period, the legal régime of water ownership was extended from Italy to the provinces, following the increasing allocation (*adsignationes*) of newly conquered lands (*ager publicus*).

3.2.2.2 The right to use water

Direct sources of information are available regarding the legal régime of urban water supply and the right to use private water in the Republican period.

Water was supplied to towns by means of public aqueducts emptying directly into public tanks and reservoirs (*castella, lacunae*) and thence into public baths, public fountains and public washhouses. Private connections with public aqueducts or reservoirs were strictly forbidden; only excess water (*aqua caduca*) could be diverted exceptionally from public mains, under a perpetual concession from the administrative authorities, and subject to payment of a rent (*vectigal*).[16]

From early in the Republican period, water use rights could be purchased or otherwise acquired from the owner of a private water in the form of *ad hoc* servitudes, the most important of which being the water conduit (*aquae ductus*) and water hauling (*aquae haustus*). *Aquae ductus* conferred the right to divert water from a private water source and to convey

---

11 Hyginus, *De condicione agrorum*: 112, 114, 120, 125; Siconius Flaccus, *De condicione agrorum*, 150–151, 157; Pomponius, in the *Digest*: 41,1, 30, 3.
12 Ulpian, in the *Digest*, 43, 14, 1, pr. and 6–7.
13 *Ibid.*, in the *Digest*, 39, 2, 24,12; 39, 3, 1, 12; 39, 3, 8; 43, 20, 1, 6 and 28; 43, 22, 1; 10, 3, 4, 1; Pomponius, in the *Digest*, 39, 3, 21; 8, 1, 15.
14 *Institutiones*, 2, 4; Astuti, G., *Acque-Storia*, Enciclopedia del diritto, I, 356–357; Girard, *op. cit.*, 237–238.
15 Girard, *op. cit.*, 259–61, 351.
16 Frontinus, *De aquaeductibus urbis Romae*, 94; Costa, *Le Acque, op. cit.*, 38.

it to one's own land by means of suitable water pipes and fittings; *aquae haustus* conferred only the right to draw water from a private source and the additional right of access thereto.

The distinction between the *aquae ductus* and the *aquae haustus*, which constitutes a fundamental of classic Roman water law and which dates as far back as the Law of the Twelve Tables (*rivus = aquae ductus, aqua = aquae haustus*),[17] was definitely fixed during the Principate.

In the earliest stage the acquisition of servitudes probably followed the rules governing the acquisition of land ownership. Later a distinction appeared between the legal status of Italic land and that of provincial land. As a consequence, the modes of acquisition of a servitude on Italic land were (i) alienation (*mancipatio, in jure cessio*), (ii) magistrate's judgement (*adjudicatio*), (iii) last will; servitudes could be acquired on provincial land only by an agreement (*pactum, stipulatio*).

3.2.2.3 Protection from harmful effects of water and control of waterworks and structures

The earliest concern of Roman law-makers with regard to the harmful effects of water was for the damage caused by the flush of torrential rains.

In fact, the Law of the Twelve Tables (Table VIII, 9) stated that either rainwater was to be retained upstream or, in default, compensation was to be paid for the damage occurring downstream. Presumably the action for the protection against rainwater (*actio aquae pluviae arcendae*) originated as a means of judicial redress of this principle. This *actio* seems to have had regard only to each family's small holding of public land, constituting its homestead (*ager* of *heredium*), with the exclusion of urban land (*fundus urbanus*).

As to the legal aspects of overflow protection, a permissive régime probably governed the construction of waterworks by riparian landowners for protection purposes. Control by water authorities probably took place, either directly or following a complaint, only when navigation and/or community uses were impeded.

3.2.2.4 Water administration[18]

In early Republican times water administration in the matter of waterworks was probably part of the consuls' financial and administrative competence.

Later, in the middle of the fifth century BC, water administration duties passed to the censors. Censorship was an extraordinary five-year post, the creation of which tradition places in 443 BC; it was first reserved for patricians, later (from 351 BC) open to plebeians. Censors were appointed in the number of two by the assembly (*Comitia Centuriata*). Censors, whose functions included matters of civil status, census, survey, finance and public

---

17 Table V, 3 in the interpretation of Padelletti, G. & Cogliolo, P. (1886) *Storia del Diritto Romano*. Florence, Cammelli, 253.
18 Information on water administration in the City of Rome is drawn from Homo, L. (1929) *Roman Political Institutions*. London & New York; Information on water administration in Italy and in the Provinces from Costa, E. (1927) *Cicerone Giure Consulto*. Vol. I. Bologna, Zanichelli, 399 and ff.; also, Lanciani, R. (1880) *Topografia di Roma Antica: I commentarii di Frontino intorno alle acque e gli acquedotti*. Atti dell'Accademia dei Lincei, Serie 3a, Vol. IV. Rome.

works, were responsible for financial aspects of water administration, with particular reference to the construction, maintenance and management of Rome's aqueducts and for policing the bed and banks of the Tiber and its tributaries. Furthermore, they had judiciary powers for the settlement of water disputes and responsibilities relating to public waterworks and inland navigation throughout continental Italy.

Towards the end of the Republican period, water administration became the responsibility of the aediles; the control and inspection of urban aqueducts, fountains, public baths, the allocation of surplus water and water concessions were taken over by the quaestors. Both aediles and quaestors were responsible for ensuring an uninterrupted water supply. Plebeian and curule aedileships, both sharing responsibilities in water administration, were ordinary annual magistracies, heading the great urban services: cleaning and upkeep of public buildings, fire brigade, police, food supply, public games. They were created in the middle of the fifth century BC.

Quaestors were generally in charge of the treasury and finances. Censors, aediles and quaestors had administrative, judicial and religious powers and functions in the civil domain, since a clear separation of powers was alien to Republican Rome.

Italic burgs (*civitates*: *municipia* and *coloniae*) autonomously owned and administered public waters existing within their territories. Local magistracies, which were modelled on Roman magistracies,[19] were responsible for local water resources with regard to irrigation, urban water supply and water police.

Water administration in the territories outside Italy conquered by Republican Rome varied according to the particular legal status existing between Rome and those territories or burgs. When they were entirely subject to direct Roman rule (*civitates vectigales*), water administration fell under the responsibility of the Roman governor (a magistrate or a promagistrate); when a treaty (*foedus*) governed the relationships between Rome and the conquered burgs (*civitates liberae et foederatae*), a large degree of autonomy was left to the local authorities, also in the field of water administration.

Finally, the Senate of Rome also took an indirect share in the administration of water throughout the Roman world. In fact, besides supervising the conduct of all Roman magistrates and provincial governors, the Senate had supreme control of state finances from the earliest stages of the Roman constitution, both with regard to expenditures (including public works) and to revenues (including water rates – *vectigalia*).

3.2.3 The Principate (27 BC–286 AD)

After the battle of Actium (31 BC) Octavian stood alone as consul: in 28 BC he resigned his command and immediately afterwards was bestowed consulship with the command of a proconsul (*proconsulare imperium*) for ten years and the title of Augustus; he was also called Emperor (*Princeps*). In 24 BC, Augustus was bestowed tribuneship and, in 19 BC, consulship for life.

Augustus retained the forms of the Republican constitution, but the duplication of his offices and the exercise of the proconsular military command and of the tribuneship's right

---

19 A double college of *duoviri iuri dicundo* and *duoviri aediles* in the *coloniae*; and a single college of *quattuorviri iuri dicundo* and *aediles* in the *municipia*; from Costa, *Cicerone Giure Consulto, op. cit.*, 401–402.

of veto (*tribunicia potestas*) were grave breaches of the true principles of the former Republican constitution.

The new constitutional order which Augustus gradually shaped was a kind of diarchy, as power shifted to the hands of the emperor and the Senate. The diarchy further evolved, starting from the beginning of the third century AD, toward absolute monarchy, as all powers were ultimately concentrated in the hands of the emperor alone.

As a result, during the Principate sovereignty passed from the people to the emperor, and, with the decline of the old popular assemblies, the old Republican magistracies also declined. They were gradually deprived of all authority in favour of an entirely new internal administrative machinery.

During this period the sources of law were:[20]

(i) custom;
(ii) statutes (*leges*);
(iii) edicts of the magistrates (until revision of the praetor's edict by Salvius Iulianus, cir. 131 AD);
(iv) decisions of the Senate (*Senatus Consulta*);
(v) imperial constitutions (*edicta, decreta, rescripta, mandata*);
(vi) opinion of the jurists (*responsa prudentium*), which during this period became a direct source of law.

While C. Cassius Longinus (cir. 60 AD), was unanimously deemed by his contemporaries to be the most skilled in water law, Sextus Iulius Frontinus (40–103 AD) wrote a comprehensive treatise on Roman aqueducts, including water law and administration.

The major legal enactments relevant to water law during this period are the following:

(i) a *Senatus Consultum* of 10 BC which directed that private connections with public water supply installations had to be made only at reservoirs and tanks (*castella*);[21]
(ii) the *Lex Quinctia de aquaeductibus* (8 BC). The law forbade abusive private connections (*nova foramina*) with public mains, regulated the distribution of water according to the terms and conditions of the existing administrative concessions, and set forth penalties;[22]
(iii) the Augustian Edict on the Venafrano Aqueduct (*Edictum Augusti de Aquaeductu Venafrano*) (cir. 10 BC);
(iv) a law of the college of laundrymen and dyers (*Lex collegii fullonum*), first century AD, directing that a water right over at least two intakes from public mains, (*lacunae populi Romani*) was a condition for admission to the college of laundrymen and dyers, or for the exercise of laundry and dyeing industries (*fulloniam facere*);[23] and
(v) the Agrarian Law of Nerva (*Lex Nervae Agraria*) (96–98 AD), which dealt with the connection of urban private dwellings with public aqueducts and reservoirs for domestic and/or industrial (mainly washing and dyeing industries) purposes.

---

20 Girard, *op. cit.*, 49–69; Turner, *op. cit.*, 91–92.
21 Reported by Frontinus, *De Aquaeductibus; op. cit.*, 106.
22 *Ibid.*, 129.
23 Reported by Bruns, *Fontes*, 395–396.

Furthermore, we have:

(i) a *Lex rivi* which forbade bathing and the washing of clothes in a stream of Savoy;
(ii) the perpetual Praetor's Edict as revised by Salvius Iulianus cir. 131 AD (*Edictum Perpetuum*), many sections of which expressly dealt with waters;
(iii) a *Rescriptum* by the Emperors Antoninus and Verus (cir. the end of the second century AD) dealing with private diversions of water from public rivers. This *rescriptum* sanctioned the opportunity of an apportionment of public water for irrigation, proportional to the needs of lands. This principle would not have applied with respect to those water rights which had been granted by an administrative concession. In any case, existing third-party rights had to be safeguarded.

It is to be noted that the first two centuries of this period, together with the later part of the Republican era, constitute the period of Roman classical law.

3.2.3.1 The classification and ownership of water

During the Principate, the public or private nature of water continued to derive from the legal status of land. In addition, the concept of public river (*flumen publicum*) developed and included all perennial rivers.

Non-perennial watercourses such as torrents (*flumina torrentia*) were private; however, whenever the law under which an *ager publicus* had been allotted, or custom had considered such torrents as public, these continued to be regarded as public. It seems, therefore, that torrents could be either publicly or privately owned.

Classification of water as public or private continued to entail consequences with regard to their ownership and right of use. It is to be noted that the concept of *res publicae* as the people's property (*res populi*) gradually faded during this period, and, after the rule of the Severi emperors (cir. the middle of the third century AD), i.e., with the completion of the transfer of sovereignty from the people to the emperor, a new concept evolved whereby *res publicae* came to mean 'people's right of use.'

Running water (*aqua profluens*) ranked as thing common to every one (*res communis omnium*), as any man was entitled to make use of it, and no one could claim ownership thereof. This classification had, however, the value of a statement of principle, as the effectual usage of running water followed the public or private condition of the watercourse it belonged to.

3.2.3.2 The right to use water

During this period a very permissive régime developed with respect to the right to use public watercourses, both for community (or public) and private purposes. Citizens (*cives*), but also aliens (*peregrini*), were entitled to community uses of public watercourses for drinking and domestic purposes, watering of cattle, fishing and transportation, without the requirement of an administrative concession or permit.[24] Similarly, administrative concessions were

---

24 Ulpian in the *Digest*, 39, 2, 24; Gaius in the *Digest*, 43, 14, 1, 1, 41, 1, 7, 5; *Institutiones*, 2, 1, 1–2 and 4. It is with regard to the basic uses of water for watering of animals, drinking and domestic needs, that running water

not needed for water diversions for irrigation or industrial purposes, provided that neither community uses nor existing diversions were impeded (*sine iniuria alterius*). In the case in which a water diversion hindered community uses or existing diversions, the water authority had the power to interfere and either terminate or modify such diversion.[25]

From the period of the late Republic, private rights to divert water from public watercourses and also exclusive fishing rights could be secured through an administrative concession which protected water users against any subsequent intervention of the water authorities.[26]

With respect to the right to divert water from a perennial river for irrigation and/or industrial purposes without administrative permission or concession, Principate lawyers upheld that such rights could either relate to a given land (real rights) or to a given person (personal rights). The holder of a real water right was entitled to divert as much water as was needed for a particular use.[27] In the case of a personal water right separated from land (*ius aquae separatum a fundo*), which was a perpetual and transmissible water right, the holder was entitled to divert and utilize water wherever possible at his discretion (*quocumque aquae duci possint*).[28] In both cases a right of way was included for crossing other people's lands with pipes and for repairing them.

The legal régime of urban water supplies underwent a remarkable change: connecting private dwellings with public aqueducts and reservoirs became normal practice, subject to the terms and conditions of an administrative concession (*epistulae caesaris*). The right to divert water from a public water supply installation (*ius aquae ducendae ex castello, ex rivo*) could be granted for a given place or to a given person; such right could be either perpetual or temporary and was subject to the payment of an annual rate (*vectigal pro aquae forma*) and to the possibility of cancellation at any time.[29] A water concession ended with the death of the concessionnaire.

With regard to the use of private waters, the owner could enjoy his own water, both surface and underground, up to its total depletion, irrespective of the damages which such conduct could cause to neighbouring landowners (*ius utendi et abutendi*).

A landowner was further entitled to cut aquifers springing on his land, to draw to his land groundwater streams flowing below his neighbour's land, and even to cut aquifers feeding his neighbour's wells.[30] Towards the end of the Republican period, however, this principle

---

(*aqua profluens*) is classified as *res communis omnium, iure naturali* (thing common to every man according to natural law): Marcianus, in the *Digest*, 1, 8, 2, 1; also *Institutiones*, 2, 1, 1); it is with regard to fishing and navigation that public water is classified as *res in publico usu* (thing for every man's use).

25 Pomponius, in the *Digest*, 43, 12, 2, known as *lex quomimus*. Ulpian in the *Digest*, 43, 12, 1, 12, and 39, 3, 10, 2, with regard to diversions of water from navigable rivers; see also Costa, *Le Acque*; *op. cit.*, 37–38.

26 Ulpian, in the *Digest*, 39, 3, 10, 2, and 43, 13, 1, 1 and 9; Costa, *Le Acque*; *op. cit.*, 16, 18. See also a Constitution of the Emperors Arcadius and Honorius, which aimed at preserving an old water concession (Theodosian Code, 15, 2, 7; Justinianean Code, 11, 43 (42) 4).

27 Papirius Justus, in the *Digest*, 8, 3, 17; also see Pomponius, *ibid.*, 43, 20, 3, 1; and Diocletian and Maximian Emperors, in the Code of Justinian, 3, 34, 7.

28 Ulpian in the *Digest*, 43, 20, 1, 12; Paulus, *ibid.*, 10, 3, 19, 4; Julianus, *ibid.*, 43, 20, 5, 1.

29 Ulpian, in the *Digest*, 43, 20, 1, 38–9; 43; 7, 1, 27, 3, and 30, 39, 5; Papinianus, *ibid.*, 19, 1, 41; Frontinus, *De Aquaeductibus*; *op. cit.*, 107, 109. As a consequence, those uses of public water granted under payment of a rate to the state or to a township were classified also as *res in patrimonio populi* (the people's patrimony).

30 Ulpian, in the *Digest*, 39, 2, 24, 12; 39, 2, 26; 39, 3, 1, 12; Pomponius, *ibid.*, 39, 3, 21 and 8, 1, 15, pr.

suffered restrictions with respect to the drainage of rainwater. The upstream landowner was no longer free to interfere with the natural flow of rainwater, as he could be subject to the downstream landowner's civil action (*aquae pluviae arcendae*).

As a consequence of the extended use of private waters for irrigation, industrial and domestic purposes, this period witnesses a clearer definition and increased use of water servitudes.

A right to use private water could be purchased or otherwise acquired[31] from the owner under the form of a servitude; this was either for a given usage, such as watering of animals (*servitus pecoris ad aquam adpellendi*, or *appulsus*)[32] or for a given amount of water. In the latter instance, in order to divert water from a private source of perennial running water and convey it to the user's land, a servitude for conveying water (*servitus aquae ductus*) could be acquired; this servitude included the right to lay water pipes and fittings through intermediate lands and maintenance rights (*reficere*, *purgare*).[33] A striking feature of the *aquae ductus* was that, in the opinion of Principate lawyers, the diversion could be made only at the source (*caput aquae*) of a perennial running water.[34] This limitation, which disappeared in the Late Empire,[35] entailed that any additional intake from a channel carrying water under an *aquae ductus* servitude could not take place under a similar *aquae ductus* servitude, but only under the form of a mutual obligation.

In the case of a non-permanent water intake, such as drawing water from a private well, tank or pool, a servitude for drawing water (*servitus aquae haustus*) could be acquired; such servitude would entail the right of access to the water point.[36]

Under classical law, these water servitudes were considered real rights and appurtenances to a given land (*iura aquarum*).[37] In addition, private water use rights could be acquired as personal rights by usufruct (*usus fructus or usus*).[38]

3.2.3.3 Protection of existing water rights: the interdicta

The legal system of protection of existing water rights, as developed in classical law is briefly outlined herebelow.

This system was largely based on the so-called *interdicta* (literally, an interdiction to do). *Interdicta* were the magistrate's (the praetor, in Rome) binding injunctions, under

---

31 Under classical law the *mancipatio* and the *in jure cessio* disappeared and the *pacta et stipulationes* (agreements) took their place as a mode of acquisition of a servitude on both Italic and provincial land (from Girard, *op. cit.*, 370).
32 Papinianus, in the *Digest*, 8, 3, 49 (probably amended); the holder of this servitude was entitled to water only so many head of cattle as agreed upon with the owner of the water (Ulpian, in the *Digest*, 43, 20, 1, 18).
33 Ulpian, in the *Digest*, 43, 21, 1, pr.; with regard to the terms and conditions of the rights of maintenance, see Costa, *Le Acque; op. cit.*, 51–54.
34 Ulpian, in the *Digest*, 8, 5, 10, 1; Paulus, *ibid.*, 8, 3, 9; Pomponius, *ibid.*, 8, 4, 11, 1. Water could be conveyed through uncovered channels (*rivi*) Ulpian, in the *Digest*, 43, 21, 1, 2 or by water pipes (*fistulae*), Paulus, in the *Digest*, 39, 3, 17, 1; it could be diverted continuously or at intervals, depending on the deed establishing the servitude; Ulpian, in the *Digest*, 43, 20, 1, 22.
35 See, for instance, a rescriptum by Caracalla, reported by Ulpian, in the *Digest*, 8, 4, 2.
36 Marcianus, in the *Digest*, 8, 2, 10; Ulpian, *ibid.*, 8, 3, 3, 3; Pomponius, *ibid.*, 8, 6, 17; Africanus, *ibid.*, 8, 3, 33, 1.
37 Ulpian, in the *Digest*, 8, 3, 1 pr.; and 43, 20, 1, 16; Pomponius, *ibid.*, 8, 3, 24 and 25; Paulus, *ibid.*, 8, 3, 23, 3.
38 Paulus, in the *Digest*, 8, 3, 37; Modestinus, *ibid.*, 7, 8, 21; Papinianus, *ibid.*, 8, 3, 4.

which, upon the mere claim of an alleged offended party or, in some instances, of any citizen (*quivis de populo*, popular action), the alleged offender was summoned to stop the nuisance and restore things to their former state. Investigation concerning the grounds for the injunction would only take place following any action which the claimant could bring against the alleged offender, on the offender's failure to comply with the *interdictum*.[39]

This speedy and effective means of judicial protection applied extensively to water rights. Water *interdicta* included a series of *interdicta de fluminibus* which protected common use rights of perennial rivers (for transportation, fishing, etc.);[40] the *interdictum de aqua castellaria*, which protected existing rights to divert water from public aqueducts or reservoirs; the *interdictum de aqua cottidiana et aestiva*, which probably protected both rights to divert private water under a servitude and rights to divert water from a public watercourse. Other specific *interdicta* could be issued for the protection of water servitudes and of other rights relating to water diversion from public and private waters (*interdictum de rivis, de fonte, de fonte reficiendo*).[41]

Other *interdicta*[42] included the following:

(i) The *interdictum* '*ne quid in flumine publico ripave eius fiat, quo filus navigetur,*' which aimed at the conservation of the bed, the banks and the regular flow of public navigable rivers. It applied whenever navigation was impeded or threatened by the widening or the restricting of the riverbed, or by too many diversions of water from the river or by any work and structure being carried out or existing thereat. On the jurist Labeo's authority this *interdictum* was extended to public non-navigable rivers.

(ii) The *interdictum* '*ne quid in flumino publico ripave eius fiat, quo alite aqua fluat atque uti priore aestate fluxit;*' it aimed at the conservation of the quantity, the quality, the level and the speed of public rivers, with regard to the inter-equinoxial period of the year. This interdictum was a popular one.

(iii) The *interdictum* '*ut in flumino publico navigare liceat,*' which protected the freedom of navigation on public rivers, lakes, pools and basins, including the freedom of loading and unloading on the banks or shore thereof. This *interdictum* was extended to the protection of existing water rights on a public lake or pool, under lease of the state or of a township (*municipium*). It also covered the freedom of fishing in public lakes and pools, and the freedom of watering one's cattle at public rivers.

(iv) The *interdictum* '*de ripa munienda,*' which protected the riparians' right to build protective works and structures on the banks of public rivers or on the shores of public lakes, ditches and pools for overflow-protection purposes.

---

39 Serafini, F. (1888) *Istituzioni di diritto romano comparato al diritto civile patrio*. Firenze, G. Pellas, 185–187; Girard, *op. cit.*, 1039–1047; Costa, *Le Acque, op. cit*, 49.
40 *Edictum Perpetuum*, XLIII, 241–244.
41 The *interdicta 'de fonte'* and '*de fonte reficiendo'* protected the servitudes '*aquae haustus*' and '*pecoris adpulsus*' (Ulpian, in the *Digest*, 43, 22, 1). The *interdicta 'de fonte'* and '*de rivis'* protected the rights to keep the *fons* (source of water), the intake point and the water-channels, pipes and reservoirs in good condition (Ulpian in the *Digest*, 43, 21, 1, pr., and 43, 22, 1, 6–7).
42 Ulpian, in the *Digest*, 43, 12, 1, 15–22; 43, 12, 1, 17–18; 43, 13; 43, 14; 43, 14, 1, 7–9; 43, 15.

Recourse to the procedures of the *interdicta*, from extraordinary, as it was under the late Republic, became, during the Principate, an ordinary means of judicial remedy for the protection of existing water rights against any nuisance.

In addition to the procedure of the *interdicta*, other ordinary judicial means protected the ownership and possession of things or of any servitude from nuisance,[43] and the individual's freedom to use public things.[44]

3.2.3.4 Protection from harmful effects of water and control of waterworks and structures

Principate lawyers gave extensive consideration to the legal aspects of the prevention of overflow, then considered the most harmful effect of water.

Riparian landowners were entitled to build reinforcement structures (*munitio*) on the banks of both perennial rivers, whether or not navigable, and of torrents, without the need of an administrative permission. However, in the case of a perennial river, navigability and/or suitability for public utilizations other than navigation had to be safeguarded.

The aforementioned *interdicta de fluminibus* could be utilized as a means of redress for the restoration of a river back to its original state. In addition, a ten-year guarantee had to be given to the other riparian landowners against any possible damage arising from the construction of waterworks (*satisfactio damni infecti nomine*). The riparian landowners' right to build reinforcement structures was protected from private interference by an ad hoc interdictum '*de ripa munienda;*' this *interdictum* also applied to the right to build such structures on the shores of public lakes and pools.[45]

For the purpose of protecting his land from the rush of torrents, a landowner was entitled to undertake all necessary works on his own land; however, the trespassing on upstream or downstream lands had to be expressly permitted either by the local law of public land allotment (*lex agri*) or by custom (*vetustas*). As a consequence, the disadvantages or advantages of possible overflow were the landowner's burden.[46]

Thus, administrative control on waterworks and structures affecting banks of perennial rivers followed the issuance of an *interdictum*, and the restoration of a river to its former state would take place if the works under accusation had proved harmful. These *interdicta* '*de fluminibus*' protecting existing rights of use from private nuisance may also be regarded as judicial instruments whereby private citizens could induce the administrative authorities' intervention for the control on waterworks and structures.

Legal attention was also given to the harmful effects of torrential rain. Under classical law, an action for protection against rainwater (*actio aquae pluviae arcendae*) was a judicial

---

43 The servitudes were generally protected by the *confessoria in rem actio* (*Institutiones*, 4, 6,; *De actionibus*, 2; Girard, *op. cit.*, 376 and ff.). However, a specific *actio confessoria de aqua* protected the *servitus aquae ductus* (*Edictum Perpetuum* XXX, 176).
44 Such as the criminal *actio iniuriarum* which protected human beings' freedom to perform lawful activities (Ulpian, in the *Digest*, 43, 8, 2, 9; *ibid.*, 47, 10, 13, 7, partly amended).
45 Ulpian, in the *Digest*, 39, 2, 24, pr.; 39, 2, 7, pr.; 46, 5,1, 7; 43, 15, 1, pr. 6.
46 Paulus, in the *Digest*, 39, 3, 2, 9; Gordianus, in the *Code*, 7, 41, 1; Ulpian, in the *Digest*, 39, 3, 1, 23.

means whereby the downstream landowner could request the destruction of those upstream works which had changed the natural drainage of rainwater.[47]

Provisions concerning the protection of waterworks and structures included, first of all, the possibility of expropriating lands for the construction of waterworks. Servitudes for crossing private lands with pipes had to be acquired from the landowner who, sometimes, gave them free as a gesture of generosity. Public lands could not be crossed with pipes without a permit of the responsible municipal or governmental authority.[48]

For the protection of aqueducts, a strip of land on both sides was to be left free from any encumbrance (*vacuum agrum*). The total width of such protected area could vary between 20 to 30 feet, including the width of the aqueduct itself. Within its limits it was prohibited to build, plant, let vegetation grow, or do anything likely to cause damage to the aqueduct.[49] Offenders were punished with a fine of 10.000 sersterces, of which half was rewarded to the delator.[50] The planting of trees in this area could be punished with the confiscation of the offender's land.[51]

3.2.3.5 Water administration[52]

As far as the City of Rome was concerned, in the early organization of the imperial administrative system the Republican magistrates kept the whole of their urban attributions; only the censorship was abolished.

The curule and plebeian aediles were maintained at the head of the great urban services which they ran in Republican times, such as the water service (supply, upkeep and distribution).

But, although the magistrates of Rome kept the whole of their urban functions, they didn't exercise them alone, as they were now confronted with the competition of the Emperor and his agents. With respect to water administration two urban executive commissions, one for water, and another for the bed and banks of the Tiber and drains were created.

The Water Commission (*Cura or Statio Aquarum*), instituted in 11 BC and headed by a water commissioner (*curator aquarum*) assisted by two praetorians (*adjustotes*), was in charge of everything connected with the water supply of Rome. The powers of the Water Commission, reasserted by a *Senatus Consultus*,[53] were gradually extended to the whole system of existing aqueducts and therefore to a large area outside the City.

The water commissioner was appointed for life by the emperor, with the approval of the Senate. He was of senatorial rank. In addition to two assistants, his office included one general inspector (*procurator libertus Caesaris*), one *tribunus aquarum*, one or more water engineers (*architecti*), and a large number of specialized workers (*familia aquaria publica* and *familia aquaria Caesaris*), including one chief (*prepositus aquariorum*), one registrar of water sharing (*tabularius rationis aquariorum*), measurers (*libratores*), plumbers (*vilici*

---

47 Girard, *op. cit.*, 623.
48 Lanciani, *op. cit.*, 597–598.
49 *Ibid.*, *op. cit.*, 600–601.
50 Frontinus, *op. cit.*, 127.
51 Lanciani, *op. cit.*, 605.
52 Homo, *op. cit.*, 311–320; Frontinus, *op. cit.*; Lanciani, *op. cit.*
53 Mommsen, staatw. 2, 416, quoted in Lanciani, *op. cit.*, 592.

acquarii or serf class), those in charge of reservoirs (*castellarii*), inspectors (*circitores*), stone and other workers (*silicarii, tectores*).

The Water Commission maintained either one double entry or two water registers. One of them contained information on the permanent inventory and availability of water resources, with an indication of their origin, sources, springs, volume, water courses, reservoirs, water fountains and monuments. It also contained information on water distribution, which could be in favour of the emperor (*nomine Caesaris*), in favour of public uses (*usibus publicis*) and in favour of individuals (*nomine privatorum*). The other part of the water register, or the other register, contained information on modifications of water rights, water users and water distribution. As soon as a concession came to an end, this was recorded (*in actis*) and the water returned to the administration for reallocation to a new concessionnaire.

It would appear that, particularly at the level of the plumbers of the Water Commission, there was a great deal of corruption. The purest water was sold privately and replaced by more ordinary water; during the interval between a terminated concession and a new one, the water was sold privately, surplus water was sold, etc. As one author put it, 'It is in the instinct of both ancient and modern water guards to commit such frauds . . .'[54]

The Commission for the Bed and Banks of the Tiber (*Cura alvei et riparum Tiberis*), created by Tiberius in 15 AD, was responsible for keeping the channel of the Tiber in good condition and for the prevention of floods. It was composed of five members of consular rank, one of whom was the president. First chosen by lot, later they were appointed directly by the emperor, as were the members of other similar commissions. Emperor Trajan added the maintenance of drains to its duties, and its name expanded accordingly (. . . '*et cloacarum urbis*).

In accordance with the general principles of Roman public law, the water commissioners had judicial powers within their own sphere in addition to their administrative competence. Furthermore, they were made responsible for grain distribution (*Curatores aquarum et Miniciae*)[55] cir. 200 AD.

Curule and plebeian aedileships, gradually drained of their real powers, became useless and superfluous offices, until they disappeared before the middle of the third century AD.

Outside the City of Rome a gradual concentration of the water administration of Italy in the hands of the emperor took place as the territorial jurisdiction of the imperial executive water commissions extended. Italy was divided into administrative districts (*regiones*), each governed by an imperial representative (the consulars under Hadrian, the juridici up to Diocletian), vested with administrative and judicial powers, who acted as intermediaries between the central government and the traditional autonomies enjoyed by Italic burgs.

As during the preceding Republican period, water administration in the Italic burgs (municipalities and colonies – *municipia* and *coloniae*) continued to be the responsibility of the local magistracies (*aediles* or *quinquennales*).[56]

In the provinces, the surviving administrative autonomies of free provincial burgs (*civitates liberae et foederatae*) continued. In those provinces which were subject to direct

---

54 Lanciani, *op. cit.*, 530–543.
55 From the name of the place where grain distribution cards (*tessere frumentarie*) were issued; in Lanciani, *op. cit.*, 524.
56 *Quinquennales* corresponded to the Republican censors. Lanciani, *op. cit.*, 522.

imperial rule (*civitates vectigales*), the imperial or senatorial governor's administration was in charge of major public works and of the licencing and control of private connections with public water supply installations.

Within the new administrative framework, the Senate was doomed to play a role of ever-decreasing importance, its competence confined more and more to the domain of urban administration. The Senate had supreme control over public property in the City and directed and supervised the magistrates in the exercise of their annual functions. The senatorial budget, with respect both to expenditure (public works, roads, food-supply), and to revenue (customs, water, various taxes), was largely a local budget. This evolution, which began during the reign of Augustus and continued for two centuries, accelerated under the military anarchy and culminated under Diocletian and Constantine. By the third century AD, in spite of certain prerogatives, chiefly honorary, the Senate was little more than the town council of Rome, and even in that limited sphere it had to compete with the emperor's authority as exercized through his agents.

3.2.4 The Absolute Monarchy or Late Empire (286–565 AD)

The period of the Absolute Monarchy begins with the accession of Diocletian in 286 AD and ends with the death of Justinian in 565 AD. During this period all ultimate powers (legislative, executive and judicial) were concentrated in the hands of the emperor, regarded more as a lord (*dominus*) than as an emperor (*princeps*). The main seat of government shifted from West to East, with the consequent increasing reception of oriental notions in all sectors of public life and in the evolution of the legal system.

The division of the whole Roman world, including its administration, into two parts, the Eastern and the Western, virtually gave rise to a federal system. Legally, however, these two parts were regarded as fractions of a single whole.

During this period the sources of law were:[57]

(i) custom;
(ii) the Imperial Constitutions (*edicta, rescripta, decreta*, a few *mandata*) eventually known as *Leges*.

It became customary to cite former legal enactments not by reference to the original text (e.g. *Leges, Senatus Consulta*, praetor's edict), but by reference to the works of the commentators, particularly of the later Empire, since the quality of the living jurists deteriorated.

For practical purposes, the countless constitutions delivered by the emperors in the exercise of their legislative power (*ius edicendi*) were collected and systematized first by private, later by official codifications.

The major codifications of this period were:

(i) the *Codex Gregorianus*, drawn up in 294 AD, a private code which collected the Imperial Constitutions from S. Severus to Diocletian (196 AD to 295 AD);

---

57 Girard, *op. cit.*, 70–81; Serafini, *op. cit.*, 22–26, 38–47, 54–56; Turner, *op. cit.*, 102–103.

(ii) the *Codex Hermogenianus*, drawn up in 334 AD was also a private code, collecting the Imperial Constitutions from Diocletian to Constantine (291 AD to 378 AD);
(iii) the *Codex Theodosianus*, which was an official code promulgated in 438 by Emperor Theodosius II in the East and accepted also in the West in the same year by Emperor Valentinian III. This code collected the Imperial Constitutions from Constantine onwards (312 AD to 436 AD).

In the middle of the sixth century AD, the Eastern Roman Emperor Justinian (527–565 AD) conceived a consolidated system of law. This codification combined and systematized the large number of laws in force and the thousands of scientific law works which had been drawn up by the jurists of the past. The result of this huge work (*Corpus Iuris Civilis*) consists of four parts:

(i) the Code (*Codex*), containing the Gregorian, the Hermogenian and the Theodosian Codes combined, edited and modernized. It was promulgated in 529 AD and replaced five years later by an entirely new second edition (*Codex repetitae praelectionis*);
(ii) the Digest (*Digestum* or *Pandectae*), which systematized Roman law contained in the writing of jurists of the past who had been vested with the right to deliver official opinion (*responsa*). The Digest was promulgated in 533 AD;
(iii) the Institutes (*Institutiones*), which was an official law-book (533 AD);
(iv) the New Constitutions (*Novellae Constitutiones*), which was a new code collecting the Imperial Constitutions from 535 to 565 AD.

The compilation of Justinian constitutes by far the major available source for enlightenment on both the classical and the post-classical systems of Roman water law.

3.2.4.1 The classification and ownership of water

The former distinction between public and private waters was maintained, but an extension of publicity to non-perennial rivers (*flumina torrentia*) seems to have taken place, thus increasing the category of waters ranking as public.[58]

Res publicae, which included public waters, no longer were the equivalent of the people's property (*res in patrimonio populi*), but seem to have become the synonym of the people's right of use (*res in publico usu*).[59] The former people's property (*res in patrimonio populi*) came to be known as Caesar's or fiscal property (*res in patrimonio Caesaris* or *res fiscales*). Accordingly, public water seems to have maintained the qualification of state's public property, separate from the emperor's own property (*res fiscales* or *res in patrimonio Caesaris*).

---

58 This may be inferred from the Institutes 2, 1, 2, in contradiction to Marcianus, in the *Digest*, 1, 8, 4, 1 (See Astuti, *op. cit.*, 350–351). With regard to the legal regime of private land ownership, the distinction between Roman tax-free land ownership (*dominium ex jure Quiritium*, which applied only to Italic soil) and taxable provincial land ownership (*dominium ex jure gentium*) disappeared as Italic soil was gradually subjected to land tax. Finally, this distinction was abolished by Justinian, and all private land ownership became subject to land tax (from Girard, *op. cit.*, 260–261).
59 See Astuti, *op. cit.*, 356–357; Costa, *Le Acque, op. cit.*, 13ª14; Vassalli, F. (1908) Sul rapporto fra le *res publicae* e le *res fiscales* in diritto romano. *Studi Senesi*, Vol. XXV, 7.

The further classification of running water (*aqua profluens*) as a thing common to everyone (*res communis onmium*) continued to keep its full significance as a statement of principle; *res communis* included those things which could not be the object of ownership in view of their nature.

3.2.4.2 The right to use water

The *Corpus Iuris* introduced some important innovations with respect to the legal régime of the right to use public and private water.

The general freedom of use of public watercourses remained unchanged, particularly with regard to drinking and domestic purposes, the watering of animals, transportation and fishing. However, the ever-increasing granting of concessions and leases, as well as the acknowledgement of water usurpations and priviliges for fishing, fish breeding and other purposes, tended to infringe upon this principle.

Justinian lawyers even stated that if any person had been fishing in a given stretch of a public river or lake for a long time, he would acquire an exclusive right to fish therein (*ius praeoccupationis*), on the grounds of prescription (*longi temporis praescriptio*).[60] This late Roman mode of acquisition of water rights continued throughout the Middle Ages and became a mode of acquiring water rights for the utilization of mills.[61]

From the Late Empire (fourth century AD), the traditional permissiveness of classical Roman law was superseded by an entirely different system, which made the administrative concession the only legitimate mode of acquisition of a right to divert water from public watercourses for irrigation and/or industrial purposes. A further blow to the ancient permissive system was brought about by a general prohibition to divert water from navigable watercourses and from affluents therof, which seems to have been introduced under the law of Justinian.[62]

The system of administrative concessions to make private connections with public aqueducts and/or reservoirs for household or industrial purposes which had existed ever since the age of the Republic continued. Concerning this use and those other uses of public water which were granted under payment of a rate to the state, public water seems to have been classified henceforth as things in the emperor's patrimony (*res in patrimonio Caesaris* or *fisci*, or *res fiscales*).[63] These concessions were called *sacrum rescriptum* or *divini apices de sacrum epistolarum scrinio*.[64]

Administrative control was increased, however, because of ever-increasing usurpations (illegal connections to aqueducts). A pertinent *interdictum* ('*de aqua castellaria*') could be issued only to the holders of a regular concession for connection to an aqueduct.[65] In addition, the central water authority exercized strict control over the existing connections and legalized only those of long-lasting usage (*usus vetus*).[66]

---

60 Marcianus, in the *Digest*, 44, 3, 78, (as probably amended), in opposition to Papinianus, *ibid.*, 41, 3, 45.
61 See Bertolo from Sassoferrato (1570) *Primum Digesti novi partem*.Venice, 148.
62 Pomponius, in the *Digest*, 43, 12, 2, as amended (*lex quoimus*).
63 Astuti, *op. cit.*, 356–357; Costa, *Le Acque; op. cit.*, 13–14.
64 Frontinus in Lanciani, *op. cit.*, 594.
65 Ulpian, in the *Digest*, 43, 20, 1, 44, probably amended (Costa, *Le Acque; op. cit.*, 55)
66 See the Constitutions issued by the Emperors Arcadius and Honorius (in the Theodosian Code, 15, 2, 7, and in the Justinian Code, 11, 43, 4); Valentinian and Theodosius (in the Teodosian Code, 15, 2, 4, and 5–6; and in the

The practice of legalizing unlawful diversions of water from public aqueducts or reservoirs on the grounds of long-lasting usage had interesting developments in the Middle Ages. While under late Roman law (post-classical law) long-lasting usage served only as a presumption that an administrative concession had been issued, the same long-lasting use (*usus vetus*) later evolved into a mode of acquistion of a right to use public water.[67]

As to the unlimited right to use private water, the law of Justinian introduced some restrictions: a water right holder could no longer use his water for the sole purpose of damaging his neighbour.[68]

Furthermore, the law of Justinian abolished the restriction whereby a right to divert water from someone else's private source of perennial running water (*ductus aquae*) could be acquired only from the source (*caput aquae*). As had been practiced during the Late Empire, private running water which sprang from a perennial source could thenceforth be diverted from any point (*ex quocumque loco*) under a regular water right separated from land (*ius aquae separatum a fundo*), no longer necessarily a real right.[69] These water rights separated from land were those under which public and, thenceforth, private running waters were diverted and were the origin of the so-called perpetual water rights, largely applying to irrigation, industrial and miscellaneous uses of water throughout the Middle Ages.[70]

3.2.4.3 Protection from harmful effects of water and control of waterworks and structures

Under the law of Justinian, legal action for defence against damages caused by rainwater (*actio aquae pluviae arcendae*) changed its meaning. In parallel with the shifting of relevant concern, the former significance of 'preventing harmful effects of rainwater downstream,' evolved to that of ensuring the natural drainage of rainwater for 'protecting beneficial uses downstream.'[71]

No other considerable change seems to have occurred with respect to this topic, as the regulation which had developed under classical law was incorporated into the *Corpus Iuris*.

3.2.4.4 Water administration[72]

The bureaucracy which developed in the last period of the Republic and was extended during the Principate reached its height in the Late Empire. From Diocletian and Constantine onwards (fourth century AD), the City of Rome's water supply, the Tiber and the drains continued to be the responsibility of the specialized technical water service (*Curatelae*

Justinianean Code, 11, 43, 2–3 -5).
67 Pomponius in the *Digest*, 43, 20, 4, 4; and the Code of Justinian, 11, 66 (65), 5; Astuti, *op. cit.*, 367.
68 Ulpian, in the *Digest*, 39, 3, 1, 12, as amended.
69 Paulus, in the *Digest*, 8, 3, 9, as amended. Under the law of Justinian, the mode of acquistion of servitudes came under five headings: (i) agreement, (*pacta et stipulationes*); (ii) last will; (iii) the law; (iv) the magistrate's judgement (*adjudicatio*); and (v) long-term use (*longi temporis praescriptio*); from Girard, *op. cit.*, 370–373; Serafini, *op. cit*, 313–317.
70 Costa, *Le Acque*; *op. cit.*, 48, 58.
71 *Digest*, 39, 3; Astuti, *op. cit.*, 349.
72 Homo, *op. cit.*, 371; Lanciani, *op. cit.*, 524.

aquarum), which became a strictly graded department under a great imperial dignitary, the Praefect of the City, with a staff composed entirely of imperial officials (*aquarii*, at the lower grade). The name of those responsible for water administration changed from Water Commissioner (*Curator aquarum*) to Water Consuls (*Consulares aquarum*).

The organization of water administration continued as during the Principate without major changes. In Italy and in the provinces water administration reponsibilities presumably passed entirely to the emperor's vicars (consulars, praesides and correctors both in Italy and in the provinces, proconsuls also in the provinces) in parallel with the gradual suppression of the surviving local autonomies.

3.3 Intermediate water law principles in Europe (565–1812)

3.3.1 First period (fifth century AD–1158)

3.3.1.1 Introduction

The legal history of this period, which goes from the fall of the Western Roman Empire to the Diet of Roncaglia (565–1158), can be summarized as follows.[73]

Spain became the homestead of Visigoths or West-Goths. An impressive fragment survives of a Code of laws of Enric, king of the Visigoths (466–485 AD). These laws applied to the relationships between Visigoths and Romans, as well as to those among Visigoths. In 506 AD King Alaric II promulgated a 'Breviarium' of Roman law (*Lex Romana Visigothorum*) for his Roman subjects, which was not applicable to Visigoths. In the middle of the seventh century AD, King Recceswinth enacted laws for all his subjects, irrespectively of their belonging to one group or to the other (*Lex Visigothorum*).

Italy was occupied by Ostrogoths, or East-Goths. King Theodoric (493–526 AD) enacted a law for both Goths and Romans (*Edictum Theodorici*), which was a condensed outline of Roman legal principles with no reference to original texts. During the sixth century the Ostrogoths were driven out of Italy by the Byzantines, and the Compilation of Laws of Justinian was extended to the country (554 AD). From 568 AD, Italy was invaded by the Longobards. Thus, while that small part of Italy which remained under the rule of the Byzantine emperors (southern Italy) continued to follow Roman-Justinianean law, the Longobard kings from Rothar onwards produced a flow of new legislation (643–755 AD). In 774 AD northern and central Italy were conquered by the Franks of Charlemagne, while the south remained under the Byzantine rule until it was conquered by the Normans (eleventh-twelfth century).

France became the home of the Franks and Burgundians. Burgundians settled in Burgundy. At the beginning of the sixth century King Gundobald appears to have enacted a code for his people (*Lex Burgundiorum*), of which a ninth-century version containing later additions is extant. A *Lex Romana Burgundiorum* was subsequently promulgated for Roman subjects only.

The Franks were divided into three tribes: the Salic, the Ripuarian and the Chamavian Franks. The Salic Franks were little affected by Roman civilization. When they united with

---

73 For the texts of the legal enactments of this period, see: Zenmer (1894) *Leges Visigothorum Antiquiores*, and Pertz (1878–1889) *Monumenta Germanicae Historiae*. Hannover; Leges, Tomes 1, 3, 4 and 5.

the Ripuarian Franks under King Clovis (480–510 AD), this latter appears to have produced primitive legislation unaffected by the law of Rome. During the sixth century, the Franks subjected the Burgundians and the Bavarians to their rule; as a consequence, the *Lex Romana Burgundiorum* was replaced by the *Breviarium Alaricianum*, which had been retained by the Roman subjects of Frankish Gaul as their own native law. Between the sixth and the eighth centuries, the Frankish rulers issued, with the consent of their subjects, sporadic legislation in the form of *Capitularies*. In the late part of the eighth century, the *Lex Salica* or law of the Salic Franks was enacted, being a mixture of additions to the original Clovis legislation of 200 years earlier. The seniors of Charlemagne united the Empire of the Franks, and in this period the *Lex Ripuariorum* or Code of Ripuarian Franks was enacted. At the beginning of the ninth century a *Lex Francorum Chamavorum* appeared, and a collection of *Capitularies* was made in 827 by Ansegisus, abbot of St. Wandrille. In the year 800 the Frankish king Charlemagne was crowned Holy Roman Emperor. The new political power, known as the Holy Roman Empire, included the Kingdom of West-France, Aquitaine and East-France (Bavaria), and the Kingdom of Italy (north and centre).

In the middle of the fifth century AD, England was invaded by the German tribes of the Angles and the Saxons; their settlements gave rise to seven small kingdoms. Written law began with Aethelbert of Kent (about 600 AD), and a collection of laws bearing his name but written much later commenced the series known as 'Anglo-Saxon laws.' The collection that followed, called 'Laws of Hlothhere and Eadric,' shows a process of deterioration. In Ireland, the so-called 'Brehon Law Tracts,' which appeared much later, gives evidence of an age-long development of legal thought. In Wales, the Codes of the 'Ancient Welsh Laws' (twelfth century) have many features in common with the Brehon Laws.

Other barbaric peoples which produced some written legislation during this period were:

(i) the Alamanni, who settled in the upper valleys of the Rhine and Danube, formerly known as the Roman province of Retia. During the late eight century, a *Lex Alamannorum* was enacted, being a document of ecclesiastical authorship.
(ii) the Bavarians, who settled first in Bohemia, then in Bavaria; they enacted a *Lex Baiuwariorum* which is largely a copy of the Code of Enric, king of the Visigoths (466–485 AD), and of the *Lex Alamannorum*. The *Lex Baiuwariorum* is contemporary with the *Lex Alamannorum*.
(iii) the Saxons, who produced the *Lex Saxorum* at the beginning of the ninth century.
(iv) the Frisians, who populated the inner regions of Germany up to the North Sea and produced the *Lex Frisonum* during the same period.
(v) the Danish Angles, who made the *Lex Anglorum et Werinorum* during the same period.

Throughout this time, the study of Roman law had never completely ceased. However, Roman law appears to have been largely overwhelmed by the competing Germanic customs and traditions.

All the above-listed codes or national laws show a deterioration with respect to the preceding Roman period. The growth of law was hindered and distorted by the decay of Frankish authority, the widening influence of the Church in the field of law, the growing independence of the nobility, the extension of trial by oath of purgation and the introduction of trial by battle, the proliferation of local technicalities, and the increasing importance of procedure.

After Charlemagne, the process of legislation ceased on the continent.

3.3.1.2 Principles of water law under Roman-Barbaric rule

The laws of the German people contained a very rudimentary legal discipline for water, which concentrated on the use thereof.

This discipline was based on a series of penalties, some of which aimed at preserving the freedom of navigation, fishing, drinking and domestic uses, and watering of animals. Other penalties were aimed at protecting existing water rights from nuisance, such as water pollution, excess diversions or water hoarding (*aquas concludere*), withdrawal of other people's irrigation water, destruction of watermills, construction of waterworks for purposes of protection from overflow or water diversions.[74] In addition, the legal distinction between public and private water became obsolete.

However, it is worth pointing out that a few water law provisions contained in the above-mentioned Germanic laws appear to have been influenced by Roman law or by the Barbaric-Roman codes.

For instance, the *Lex Visigothorum* made a distinction between major rivers (*flumina maiora*) and minor watercourses. With regard to *flumina maiora*, the law provided for freedom of navigation, fishing and other community uses of water by directing that riparians could not build up weirs which would prevent the flow of at least one half of the running water, or which would prevent navigation or fishing by nets. In addition, the law prohibited breach of existing rotation-turns in withdrawing irrigation water from other people's diversion channels. Different penalties, which appear to have been influenced by the Roman action against damages (*actio iniuriarum*), applied in respect of unlawful diversions from major watercourses (*maiores aquae*) and unlawful diversions from minor watercourses (*minorum derivationes aquarum*).[75]

3.3.1.3 The feudal system[76]

Before describing the water legal system which developed after the fall of the Roman Empire and the barbaric invasions, and to allow a better understanding of the basic principles of water law, it may be useful to give a brief outline of the political and institutional system of the feudal period.

Starting from the Longobard rule up to the French Revolution at the end of the eighteenth century, West Europe experienced the feudal system, which was at the same time a social and a political system characterized by the concession of benefits, vassalship and the usurpation of public functions.

---

74 *Lex Salica*, XXVII: *de navibus*; XLIV: *de retibus*; *Lex Ripuariorum*, XLVI (XLII): *de venationibus*; *Lex Burgundiorum*, XCIV: 1–2; *Lex Alamannorum*, LXXXIII (LXXXIV): *de eo qui aliquam clausuram in aquam fecerit*, 1–2; *Lex Baiuwariorum*, X: *de fonte*, 22–23; XIX: *de navibus*, 9; *Lex Frisonum*, XXVIII: *de flumine obstruso*; *Lex Visigothorum*, VIII: *ut qui in transitu fluminis culturam facit laborem sepe circundet*, 4. 28; *de discretione concludendorum fluminum*, 29; *de confringentibus mulina et conclusiones aquarum*, 30; *de furantibus aquas ex decursibus alienis*, 31; Rothar's Edict (Longobards), c. 150, 265–268, 299, 306.

75 See, among others: *Lex Romana Burgundiorum*, 17, 2; *Lex Romana Visigothorum, Pauli Sententiae*, 1, 17, 2; *Lex Romana Utinensis*, 23,23 (20); *Lex Romana Visigothorum, Codex Theodosianus*, 15, 2, 1; *Epitome Gaii*, 2, 1, 3; *Pauli Sententiae*, 5, 4, 13; *Lex Romana Utinensis*, 17, 10; *Lex Visigothorum*, VIII, 4, 29–31.

76 De la Torre Ruiz, H. (1981) *Regimen Juridico de las Aguas*. Bogota, 17–19.

The feud was the smallest unit of the feudal system and consisted of land or a real right granted by the feudatory to another man, the vassal, together with the duty to perform certain services. Thus, the feud was a contract which had both a real and a personal element. The real element was normally represented by the concession of land, including all watercourses and water sources located within the perimeter therof. Such concession was often mere fiction, as many landowners, desiring to participate in the feudal organization, which implied solidarity and mutual assistance, conferred their land to a feudatory, who immediately returned it to them. The personal element consisted of the promise by a vassal who had received a feud to serve his feudatory. A vassal could himself be a feudatory by granting land to another less powerful individual in exchange for services. Thus the feudal system was a hierarchical/ pyramidal organization, at the top of which was the king, feudatory of all feudatories, holder of all rights.

The régime of the feuds varied in time and space. The feud was granted administration rights, the right to levy dues for the transit through lands and water, a portion of fortification, churches and ecclesiastical rents. Starting fom the eleventh century, a feud could also receive sums of money. Important feudatories and clergymen were also granted immunities and privileges by the king, which had the purpose of rendering the property or monastery they controlled inaccessible to the agents of the public power. There were two types of immunity, i. e., the donation of fiscal land and the title of immunity. In both cases the immunity was not a personal right, as it was acquired together with the land and necessarily linked thereto. The owner could dispose of the immunity only together with the land, by special king's authorization.

During the feudal period, the rural populations were grouped around the feud, from which they received protection in times of peace and war in exchange for limitations of their personal freedom. Many peasants became servants and lost their rights.

In parallel with the development of the feudal system, the dismembering of the state took place, and other social groups, the seignories, came to fore. Each of these social groups was headed by a senior who ordered the military service of his subjects, administered justice and levied taxes and duties. During the Carolingian period many king's officers, aristocrats and personalities of the clergy (bishops and abbots), having received certain immunities, usurped public functions.

3.3.1.4 Water law principles under the feudal system

Under the rule of Charlemagne, navigation in inland waterways and the connected operations of mooring and docking were subject to various royalties (*thelonea, transitura, palifictura, ripaticum, portorium*), in recognition of the emperor's right of ownership (*demanium*) of all non-private waters.[77]

Under Charlemagne's successors, and parallel to the extension of the feudal system, a process of breaking up and alienation of, and encroachment on, the emperor's water domain and the connected water royalties took place. As a consequence, townships, feudatories, monasteries and individual riparians vested themselves with the ownership and/or with the

---

77 Many capitularies issued by Charlemagne give evidence of the growth and diffusion of water royalties. See, for instance, the *Capitulare Italicum* (Pipinus, 19 (23); Ludovick, 23 (25), 34 (37), etc.).

right of exclusive use of waters.[78] The difference between public or private, navigable and non-navigable waters disappeared.

Accordingly, various taxes and tolls were levied with respect to inland navigation, milling (*ius molendini*) and fishing (*redditus piscationum*, consisting of a share of the fish). Diverting water for irrigation purposes was subjected to a concession, which took the form of either alienation or lease of given amounts of water. However, free water use rights survived whenever either the custom or the tacit consent of the holder of a water royalty (*Wasserregal*) made it possible.[79]

3.3.2 Second period (1158–1812)

3.3.2.1 Introduction

This period from 1158 to 1812 goes from the Diet of Roncaglia to modern codifications. Before the twelfth century, law had been recreated in Europe. Canon law developed from the eighth century onwards. In the eleventh century the study of law had spread among laymen, and the study of the Institutes and the Digest of Justinian, newly discovered, ushered in the systems of law of the modern world.

During the Middle Ages the basic principles of Roman water law continued to prevail together with some German traditional law and were adapted to the historical and political times. Thus the Constitution of Frederick I includes navigable waters among public things which have become the property of the prince.[80]

The study of Roman law, which had never completely ceased during the preceding period, flourished in Italy at Bologna under Irnerius who taught there from 1088 to 1125. He founded the celebrated law school of Glossators (Martin, Bulgarus and others), who commented the compilation of Justinian; a systematized *summa* of their work was drafted by Accursius at the beginning of the thirteenth century.

All subsequent developments of Roman law were based on Accursius's compilation. As a consequence, the lawyers who lived between the fourteenth and the sixteenth centuries, and who are known as Post-Glossators (Bartolus, Cinus, Baldus, etc.), could develop a good deal of entirely new law under the shield of Roman law.

This Roman-derived law, which spread from Italy to France, Germany and thence to the rest of Europe, was the law in force up to the seventeenth century (*ius commune*); the legislations in force included it or referred to it, and in both cases adjusted it to the conditions peculiar to each state. However, it is worth pointing out that some continental states, such

---

78 See, among others: a *diploma* delivered by the Emperor Ludovick II to the Monastery of St. Sixtus (Piacenza, Italy, 862 AD); a *placitum* of the Bishop of Cremona (Italy, 998 AD); a donation of King Liutprand to the Monastery of St. Peter (Pavia, Italy, 712 AD); King Adelchi's confirmation of previous donations to the Monastery of St. Salvatore (Brescia, Italy, 773 AD); donations of the Emperor Ludovick II to the Bishop of Padua (Italy, 866 AD) and to the Church of St. Regina (Mantua, Italy, 870 AD); a donation of Countess Matilde to the Church of St. Cesario (Italy, 1114 AD). All these deeds give evidence of the fact that local authorities claimed for the very ownership of, or for exclusive rights of use of, domain waters.
79 Astuti, *op. cit.*, I, 373–376.
80 *Flumina navigabilia et ex quibus fiunt navigabilia*, in *Corpus Juris Civilis*, Krigelii, part III, *de feudis*, 871, title LVI, *Quae sunt regalia*.

as the Republic of Venice, developed a water legislation which disregarded the dominant principles of Roman common law.

In Germany, Holland and Austria, Roman law principles influenced the remodeled legislation adopted in the fifteenth and sixteenth centuries, and together with local customs, gave birth to the German Civil Code of 1900 and to the Austrian Civil Code of 1869. In France, a similar fusion of Roman and customary law gradually developed during the Middle Ages and served as the basis for the French Civil Code of 1804.

In the Eastern Roman Empire, the Roman system of law continued to operate not only up to 1493, the date of the fall of Constantinople to the Turks, but, with minor modifications, up to modern times. The Turks introduced Islamic law; however, the Justinian codifications were translated into Greek and other Greek compilations of law were made.[81] Roman-Byzantine law continued to govern the relations of the Christians within the Ottoman Empire until the Balkan states achieved their independence. Thus, the modern codifications of Greece (Civil Code of 1827), Serbia (Zarkan Code based on the Ecloga), Bulgaria (Zakon Soudni Lioudem, also based on the Ecloga), Romania[82] and Russia[83] all had, until recently, basic principles of Roman-Byzantine water law.

3.3.2.2 The classification and ownership of water

The concept of the distinction between public and private waters was reintroduced, on the grounds of both Roman law and of original Medieval law.

Glossators classified all navigable watercourses and the tributaries thereof as public.[84] Post-glossators added all perennial watercourses, both navigable and non, to that classification.[85] Other watercourses were considered private. Classification of watercourses as public entailed the vesting of the ownership thereof in the emperor (*in domino et patrimonio principis*).

However, parallel to the shifting of the real power from the Holy Roman Emperor to the local political entities, the ownership of public watercourses shifted from the emperor's domain to the domain of national or sub-national kingdoms and of the autonomous burgs, seignories and principates.[86]

---

81 *Ecloga* of the Isauri, 726 AD; *Procheiros Nomos*, or Prochiron, 886 AD; *Basilica*, 888–911 AD; Eparchicon Biblon and Novellae; *Hexabiblos* of *Harmenopoulos*, a private collection, 1345.
82 Photinopoulos (1765), Hypsilanti (1780), Caradja (1818), Callimachi (1816) Codes and the Civil Code of 1864 (based on the French Code).
83 Cormtsaja Kniga Code (1284) and Georgian Code (1723) based on the Ecloga; the Hexabiblos, translated in Russian, remained in force in Bessarabia until 1917.
84 See the *Constitutio de regalibus*, enacted by Emperor Frederick I at the Diet of Roncaglia (1158), reported by Pertz, *Monumenta Germanicae Historiae*, Leges, II, 110. The text of the *Constitutio* was later inserted in the *Libri feudorum* (ibid., II, 55). Also see a *diploma* of Frederick I (1159), which is reported by Ughelli, *Italia sacra*, IV, 524.
85 See Bartolus of Sassoferrato (1570) *Primam Digesti novi partem*. Venice, *Ad lex quominus*, 148.
86 See: the Peace of Constance (1183), c. 1; Caepolla (1552) De servitutibus rusticorum praediorum. *Varii tractatus*. London, c. 31, *de fluminibus*, n. 14; Bragiotti (1669) *Epitome iuris viarum et fluminum praxim rei aedilis comprehendens*. Rome, 14 and ff.

In Italy and in other countries, from the fifteenth century onwards, new classifications of waters as public or private were introduced by written legislation which departed from Roman common law and considerably widened the category of public waters.

At this point, it is worth recalling the special water legislation enacted by the Republic of Venice (Decree of 6.2.1556), whereby all waters, none excepted, were classified as the Republic's public domain.

Under the subsequent legislation of many states, a general trend to include torrents in the category of public waters strengthened.

3.3.2.3 The right to use water

During the second period, the right to use public water was based on the system of royalties (*iura regalia*). This system consisted of a general water royalty (*Wasserregal*), and of several specific water royalties peculiar to single utilizations, such as the *ius piscandi*, or *Fischereiregal*, which applied in respect of fishing.

Accordingly, the right to use public water both for communities and privately was subject to an authorization, concession or permit granted by the holder of the relevant water royalty, having the power to levy various taxes in respect thereof.[87]

Water royalties were considered an appurtenance of the ownership of public water; they therefore shifted from the emperor's domain to the domain of national or sub-national kingdoms and of the autonomous burgs, seignories and principates.[88]

However, the above-outlined system of *iura regalia* had different applications in the legislation of any new sovereign political bodies which attained independence from the imperial authority.

In Italy, for instance, the statutes of some burgs, such as Turin, granted the freedom of diverting water from public watercourses; others, such as Milan and Padua, subjected the use of water to an administrative concession or authorization and to the payment of a water rate.[89]

The trend to subject all utilization of public water to a concession, authorization or permit strengthened under the legislation of principates and absolute monarchies. Thus, under the water legislation of the Republic of Venice which was in the van in this field, all uses of water were brought under administrative concession, and all existing water use rights were subjected to an administrative recognition. Water rates were due in respect of water uses. In addition, the law created a register of waters and of diversions under the control of a special water magistracy (*Provveditori alli beni inculti*), and it prohibited any alienation of water among water users. Venice, a town 'born out of water,' recognized early the importance of water as a social commodity which the state had the duty to conserve for ensuring its fullest and equitable utilization for the benefit of all.

Some of the water legislation presently in force in Italy (Law of 1933) appears to have been largely influenced by the Venetian legislation of 400 years ago.

---

87 Astuti, *op. cit.*, 376.
88 See the *Constitutio de regalibus* (1158) and the Peace of Constance (1183); Caepolla, *op. cit.*; Bragiotti, *op. cit.*, loc.cit.
89 Astuti, *op. cit.*, 381.

The right to use private waters was largely regulated by Roman common law. However, new legal institutions were introduced by positive legislation. For instance, the statutes of most Italian burgs developed a compulsory servitude of aqueduct, whereby the right to convey water through other people's land could be established by authority, under payment of an indemnity.

3.4 Intermediate water law principles in Ibero-America

Spanish legislation, based on Roman law principles with some influence of Moslem law in southern Spain, particularly concerning irrigation uses, was instrumental in shaping the laws of all Latin American countries.[90] On May 4, 1493, Pope Alexander VI issued a Papal Bull by which he gave the Catholic kings all newly discovered lands, including waters. Water use became subject to a special king's permit (*merced especial de los Reyes*).[91] Roman water law principles are to be found in the modern codifications of Latin American countries.

The Laws of the Indies (*Leyes de Indias*), promulgated by Spain in 1550 for her American colonies, consisted of a synthesis of ancient Spanish legal principles and local customs, as adapted to the needs of these colonies.

3.4.1 Early legal principles

The ancient legal principles in force in Spain, contained in the *Siete Partidas*, linked the use of water to the community. So, the *Partidas* directed that rivers, ports and the public ways belonged to all men jointly, so that any person coming from a foreign land might use them in the same way as those living in their vicinity.[92] These common goods were attributed to the crown and their ownership vested in the prince as the representative of the community. Due to their nature of common wealth, they were inalienable.

These principles were combined in the Laws of the Indies together with the existing local customs which were not contrary to them. At the time of colonization, in fact, there was in South and Central America a strong sense of community in all aspects of the organization of indigenous societies. Recent studies show that the main Meso-American societies were based on clans, or groups of individuals having the same roots, having collective rights on land. In the indigenous agricultural practice, the collective use of land by the clan necessarily implied a collective use of water.

3.4.2 Water regulations

The Laws of the Indies gave this practice a legal shape, and extended it to Spanish subjects. Law No. 11, Title 17, Book 4, promulgated under Emperor Carlos on November 20, 1536, and concerning irrigation, directs that the order observed in the distribution of water among the native population be maintained with regard to the distribution of land to Spanish subjects. Thus the Laws accepted the concept that water is a common good which must be

---

90 De la Torre Ruiz, *op. cit.*
91 Leyes de Indias, Titulo XVII, Libro IV, 1550.
92 Title 28 of Law No. 6 of *Partida* No. 3.

distributed within the community for the benefit of its members, but vested its ownership in the crown, and entrusted its administration to the Spanish authority, considered as the representative of the community.

Water uses became the object of special king's permits (*mercedes*) granted by the Spanish government authorities for certain purposes, such as domestic, drinking needs and irrigation. Such permits could be revoked. The acquisition of water rights by way of prescription was not envisaged, the concept of individual use not being compatible with community interests. The violation of permit requirements could be punished with a fine.

The Laws of the Indies also made special provisions concerning the distribution of water to the indigenous population for irrigation and livestock watering purposes. To this end, water judges were appointed.

References

Costa, E. (1919) *Le Acque nel Diritto Romano*. Bologna, Zanichelli.
Frontinus, *De Aquaeductibus Urbis Romae*, of which there is an edition with a full commentary by R. Lanciani (1880) and a critical edition by F. Krohn, Leipzig, 1922.
Herschel, C. (1899 and 1913) *Frontinus and the Water Supply of Ancient Rome*. Boston and London.
Turner, J.W. Cecil (1953) *Introduction to the Study of Roman Private Law*. Cambridge, Bornes and Bowes Publ. Ltd.

Chapter 4

Definition and sources of water law

4.1 Introduction

For the benefit of those interested in water law and administration and also to make the concept of water law understandable to non-legally trained people, it is necessary to define it and to follow a practical approach, in order to stress the multiplicity of its sources in as simple a manner as possible.

Water law is something more than just a law governing water. Likewise, from the institutional standpoint, there is not one single agency responsible for water management. Thus, as regards legislation, we have a multiplicity of sources with all their implications; in the case of institutions, there can be conflicts of jurisdiction and rivalries between them, which can have effects on the rational use of water. Also, it is important to know the interconnections between 'water law' and other legal disciplines. The subject here is treated from a comprehensive and comparative standpoint, taking into account the situation as it presents itself in most countries, rather than be limited to a single situation in a particular country.

4.2 The content and concept of water law

In defining the concept of water law, the jurist first draws a distinction between 'law' and 'legislation.' 'Legislation' usually refers to positive, written law as promulgated under procedures defined in the constitution of a country, while the concept of 'law' covers not only the written laws, but also all the regulations of human activities, of which legislation will be one, but only one. Here, the sources of law, and not those of legislation, are considered, the latter yielding only part of what we are seeking.

Water law is made up of all the provisions which in one way or another govern the various aspects of water management, i.e., water conservation, use and administration, the control of the harmful effects of water, water pollution and so on. Because water is a basic natural resource for every country, the latest theories and the practice of states tend to treat the juridical aspects thereof in a comprehensive fashion.

Water law is thus a novel legal discipline which encompasses a wide area of administrative law (i.e., governing administrative activities), civil law (which governs activities between men and men and men and things), criminal law (which punishes actions contravening the laws in force) and, to complete this list, other branches of law – commercial and industrial law, the law of communications – every time water is involved. The complexity of water law is the result of its interconnection with many other branches of law and also of its influence on a country's economic and social development. Its content is vast as a consequence of the diversity of the subject matter and questions it is called to deal with.

Accordingly, the definition of water law derives from its content, and it will serve some purpose to enumerate the different problems to which this branch of law is directed: water resources policy, inventory of (or information on) water resources, laws governing surface water, groundwater and atmospheric water; ownership rights, priorities among the various uses, existing rights, zoning, planning at the national, basin and local levels; procedures for acquiring ownership rights or use rights, servitudes, i.e., rights of passage for water over land belonging to others, the regulation of all beneficial uses such as domestic uses, municipal supply, irrigation, hydropower production, industrial and mining uses, navigation; control over the harmful effects of water, such as floods, drought, poor drainage and the protection of the banks of watercourses; rules governing the financial aspects of water, i.e., taxes, water rates and fees; the safeguarding of water quality and pollution control; provisions regarding the interdependence of water and other natural resources in their relationship with the environment.

Water law therefore encompasses all rules organizing or otherwise governing the administration of water as such. The authority responsible for water administration, as established by the law, may have comprehensive jurisdiction for the entire resource, or it may be concerned with one specific economic sector governing one or more of the different uses to which water may be put. The administration concerned will have a territorial jurisdiction, which may be national, regional or basin level, or international or regional, as in the case of the Danube, Indus, Mekong, Senegal, Chad, Nile, Rio de la Plata and other international basin commissions.

Water law so conceived must address itself to such conflicts of jurisdiction as may arise between different government departments territorially: regions, districts, provinces; or sectorially: irrigation, water supply, hydropower, navigation. Thus, it can be resorted to as an instrument for attaining different objectives, be they land reform, land settlement or land use planning, to the extent that these are conditioned by the availability of water. It may also have an influence on environment protection activities, mining development, petroleum prospection and exploitation, fisheries, agricultural development, regional integration or decentralization, energy, public health, leisure activities and weather modification.

It should not be forgotten that water is limited in both quantity and quality and yet must satisfy ever-increasing needs. The development of novel technologies and the introduction of planning and of the concept of water resources management by river basin have provoked interest in the study of water law, both in its national and international implications. In the last few decades, most countries have faced the task of bringing their laws and institutions up to date and have introduced new legislation or new water codes, as well as changes in their relevant administrative structures.

Similar interest has been evoked in international water law, i.e., the law which regulates the use of rivers, lakes, or drainages basins, both surface and underground, which are shared between two or more countries. Following the Stockholm Conference of 1972 on environment protection, international water law has dealt with the question of the environmental aspects of international rivers. Then, the United Nations Water Conference held in 1977 at Mar del Plata, Argentina, affirmed the importance of national and international water law and the need for states to set up an appropriate institutional framework to secure efficient planning and use of available water. These concepts have been reiterated in the course of the Dublin and Rio de Janeiro Conferences, which were held in 1992, and will be covered more in detail in following chapters.

4.3 The relationship between water law and other legal disciplines

Water law is closely related to other legal disciplines, among which the following may be mentioned:

4.3.1 Constitutional law

Since the constitution is the basic public law regulating the institutions of a country and the legal régime of its natural resources, water law cannot ignore its fundamental principles, and, indeed, must conform to and not derogate to its precepts. In countries which do not have a constitution, the basic precepts must be included in some other legally binding enactment or overall policy-declaratory instrument as a fundamental statement of principles which establishes the framework within which a water law is to be enacted and subsequently implemented.

4.3.2 Administrative law

Administrative law regulates the relationship between the public administration and the people. The relationship between water law and administrative law is therefore evident. Before water law became an autonomous legal discipline, many of its provisions were included in the administrative law. In fact, a number of institutions regulated by administrative law are of direct interest to water law.

4.3.3 Civil law

In many countries, water ownership, utilization and conservation are regulated by the civil law, particularly through provisions concerning ownership of land and waters, riparian relationships and rights. Since modern water legislation tends to regulate water separately from the civil code, the provisions of these two legal instruments may conflict, particularly as regards groundwater ownership, which the civil code often attributes to the owner of the overlying land. In this connection, it is worth noting that the concept of complete ownership of anything located below one's land has already been infringed upon in the case of other natural resources, such as gold, oil or archaeological findings, which are usually considered state property.

4.3.4 Criminal or penal law

Criminal or penal law generally contains provisions which provide for punishment in the case of water-related offences, which may include the pollution of water, the destruction or damaging of hydraulic works and structures, etc. As a consequence, water law must refer to these provisions and interrelate some of its contents to specific articles of the criminal or penal code.

4.3.5 Agrarian law

This legal discipline, which deals with the management of agricultural land, is closely interrelated with water law in view of the fact that many situations are considered by both water law and agrarian law. However, water law tends to abstract from agrarian law those questions which are less dependent on land and geared more towards water management aspects. In some countries, water law provisions are included in an agrarian law.

4.3.6 Mining law

Mining law regulates the management and operation of mining activities, i.e., those relating to the abstraction of solid or gaseous minerals. Often the licencing authorities for minerals authorize mining operations which include the right to abstract and take as much water as is necessary. In addition, in a number of countries groundwater is considered as a mineral. Therefore, mining law may regulate groundwater extraction. This legal discipline or similar legislation may also regulate the use of thermal or mineral waters in cases where these waters are considered as mineral resources.

4.3.7 Natural resources and/or environmental law

These two legal disciplines are closely connected with water law. The theory of natural resources law was developed in the mid-twentieth century and establishes, together with that of water, the legal régime of land, air, underground minerals, fauna, flora and bioresources.

As regards environmental law, the Stockholm Conference of June, 1972, was the cradle of this new legal discipline. This doctrine purports to regulate all activities which may affect the environment and, naturally, includes water quality control.

4.3.8 Public health law

Water law is also directly linked with those provisions of public health legislation dealing with water pollution, water quality standards, waterborne diseases, urban and rural water supply and sanitation, legal requirements and other relevant aspects.

4.3.9 Other legal disciplines

There are other legal disciplines with which water law is interconnected, as they relate to specific uses of water resources. These include hydropower generation, fisheries, inland navigation, forestry and national parks legislation.

4.4 Sources of water law

The sources of water law are numerous. It can be said that all causes leading to the establishment of legal rules constitute sources of law. Scholars distinguish between material sources, whether historical, geographic, economic or other, and formal sources, the different processes leading to the making of statutory rules. Others distinguish between formal, real and historical sources. By 'formal sources' it is meant the creative process of the legal norms; the 'real sources' are those factors and elements which determine the content of those norms, i.e., the real situations which the legislator is supposed to regulate. The term 'historical source' refers to those documents (writings, inscriptions) which contain texts of laws or other legal enactments. In this connection we speak, for instance, about the 'Institutions', the 'Novels' or the 'Codes' as the historical sources of Roman law.

The formal sources of law, which are the same as those of water law, are the following:

(i) legislation in general;
(ii) international treaties, interstate agreements, the binding decisions of regional bodies such as the European Union and the binding declarations of international organizations;
(iii) customary law;
(iv) case law, i.e., the decisions of national, regional and international tribunals, arbitral awards; and, finally,
(v) doctrine or scholarly opinion.

4.5 Legislation in general

In general, legislation means written law. It is a process whereby one or more state organs formulate and enact legal rules for general observation. In this regard, law is the product of legislation and should not be confused with it.

Legislation is the first and most important of the sources of law. It is grounded in the constitution, the fundamental law of a country, from which it derives its authority. It is from the basic provisions contained in the constitution governing the use of natural resources in general, or the use of water resources, that the legislation making up 'water law' will be derived. The type of general legislation, and water legislation in particular, will derive from the type of constitution and from the doctrine informing this constitution in any one country. This is a fundamental consideration, especially as regards the right and the ability of a state to control and manage its resources. For example, if a constitution consecrates the principle of private property in relation to water, the government must take this rule into consideration before it intervenes in the water sector. But if the constitution states that all water belongs to the state, the crown or the public domain, then legislation will follow this precept. Thus, it is from the constitution that water legislation emanates.

As to form, legislation may be divided into two main categories:

(i) basic act or law, which has to be approved by the legislative body of the country and which may require lengthy constitutional procedures before adoption;
(ii) subsidiary or implementing legislation, consisting of decrees or orders issued by the ministers empowered to do so by the basic act or law or, again, regulations that are more explanatory in nature and constitute an implementation of the policy formulated in the basic act or law.

As the latest theories hold, water legislation contains all those rules which directly or indirectly govern the conservation, development, protection and administration of water. These rules may be numerous and may include texts referring to the different aspects of water policy and resource management.

Legislation, which includes both the basic law and its implementing regulations, also sets up the institutional machinery for carrying out a country's water management policy.

The 1977 United Nations Water Conference recommended that countries make a study of their respective legal and institutional framework in the part concerning water, with an eye to the experience of other countries, and that states should, as far as possible, initiate organic and unified legislation on water such as may make for rational management and planning of this resource. The conference emphasized the importance for countries, when the constitution permits them to do so, of unifying the legislation into a 'frame law' recognizing, *inter alia*, public ownership of water and of the major engineering projects, thus facilitating state control. The conference recommended the introduction of new laws allowing for public participation in planning and policy decision making. Legislation should provide for this participation as an integral part of the planning, implementation and evaluation process.

Thus, legislation is the primary source of water law, but it is also highly varied and often difficult to comprehend for the non-specialist.

4.6 International and interstate agreements

International treaties concluded between sovereign states are an important source of water law, both national and international. These treaties contain provisions binding upon the contracting parties which, in consequence, the parties are required to apply in their domestic legislation.

A distinction needs to be made between general and particular (or special) conventions. General conventions are multilateral treaties, the purpose of which is to codify rules governing a given matter. Where water is concerned, it is worth mentioning the Convention of Barcelona of 21 April, 1921, which introduced the régime of freedom of navigation on navigable waterways of international concern for the countries which accepted the régime, the Geneva Convention of 9 December, 1923, dealing with the development and exploitation of hydraulic power affecting more than one state, and The United Nations Convention of 21 May, 1997, on the Law of Non-navigational Uses of International Watercourses (UN Watercourses Convention).

Particular conventions may be multilateral (concluded among several states) or bilateral (as the name suggests, only two states are concerned). As for particular conventions affecting water resources, whether multilateral or bilateral, there are thousands of these. Among them are agreements made between several countries for the management of international drainage basins: the Rhine, Danube, Oder, Scheldt, Meuse, Moselle, Lake Constance, Mekong, Senegal, Rio de la Plata, Indus, Ganges and others.

In its national or municipal legislation, each country inherits the provisions of the international conventions to which it is a party. For instance, if a basin commission formed on the basis of a binding international legal instrument imposes certain obligations on one of the member countries, the national legislation of that country must reflect these obligations. Non-compliance of the national legislation with the obligations set under international law renders the country internationally responsible, and likely to be internationally guilty of an offence.

In recent decades, large-scale water diversions and other water-related projects have created frictions among the riparian states of international watercourses. Thus, international water law has become concerned with specific issues, such as hydropower production and distribution, irrigation and flood control. Many treaties take up the question of the sharing of the waters of international rivers in order to satisfy different demands. In past decades, emphasis has been placed on the control of pollution, particularly in Europe and, generally speaking, in the industrialized world. The provisions of these treaties must be reflected in national legislation.

Whenever international treaties have led to the establishment of joint commissions to manage shared waters, the decisions of these commissions, if binding, automatically apply to each member state. This is the case of the Danube and Sava basins, and in many other instances.

Pursuant to international treaties concluded between them, certain countries have introduced parallel or harmonized legislation governing particular aspects of shared watercourses. The Scandinavian countries provide an example of this kind, Canada and the USA another, in this case concerning the Great Lakes.

A further source of national legislation is constituted by the regulations and directives of regional organizations such as the European Union or other similar rule-making organizations, such as the Council of Gulf States, and some international basin organizations such as the Senegal River Basin Development Organization (OMVS) and the Gambia River Basin Development Organization (OMVG).

Also to be mentioned are the declarations made by intergovernmental organizations and by international conferences. While these declarations do not *per se* constitute a binding source, they nevertheless reveal trends and the positions of nations as far as water law is concerned.

For instance, Article 21 of the Stockholm Declaration of 1972 on the Human Environment has introduced the notion of state responsibility for acts or omissions causing substantial injury to the environment; the European Water Charter of the Council of Europe, 1967, has set in motion some guiding principles. Also, the recommendations on water policy and transfrontier pollution of the OECD of 1976 concluded that states are under an obligation to combat pollution and are responsible for the damages that it causes in the territory of other states.

These obligations are of an international nature and oblige the participating states to take appropriate steps within their national jurisdiction to prevent harmful occurrences.

Finally, mention should be made of the agreements concluded between states or provinces of countries which have a federal constitution. Treaties or agreements of this kind are important in that they apply principles of international law even if they are not looked upon as formal sources of international water law. Agreements of this type are seen in Europe in the Federal Republic of Germany and in Switzerland, in the USA, where they are called interstate compacts, in Argentina, in India (Chambal River) and in Australia (Murray River). The decisions of the interstate or interprovincial commissions that may be created under these agreements apply internally, at the national level, such as in the case of the *Comisión técnica interprovincial del Río Colorado* in Argentina.

4.7 Customary law

Another source of water law at the national level is custom. Custom constitutes the continuous repetition of certain actions or practices by a collectivity in the conviction that they are legally binding. These customs and practices must have been observed since time immemorial and often are not enshrined in any written text. However, this does not mean they are unknown to the beneficiaries.

Custom as a source of law includes two important elements:

(i) a set of social rules deriving from a usage of a certain duration. In Latin it is called *inveterata consuetudo* (persistent custom);
(ii) the aptitude by those who follow these social rules to consider them as binding.

Any water management plan for water resources, whether surface or underground, must recognize the existence of customary water rights. In Moslem countries, custom forms the basis of water law, especially where domestic uses and irrigation are concerned. For this reason the introduction of a novel technique must take customary practices into account.

Some examples from Europe may be cited. In the Netherlands and Belgium, the members of organizations known as *waterschappen* or *wateringues*, which have been making use of the water since time immemorial, constitute users' associations which the central administrative authorities are required to consult if it is proposed to introduce any changes in the existing situation. In the Federal Republic of Germany, the water protection associations are a force to be reckoned with as an interlocutor to the authorities of the various *Länder*. The Water Tribunal of Valencia, Spain, rules on customary rights held by users; in 1960 the tribunal, which sits outside the cathedral of the city, celebrated the thousandth anniversary of its constitution. Because of its membership made up of the users themselves, the rulings of this customary court are respected and enforced. In 2009 the tribunal was designated by UNESCO as an intangible cultural heritage.

In Italy, particularly in the Po valley and in Sicily, custom continues to form the basis of local water law, and no water manager can ignore it. In Rome, princely palaces have enjoyed water concessions granted in perpetuity since the Middle Ages for the use of water in their private fountains. The water supply agency of the capital (ACEA) has been hard put to modify these customary rights and has had to pay compensation in order to ensure that the previous rights of use granted in perpetuity be limited in order to ensure a more rational use of the water. Customs and customary water users' associations are present all over the world, in Asia, Africa, Latin America and Europe. The constitutions of some Latin American countries, including Argentina, Bolivia, Brazil, Paraguay and Venezuela, explicitly recognize the customary practices and institutions of the indigenous populations. The legislation of some African countries, also, protects pre-existing customary rights.

Also at the international level, there is an 'international customary law' governing shared water resources. Implicit in such a law are the principle of equitable and reasonable utilization, the obligation not to cause significant harm and, conceivably, the obligation for states to cooperate and to negotiate in good faith.

It follows that customs and practices, both national and international, are of the highest importance. Anyone concerned with administering water should be aware of their existence and of the fact that at times they may represent an obstacle to change when modernization or a more rational use of available water is desired.

4.8 Case law and arbitral awards

Case law refers to the totality of court rulings on given matters by way of interpretation of the law, or on conflicts of interest between individuals, or between individuals and the different administrations. Rulings of this kind are yet another major source of law, whether they be rendered within the national context or, in the case of the Common Market, by the Court of Justice of the European Communities or by the International Court of Justice at the Hague, both of which have jurisdiction in conflicts between sovereign states.

Certain countries, including Denmark, Finland, Italy, South Africa and Sweden, have set up special water courts which rule on questions arising in connection with water resources management.

In countries with a federal constitution, when the courts rule on conflicts between two or more states of the federation, they apply the principles of international water law, which then are binding at the national level.

Likewise, the International Court of Justice (ICJ) at the Hague – formerly the Permanent Court of International Justice – has dealt with conflicts of law between sovereign states, the decisions of which have become obligatory at the national level. Five cases which may be mentioned are the case concerning the jurisdiction of the International Commission of the River Oder (1929); the conflict between France and Belgium (1937), caused by the diversion of water from the River Meuse; the Gabcikovo-Nagymaros case (1997), between Hungary and Slovakia; the dispute between Costa Rica and Nicaragua regarding navigational and related rights (2009); and, the case concerning pulp mills on the River Uruguay (2010), between Argentina and Uruguay, relating to alleged breaches by Uruguay of obligations incumbent upon it under the 1975 Statute of the Uruguay River. On 6 June, 2016, Chile instituted proceedings against Bolivia with regard to a dispute concerning the status and use of the waters of the Silala River.

The Court of Justice of the European Communities, also known as the European Court of Justice, deals, *inter alia*, with matters of interpretation of European law and with claims by the European Commission that a member state has not implemented a European Union directive or other legal requirement. In 1975, this court handed down a decision concerning the pollution of the Rhine. In 2003, it imposed fines on Spain for not meeting the water quality standards set forth in the Bathing Water Directive. Between 2005 and 2006, it sanctioned a number of countries, including Belgium, Germany, Italy, Luxembourg and Portugal, for failure to transpose the Water Framework Directive into their legal systems.[1]

Arbitral awards are also a source of national water law. These awards are rendered by a tribunal which may be called upon to adjudicate a dispute over water. Within the European context there is the example of arbitration between France and Spain concerning Lake Lanoux (1957); in Asia the Helmand River case (1872 and 1905); in America the Trail-Smelter case (1938 and 1941). These cases have established principles of water law which are reflected within the national context of the countries to which these arbitral awards refer. It is to be noted that an arbitral award is binding on the states which have submitted the case to arbitration.

To sum up, anyone concerned with the management and planning of water resources should know whether there are national, regional and international judicial precedents that may have a bearing on the status of water and, by that token, on any technical operations that may be needed.

4.9 Doctrine, or scholarly opinion

Doctrine, too, is a source, but not a formal one, of water law, both national and international. The term as understood by lawyers refers to scholarly opinion to be found in learned writings on the subject. It includes studies that lawyers undertake on legal questions, either for the sole purpose of systematizing the subjects or for interpreting the rules and indicating the procedures for their application. From doctrine have been developed principles and guidelines

---

1 See Chapter 6, Section 6.4.5.2.

enabling courts and administrative authorities to adopt a surer approach to problems arising where water is concerned. This same source is drawn upon in the formulation of legislation and in jurisprudence (court decisions), within individual countries.

It should be noted that doctrine does not constitute a formal source of the law and, therefore, is not objectively binding. However, it may contribute to the preparation of legislation by the parliament or influence a judgement by a court. Naturally, the opinions of doctrine are much more forceful when they are the result of the group work of specialized lawyers.

In fact, principles of international water law and administration have been developed by international non-governmental organizations such as the International Law Association (ILA) and the Institute of International Law. In particular, in 1966 the ILA produced the well-known 'Helsinki Rules,' which introduced the principle of reasonable and equitable utilization of the waters of international drainage basins. These rules were subsequently supplemented by chapters concerning flood control (1972), the protection of water resources and water installations in times of armed conflict (1976), international administration (1976), the relationship between water, other natural resources and the environment (1980), pollution (1982), groundwater (1986), cross-media pollution (1996) and private law remedies (1996). In 2004 the ILA adopted the so-called 'Berlin Rules on Water Resources,' to revise the Helsinki Rules and place emphasis on environmental aspects.[2]

From their inception, international organizations such as the United Nations and its specialized agencies have addressed themselves to water law questions, preparing, *inter alia*, numerous publications and research data. The Food and Agriculture Organization of the United Nations (FAO) has also developed a methodology for the systematic analysis of water laws and institutions of its member nations, both at the national and the international level.

It is also worth mentioning the work done by the International Association for Water Law (AIDA/IAWL), which promotes the development of water law globally. The Association enjoys consultative status with the Economic and Social Council of the United Nations and with a number of specialized United Nations institutions. It participates actively in international meetings and conferences on water law. Members are lawyers specialized in water law and non-lawyers directly involved or interested in water law and administration who undertake technical assistance missions worldwide. The official languages of the Association are English, Spanish and French.[3]

4.10 Common law

Another source of law is common law, which developed in England and was extended to nearly all countries the legal system of which historically came under the influence of the British Empire.

Like Roman law did before it, common law holds that water cannot be privately owned, and riparian landowners have only a use (riparian) right over it. In this way, public control and administration of water is greatly facilitated.

---

2 See Chapter 10, Section 10.3.9.2.
3 www.aida-waterlaw.org. The contributions made by non-governmental organizations are discussed more in detail in Chapter 10.

4.11 Conclusion

The sources of water law are many and varied. They derive both from national law and from the obligations entered into by states under international conventions. No one who is responsible for administering water can afford to ignore the multiplicity of those sources.

References

Cano, G.J. (1976) *Derecho, politica y administracion de aguas*. Mendoza, INCYTH, Vol. 1–3.
Giner Boira, V. (1976) Las Comunidades de Regantes y el Tribunal de las Aguas de Valencia. *Annales Juris Aquarum*, AIDA II, Valencia.
Lopez, J. (1973) *Curso de Derecho de Aguas*. Buenos Aires-Mendoza, Instituto Nacional de Ciencias y Tecnicas Hidricas, Centro de Economia, Legislacion y Administracion del Agua.
Retortillo, M. (1961) *Aguas Publicas y Obras Hidraulicas*. Madrid, Tecnoc.
Spota, A.C. (1941) *Tratado de Derecho de Aguas*. Buenos Aires, Rubino.

Chapter 5

Existing systems

5.1 Introduction

To truly understand the existing systems of water law in individual countries it is essential to know to some extent their political, religious, social and legal backgrounds, together with other technical, physical, economic and related aspects. Water resources policies, administration and laws have developed through intercultural and interdisciplinary contacts and largely reflect the influence of history and physical factors. A short description of the major water law systems is a necessary starting point for a closer scrutiny of the existing situation in any one country.

In the eighteenth century, the French scholar Montesquieu stated wisely, 'Civil and political laws must be so fitted to the country for which they were enacted that it is a real coincidence if those of one nation apply to another.'[1]

While early water regulations and administrations usually related to one simple use or harmful effect of water and to penalties for offences in water activities, with the increase in different utilizations new concepts of water resources policy, administration and legislation have evolved.

Present day water administration and regulations derive from the original legal systems prevailing in any one country, together with more modern concepts or interpretations which have often been superimposed on pre-existing regulations, local uses and customs.

Throughout the world, water law and administration derive from one or a combination of more than one of the following legal and institutional systems:

(i) customary law;
(ii) Roman law, with its two main derivations, common law of England (and new USA doctrines) and civil law;
(iii) Islamic law;
(iv) Soviet law;
(v) Hindu and Buddhist law.

---

1 Montesquieu (1758) *De l'esprit des lois*.

5.2 Original Roman water law principles and their influence in subsequent legislation

The water law principles of Roman law have exerted a profound influence on the water legislation and administrations of modern nations and have, in the course of history, taken three major directives. Their original concepts are still found in the basic water law principles of the so-called civil law countries, i.e., the countries which have adopted or followed the principles established in the French Napoleonic Code promulgated in 1804,[2] the so-called common law countries, deriving their legislation from the English application of the original Roman law,[3] and the so-called water law doctrines of appropriation, beneficial use and correlative rights originating in the United States of America.

Simply stated, ancient Roman law subdivided water resources into three categories:

(i) *waters common to everybody* (*res communis omnium*), i.e., waters not capable of being the object of any ownership status. No one, whether the individual, the community or even the state or the sovereign could own these waters; together with air and the sea (shore), they could only be the object of rights of use. All flowing waters belonged to this category.
(ii) *public waters* (*res publicae*), i.e., those belonging to a community, municipality or other public institution. The use of such waters was reserved to the institutions, which had a legitimate title over them. Institutions could, in turn, grant a right of use to other users.
(iii) *private waters*, i.e., those privately owned. Only a small part of water resources were considered private: rainwater, groundwater and minor water bodies. Generally, the ownership of these waters was attached to the ownership of land. The landowner had an exclusive and unlimited right of use (and abuse) over such waters, and this right of use was without any restriction, independently of the consequences that the use could cause to neighbouring lands (*ius utendi et abutendi*).

In Roman law the right of ownership and use of water recognized to the landowner was necessarily limited both by similar rights of neighbouring landowners and by the rights acquired by or granted to third persons by the state. As to the rights of neighbours, Roman law, which regarded water as a constituent part of the land, left the owner entirely free to dispose of all the water found on his land without any consideration for his neighbours but made him liable to suffer similar treatment at their hands.

5.3 Customary water law: its importance

Customary water law and administration exists all over the world and plays an important role in water management, particularly at the users' level. Customary practices are either local, regional or tribal, and they may derive from well-established rules. Often they have persisted in spite of the introduction of subsequent water institutions and legal systems.

Custom and traditional uses have often determined, and still continue to determine, water rights and their administration. Generally, traditional customary water rights are neither written nor registered, but are transmitted orally from generation to generation.

---

2 See this chapter, Section 5.5.
3 See this chapter, Section 5.6.

Custom has been defined as the spontaneous expression of the formation of a right, while the written law represents its codified form. Historically, it is also the first and exclusive source of juridical rules; customary law has been based on the constant and uniform observance of rules of conduct accomplished by the members of a community with the conviction that it corresponds to a legal need or obligation. From early times custom has had a decisive influence in the formulation of an objective right. When legislation came into being, it considered custom as an important source, either autonomously or parallelly to other sources. As the output of written legislation increased, the state took over the power to enact legal precepts, and, as a consequence, customary law has slowly been superseded by written legislation.

While written codification has lessened the importance of custom as a source of law, the influence of custom is still felt, particularly at the local level.

Generally, customary water law deals with subjects such as:

(i) customary legal status of water;
(ii) customary water rights distribution and management;
(iii) customary procedures for the settlement of disputes among water users;
(iv) customary water administration.

Some countries have neither written water laws nor a formal water administration, but rely on local uses and customs. One advantage of customary and traditional law over written law is its flexibility, making adaptation to local needs at a particular moment easier. At the users' level it is a system generally well known, respected and followed. Sometimes it is the only system known to the population; therefore it cannot be ignored or overlooked. As an example, we may cite a particular case in an African country, in which an official of the water department of the capital went to a village in order to carry out an inspection of the water distribution system. He was badly received by the traditional customary water administration, which completely ignored his official function.

On the other hand, customary law has the character of uncertainty, since its limits are often not well defined. Furthermore, its modification is a slow process and cannot always keep pace with modern developments in water utilization and resource management needs, which often necessitate profound and rapid transformations.

Before introducing or implementing modern concepts of water resources policy, administration and legislation, a preliminary analysis of the existing legal practices, including the prevailing customs, is necessary in order to define and delimit clearly the existing customary and traditional water rights.

The advantage of a written codification is that it gives the society to which it refers the assurance of a well-defined quantity and quality of water, and a clear delimitation of individual rights.

In countries possessing written water legislation and a government water administration, local usages and customs are generally left unwritten, as they are only for minor water utilizations and for cases where the codified law lacks specific provisions relating to them. Traditional customary rights are usually recognized by subsequent water regulations on the basis of detailed procedures, including a preliminary inquiry and their subsequent recognition in the manner prescribed by the water law. The written recording or registration of existing customary water rights is one of the main characteristics of practically all modern water regulations. It is obvious that adapting customary and traditional water rights to modern requirements necessitates an adequate institution responsible for undertaking the administration of water rights.

In the arid regions of North Africa and the Middle East where water is scarce and precious, orally transmitted regulations are prevalent and strictly adhered to by the users on the basis of the principles of Islamic water law, in spite of superimposed water regulations which are often less known or ignored.

The same situation prevails, particularly at the users' level, in countries deriving their water legislation and administration from the Chinese system of law, like China and Viet Nam. In these countries, despite more recent water administrations and written legislation, sometimes copied from the west, the traditional Chinese water management practices still represent the known and respected water rules at the users' level.

To some extent, the Hindu and Buddhist principles of law have influenced and are still present in the customary water laws of Bali, Myanmar (Burma), Cambodia, Sri Lanka, Laos, Viet Nam, Thailand, and, to a lesser degree, India.

In Africa south of the Sahara, customary law still has great relevance and is respected, particularly for water use and distribution. This customary law is specifically recognized by statute in many countries, in spite of subsequently adopted legislation of a western type. Private ownership of water is unknown, and the principle of 'community of interests' exists, whereby an individual has only a right to use land and water. Traditional and customary water institutions coexist with government administrations.

In several countries of Latin America where large indigenous communities exist, pre-Colombian customary and traditional water regulations and institutions deriving from Inca, Aztec and Maya civilizations continue to govern water at the users' level.

In Europe, also, customary practices as regards water management at the users' level have survived in the written water legislation which, generally, has recognized them *de facto* or *de jure*. This is the case of the Dutch *waterschappen*, the Belgian *wateringues*, the Italian *consorzi di bonifica* or *di irrigazione* and the Spanish 'communities of irrigators.'

The difficulty of adapting modern systems of water resources policy, administration and legislation to traditional and customary regulations and institutions must not be overlooked. Since they are well known and respected at the users' level, these customs may become a major impediment when changes are required in modern water resources planning and management. An effective policy must be formulated in order to bring the customary users gradually under administrative control. As has been said, customary law cannot be ignored when introducing new legislation, and an appropriate administrative machinery should recognize it as far as possible. In a number of countries in Africa and Latin America, legislation and some constitutions explicitly recognize customary rules and institutions.

5.4 Water law principles in the Islamic system[4]

5.4.1 Introduction

Islamic water law derives from and is an integral part of the whole Islamic political and social system, which is religious in character and comprises all the rules of the Moslem faith.

---

4 For an analysis of Moslem water legislation, see Caponera, D.A. (1954) *Water Laws in Moslem Countries*, FAO Publication; Caponera, D.A. (1973) *Water Laws in Moslem Countries*. Irrigation and Drainage Paper No.20/1. Rome, FAO.

The fundamentals of Islamic water law purport to ensure that water is available to all members of the Moslem community. This is why in many modern Moslem countries water legislation considers water resources as belonging to the whole community, i.e., the state, or the public domain.

This basic principle has exerted a strong influence, particularly in those countries where the constitution provides that the state accept the principles of the Islamic doctrine, such as Afghanistan, Bangladesh, Indonesia, Iran, Malaysia, Pakistan, Saudi Arabia, Kuwait, Qatar and the UAE. However, customary water law (*'Urf*) also reflects the differences brought about by the different schools existing within Islam.

In spite of subsequent written water laws introduced by external powers or other governments, the basic principles of Islamic water law are still observed and strictly followed as local customs and usages by the population. As a consequence, any attempt to enact modern water legislation in these countries must consider both the religious and political background in order to avoid the introduction of dangerous *bid'aa* (heretical innovation), which could carry adverse social consequences.

5.4.2 The origin and sources of Islamic water law

In Arabia, before the Prophet Mohammed, in the *djahilyya* or 'period of ignorance,' there were no established water regulations. The early Arabs were bedouins who preferably set up their tents near centres where there were palm groves, olive trees, watercourses or wells. The nomad tribes were bound to the settled populations by a fraternal pact, which regulated their relations in regard to water rights. When the settled populations were strong, the nomads respected their customs. In the reverse situation, however, the wells were the cause of constant quarreling.

The teachings of the Prophet laid down the first laws with respect to water. Such teachings, divinely inspired, are known by the name of *Shari'a*, and have been collected in the *Holy Koran* (revelation), integrated by the *Sunna* (behaviour of the Prophet), as recorded in the *Hadiths* (commentaries or traditions), by custom (*'Urf*) and, above all, by the different schools of doctrine and of law.

It is important to know that after the death of the Prophet Mohammed many disputes concerning his succession and other matters led to the formation of several schools, which, by interpreting the inner meaning of the Prophet's precepts, have sought to make the principles responsive to changing and increasingly complex circumstances. Three major Islamic schools are in existence at the present time:

(i) the *Sunnites*, composed of the *Hanifites*, the *Malikites*, the *Shafi'ites* and the *Hanbalites*; they represent the largest majority of Islam (about 90%).
(ii) the *Shi'ites*, composed of many schools, among which the largest are the *Ismailites*, the *Imamites* and the *Zaidites*; they constitute about 10% of Islam. The Imamites have also political power in Iran, where the head of the state is at the same time their religious leader. As regards the Ismailites, their head is the Aga Khan. The Zaidites are mainly in Yemen.
(iii) the *Ibadites*, a branch of the ancient *Kharijites*, which are mostly in Oman.

These schools have affected not only the fundamental principles of the doctrine, but also the basic political framework, as well as the law, which consequently underwent variations and allowed for different interpretations of water legislation. Usually one or another of these schools is dominant in any one Moslem country. In addition to any modern written water legislation, it is essential to know the rite followed by the population of a specific region in order to understand the local practices and customs.

5.4.3 Water ownership and the right of use

According to the precepts of Islam, the gift of water entails a religious obligation deriving from the very nature of water, out of which 'every living creature was created.'[5] The Prophet declared that free access to water is the right of the Moslem community.

No one can refuse surplus water without sinning against Allah and against man. Also animals must not be allowed to die of thirst, and the water which remains after man has quenched his thirst must be given to them.

It would seem that the Prophet Mohammed declared that water should be, together with pasture and fire, the common entitlement of all Moslems,[6] that he prohibited the selling of it,[7] and that he had established a community of water use among men.[8] He also recognized that the ownership of canals, wells and other water sources entailed the ownership of a certain extent of bordering land or *harim* on which it was forbidden to dig a new well so as not to reduce the quantity or damage the quality of the water in the existing ones.[9] This is the concept of 'protected areas' around water points.

In addition, the Prophet advocated the practice of religious institutions and, on his advice, Othman bought the well of Ronna and made it into a *waqf*, i.e., usufruct of a collective property held in trust for religious purposes.

Islamic water law recognizes two important water rights which are regulated in detail by the different schools of law:[10]

(i) the right of thirst (*chafa*), i.e., the right to take water to quench one's thirst or to water one's animals; and
(ii) the right of irrigation (*chirb*), i.e., the right to employ water for watering land, trees and plants.

The schools have developed detailed regulations which deal with specific situations arising in connection with different types of water bodies (lakes, canals, springs, rivers, groundwaters), obligations regarding the construction, operation and maintenance of waterworks and structures, the manner and timing of uses, the procedures for the resolution of conflicts among users, etc.

Islamic water law has also developed a well-established order of priorities for water uses, both as regards the right of thirst and irrigation rights.

---

5 *The Holy Koran*, XXI, Verse 30.
6 El-Charani, Le Cheik (1898) *Kitab al Mizan* (Balance de la loi musulmane). Translated by Perron. Algiers, 388.
7 Yahya ibn Adam, Kitab al Kharadj (1896) *Le livre de l'impôt foncier*. Leiden, E.J. Brill, 55.
8 Ducaurroy, L. (1849) Legislation musulmane Sunnite, rite Hanéfite. *Journal asiatique*. Paris; Van den Berg, L.W.C. (1896) *Principes du droit musulman selon les rites d'About Hanifat et de Chafei*. Algiers, 123.
9 Yahya ibn Adam, *op. cit.*, 75.
10 For more details, see Caponera, *op. cit.*, 13–25.

Other rules regulate in detail fishing rights (complete freedom for everyone), the possibility of sale or transfer of water rights, the constitution of servitudes and rights of way, the extent of the *harim* (forbidden area), the obligations concerning the maintenance of watercourses, etc.

The interpretation of the Prophet's precepts by the different schools and the new rules they established are still followed in the Moslem world, particularly at the users' level.

5.4.4 Groundwater law

In view of the importance of groundwater in Islamic water law and of the fact that such law originally developed from the need to regulate groundwater (springs, wells), this subject deserves special attention. In general, there is no right of ownership on groundwater, which is considered to be a public good. Only wells may be owned, giving their owners exclusive or priority right of use.

As far as groundwater resources are concerned, the Sunnite and Ibadite doctrines recognize the 'right of thirst' (*chafa*) with respect to well water and water points dug on land which is not private property. Water uses are regulated and take into account the requirements and labour of the parties concerned:

(i) wells or water points dug for public benefit. The water is available to all. In case of scarcity, animals may be watered only after man has drawn his share.
(ii) wells dug by nomads. The nomads have exclusive rights to these wells for the entire length of their stay. They cannot, however, refuse surplus water to anyone who is thirsty. After their departure, the wells become public property. The rule of 'first come – first served' then prevails.[11]

In addition, under the Sunnite version of 'irrigation rights' (*chirb*), the digger of a well, whether on his own land or on unoccupied land, becomes the owner of the well water as soon as he has completed the digging.[12] Possession through use is also disputed. The owner of the well is the sole beneficiary of the right of irrigation and is not required to supply water to irrigate other land.[13] He who digs a water point on unoccupied land has, in principle, an exclusive right of irrigation.[14]

Apart from the general Islamic rule of kindness to fellow-beings, the other schools of law make scarce distinctions as regards the extent of the right of irrigation.

According to the Shi'ite doctrine, there is no difficulty when the water supply from springs and wells is sufficient to supply everyone's requirements for irrigation purposes, or when the proprietors agree on the manner of possession. However, the water is divided proportionately to the respective size of plots with due consideration for the location of the land along the canal.[15]

---

11 Ali ibn Muhammad, al-Mawardi (1901) *Traité de droit public musulman*. Paris, E. Leroux, 317–318; Khalil ibn Ishak, al Jundi (1858) *Code musulman par Khalil, rite Malékite*, Section 17.2. Algiers.
12 Ali ibn Muhammad, al-Mawardi, *op. cit.*, 321.
13 *Ibid.*, 319–320; Ahmad ibn Husain, Abu-al-Shuja, al Isbahani (1859) *Précis de jurisprudence musulmane selon le rite Chafeite*. Leiden, E.J. Brill, 90–99; Khalil ibn Ishak, al-Jundi, *op. cit.*, Sections 18 and 19.
14 Ali ibn Muhammad, al-Mawardi, *op. cit.*, 321.
15 Querry, A. (1872) *Recueil des lois concernant les Musulmans Schytes*. Vol.II, Art. 74. Paris, Imprimerie nationale.

Further, within the Sunnite doctrine, the Malikites follow the principle that the owner of a supply of water may sell and dispose of it at will, save in the case of water in a well dug for the watering of livestock.[16] However, the purpose of the sale must be exactly known and stipulated; water cannot be sold *in globo*. The Shafi'ites teach the same doctrine.[17] The Hanifites allow the sale of water in receptacles only. Any other sale of water is void.[18] The Hanbalites observe the same doctrine as the Hanifites.

Finally, in order to prevent new wells from depleting an aquifer, all schools have adopted the principle that the ownership of canals or wells entails ownership of a certain amount of adjacent land; this is the *harim* or forbidden area, varying in size according to different schools.[19]

5.4.5 The process of codification and the Ottoman Civil Code 'Mejelle'[20]

The first written laws governing water rights were issued in Algeria, Morocco, Tunisia, Arabia and Indonesia at the time these countries were under foreign control. They consisted of subsidiary legislation (ordinances, decrees, orders, etc.), issued by the governors of these territories. In general, the following procedure was adopted:

(i) all waters were declared as vested in the state, the crown or incorporated in the public domain; the state thus took the place of the Moslem community.
(ii) every use of water (other than for drinking or animal watering) left free under colonial legislation and *Shari'a* was placed under government control;
(iii) water commissions were set up to survey and recognize established water rights;
(iv) land registers were compiled in order to keep a written record of duly recognized land and water rights.

5.4.5.1 The Ottoman Civil Code 'Mejelle'

The subsequent promulgation of the Ottoman Civil Code Mejelle in the 1870's constituted an important step forward in the codification of Islamic water law.[21] A number of provisions of the code still govern water in some of the countries which had formed part of the Ottoman Empire. As these provisions reflect the old principles of Islamic water law, it is worth quoting them.

---

16 Malik ben Anas (1911) *Le Mouwatta* (*Livre des ventes*). Algiers, 122; Khalil ibn Ishak, al-Jundi, *op. cit.*, Sections 16 and 17, Article 1220.
17 Ali ibn Muhammad, al Mawardi, *op. cit.*, 320.
18 Ya'kub ibn Ibrahim, Abu Yusuf, *op. cit.*, 54; Ibn'Abidin (1869) *Al Dorr al Mokhtar*, Vol. V, 441. Beulag.
19 Ahmad ibn Muhammad, al-Kuduri, *Institut du droit mahométan* 71–72; Malik ben Anas, *op. cit.*, Vol. XV, 189; Khalil ibn Ishak, al-Jundi, *op. cit.*, 387–388.
20 For a detailed analysis of this section, see Caponera, *op. cit.*, 37–40.
21 Nicolais, D., *La Médjellé, Code Civil de l'Empire Ottoman*; Young, G. (1906) *Corps de Droit Ottoman*. Paris, Oxford Clarendon Press.

5.4.5.2 The legal status of water

Article 1234 of the Mejelle Code defines water as a non-saleable commodity to which everyone has a right (*mubah*). This definition applies to running water which has not been appropriated, to water contained in wells dug by unknown persons and to water of the sea and large lakes (Article 1236). Groundwater belongs to the community as well (Article 1235).

5.4.5.3 The right to use water

The Ottoman Civil Code Mejelle recognizes two basic water rights:

(i) water for drinking and animal watering (*hakki chefe*); everyone may quench his thirst from both publicly and privately owned rivers.

Article 1267 extends this right to animals, provided that their number is not so great as to damage the water body. Although the right to water livestock constitutes part of *hakki chefe*, it also applies to privately owned rivers. Article 1268 provides that 'anyone who has on his own property a natural stream, basin or well may prevent any person from trespassing upon his land to obtain drinking water.' However, if there is no other public water source in the vicinity, the owner is obliged to allow the drinking of his water. This also includes permitting persons to cross one's land for this purpose. Even if the water is not offered freely, anyone who wishes to drink may cross another's property on condition that he does not cause damage to the rim of the well or to the sides of the water conduit.

(ii) water for irrigation (*hakki chirb*); in the case of rivers and lakes forming part of the public domain, everyone is entitled to use water for irrigation purposes provided the rights of third parties are not infringed upon. Irrigation canals and ditches, as well as pumping installations, may be freely constructed; however, any work which causes floods, exhausts the water supply of a river or lowers the water level so as to prevent the passage of boats is prohibited.

The use of privately owned waterways is restricted to riparian landowners, unless a non-riparian owner obtains the agreement of all riparian owners.

The sale of rights of way, of irrigation rights and of overflow water from conduits is permitted as part of the sale of the land (Article 216). However, if the owner of a garden sells his land together with the right to draw water from a certain river or canal, third parties having water rights to that river or canal benefit from a right of pre-emption (Article 1015). The landowner may, however, sell his land without the water rights.

5.4.5.4 Maintenance of waterways

The maintenance of waterways is governed by Articles 1321 to 1326 of the Mejelle Civil Code.

The maintenance of rivers forming part of the public domain is incumbent upon the state. However, if the treasury is low, expenses involved may be charged to private persons (Article 1321). The maintenance of privately owned waterways is incumbent upon the owners, who are entitled to draw water for their fields or are allowed to have their animals drink therefrom. Persons having only the right to drink water are not bound to help defray maintenance expenses (Article 1322).

A distinction is drawn between cases in which several persons with *chirb* rights along a jointly owned waterway wish to undertake repair or maintenance operations and those with equal rights who oppose it. If a public waterway is involved, all persons having water rights shall be obliged to undertake maintenance operations jointly; in the case of privately owned waterways, those who do such works, after having obtained the proper authorization from the water master, may prevent the other co-owners from exercising their rights thereto until they have paid their share of the expenses (Article 1323).

In case all persons entitled to *chirb* rights on a jointly owned waterway refuse to undertake maintenance works, they may be compelled to do so for a public waterway but not for a private one (Article 1324).

The responsibility for the maintenance of watercourses begins upstream, and all co-owners must contribute to initial expenses. Provided that upstream riparians have met their maintenance obligations, they are not bound to contribute to the maintenance of the downstream sections. The principle of this provision is that everyone must contribute to maintenance expenses proportionately to the benefit he derives therefrom (Article 1326).

5.4.5.5 The harim

The Mejelle Civil Code defines the *harim* as follows.

The *harim* of a well is the protected perimeter area of 40 *arshuns* (a cubit, corresponding to 0.758 m) around the well where trespassing is forbidden (Article 1281).

The *harim* of all sorts of water bodies (springs, sources, rivers, underground conduits, etc.) is regulated in detail in Articles 1282 to 1291 of the Code.

A well dug by someone on his own *mulk* (private property) has no *harim*. A neighbour can thus dig another well near it on his own *mulk*, and the owner of the first well cannot prevent the digging of that well by saying 'it takes the water from my well' (Article 1291).

5.4.6 Islamic water administration

5.4.6.1 Customary water administration

Islamic law offers very little in terms of water organization or administration.[22] Much of the system that existed then and now was non-Islamic in origin. Islamic law either acknowledged or islamized it according to whether the administration suited to its concepts or not. At no time was there an attempt to centralize the system on a national basis, though at the local level the influence of the Islamic law was strong. Thus the organization and administrative aspects are mostly the result of either local custom or some ancient system.

At the local level, the system of water administration varies according to the locality. Traditionally, water distribution is administered by local water masters with different

---

22 Maktari, A.A. (1975) Islamic Water Law, in: *Proceedings of the Conference on Global Water Law Systems*. Valencia.

appellations (*amir al-ma/kayyal, al-ma/khabir, mirab*, etc.). Practice, however, differs widely among Moslem communities.

In towns and villages, water for domestic use is often administered by the government, but more often by the public. The scarcer the water, the more complex is the organization. The water source or sources on which the domestic supply depends is administered by an overseer who may have assistants, depending on administrative requirements. He may be the head of the community, but when the community is large he may not be so. He is paid by the community by direct contributions from heads of families. This official regulates the turns and methods of extracting water as well as the frequency. Some villages and small communities observe rules such as these without the presence of a water master, but any breach of regulations is referred to the head of the village or community.

In towns where there are abundant water sources, the government may appoint administrators. Frequently, households dig wells of their own or share a well with others. All expenses relating to maintenance are shared on a *pro-rata* basis. Where water is collected in cisterns or reservoirs, this is done on a collective basis.

Irrigation from flood water in the dry beds of intermittent watercourses or *wadis* in Arabia requires a complex organization. The *wadi* is divided into zones (upstream, midstream and downstream). Each zone is administered by a water master known, sometimes, as *shaikh* or *shaikh al-shamal* of the area. These are directly responsible to a supreme *shaikh* of the whole *wadi*. Each zone distributes water according to an order of priorities established pursuant to the following rules:

(i) nearest to the water takes first;
(ii) earlier established water rights are given preference over more recent ones;
(iii) lands located on a higher ground take water before those located at a lower level.

Water is diverted into channels. Each channel is supervised by a channel water master known, sometimes, as *shaikh al 'ubar* or *al-Sharij*, who keeps a record of all water rights of the plots of land attached to the channel. Diverted water is also deflected to plots of land according to a given sequence.

The quantity and frequency of watering depend on the volume of the water; the basic rule which entitles each plot of land to take its full share of water is determined by tradition and agricultural practices, regardless of the volume of the water or the right of other beneficiaries. Allocation of water rights in the *wadi* is determined by tradition. Sometimes this is put in writing.

When flood water gushes through in the *wadi*, the *shaikh* announces it through a traditional system of warning. Farmers are obliged by customary law to respond to this call and participate in water control and distribution. Barrages located at predetermined points are constructed well before the flood occurs. The cost of construction is borne by the beneficiaries on a *pro-rata* basis. Shares are determined by water masters, who also determine the amount of labour each landholder should contribute. Such officials also decide according to local practices the landlord's share and that of the tenants.

The payment of these officials is made by contributions known as *fruq* or *jirayah* or *ujrah*. They remain in charge as long as they act properly and are removable by a majority decision. Often offices of this nature are allocated to sons after fathers, so that for generations whole families act in this capacity, preserving tradition and ensuring conformity to the system. These same officials also act as arbitrators, and in some cases may be selected formally as judges sitting in court applying *'urf*, custom, and *'adah*, common practice. In some instances, customary law is codified.

Much of the constraints in the customary system and obstacles to rationalization are the result of resistance to change. The Islamic customary system of water administration in its various forms is doubtless suitable to the climatic and socio-economic conditions of the areas in which it is found. However, over the centuries the system has tended to become rigid and formal in nature, making change difficult.

5.4.6.2 Recent developments in the administration of water

Although the Islamic water administration and institutions are very much the result of local customs and ancient traditions and vary from area to area, many Moslem countries have felt the need for a centralized water administration.

Thus, ministries of water resources have been established in Algeria (Ministry of Water Resources), in Egypt (Ministry of Water Resources and Irrigation), in Sudan (Ministry of Water Resources and Electricity), in Iraq (Ministry of Water Resources), in Iran (Ministry of Energy and Water), in Lebanon (Ministry of Water and Energy), in Mauritania (Ministry of Water Resources and Sanitation), in Oman (Ministry of Regional Municipalities and Water Resources), in Tunisia (Ministry of Agriculture, Water Resources and Fisheries), in Somalia (Ministry of Water and Energy) and in Syria (Ministry of Water Resources). In some of these countries, centralized and strong specialized water departments and agencies have been created within the ministries. This is the case of Sudan, for instance, since the Ministry also handles electricity matters. The same applies to Tunisia, because the Ministry deals with agriculture and fishing, in addition to water. Elsewhere, water resources management has been vested in ministries with a broader mandate, such as in the case of the Ministry of Equipment, Transport, Logistics and Water of Morocco, which performs water management functions through a State Secretariat responsible for water.

As elsewhere in the world, there is a trend in Moslem countries to vest water resources management functions in river basin institutions. In Algeria and Morocco, for instance, river basin agencies have been created on the basis of the provisions of recently enacted water legislation.

5.4.6.3 Government action in Moslem countries

The Mejelle Code regulated the use of water in all countries that had belonged to the Ottoman Empire, that is, until 1911 in Libya and until 1922 in Somalia, in the Arabian peninsula, including Saudi Arabia, Jordan, Iraq, Iran, Syria, Lebanon and Turkey.

With the fall of the Ottoman Empire, Turkey became a secular republic and, in 1926, promulgated a new civil code, and, in 1956, a new groundwater law.

Some former Ottoman territories were placed under French or British mandate: Syria, Lebanon, Iraq, Trans-Jordan and Palestine. During this time, although the Mejelle Code continued and still continues, at least partially, to govern the use of water in these countries, other water laws were promulgated.

Most of these territories later became independent. In some of them, like Jordan and Iraq, new laws on water were issued to complement not only the provisions of the Mejelle Code still in force, but also the laws enacted by France and Great Britain in their capacity as trusteeship powers. These laws provided for public ownership of all waters and for existing water rights to be surveyed, recognized and registered by special committees set up for this purpose.

Other territories, such as Saudi Arabia, Kuwait, Qatar, the United Arab Emirates, Yemen and Oman, became independent Moslem states. These countries, notably Saudi Arabia, declared the provisions of the Mejelle Code to be no longer in force and re-established the sacred principles of the *Shari'a*. Except for Yemen, in this group of countries no new water laws were enacted to modify the traditional principles of Moslem customary law; however, some attempts to draft new water codes have taken place.

Egypt practically ceased to be part of the Ottoman Empire in 1830, although theoretically it remained under Turkish sovereignty until 1922. Since time immemorial Egypt has developed rules on the use of water. At the end of the nineteenth century, it promulgated numerous water regulations.

In some of the states which have since acquired independence, waters are still governed by the laws enacted during the period in which they were under foreign rule. In these states, water legislation, although occasionally respectful of local traditions and customs, was largely imported or adapted from France or Great Britain and superimposed over Moslem customary law. Basically, the public and private ownership doctrine of French continental law and the riparian doctrine of the common law of England have left their imprint which is still strong today.

In many Moslem countries, a variety of legal systems continues to coexist as regards water resources: Moslem customary law, which is followed at the local level, particularly in the oases, French or English legal provisions and more recent provisions included in post-independence water legislation.

More recently, technological development and increased demand for water have given a new impetus to the policy for a codification of water law. Comprehensive water laws were promulgated in Iran in 1968, in Tunisia in 1975, in Algeria in 1983 and 2005, in Morocco in 1995 and 2016, in Yemen in 2002 and in Indonesia in 2004. Similar developments have reached various stages of progress in Jordan, Libya, Saudi Arabia and Syria. In these and in other countries such as Bangladesh, Egypt, Iraq, Lebanon, Pakistan and Tunisia, central water administrations have been created and entrusted with the overall control of water management.

It appears that in Moslem countries modern codifications of water law aim at institutionalizing, in one form or another, the concept of community of interest in water resources, which constitutes the traditional basis of Moslem customary water law.

5.5 Water law principles in civil law countries

5.5.1 Introduction

The countries which have derived their legislation from the principles elaborated in the French Napoleonic Code, promulgated in 1804 after the French Revolution, are indifferently referred to as 'civil law' or 'code' countries. They include France, Italy, Belgium, Spain, Portugal, the Netherlands and others.

It is important to know that the feudal system, in which there was no concept of private ownership of water, and the feudal lords and the king had full control over land and water within their jurisdiction, definitively ended with the French Revolution; a new social class, the bourgeoisie, emerged with new economic interests.

As a consequence of the new economic and political order, the Napoleonic Code classified water into two categories: private waters and public waters. According to this definition, private waters were those located below, along or on privately owned land; public waters were those which were considered to be 'navigable' or 'floatable.' This is understandable, as under the new set-up introduced by the French Revolution every object necessarily had to belong to someone, either a private person, or the state or public domain. The third legal status of water, i.e., the concept of waters common to everybody (*res communis omnium*) which existed under Roman law, was eliminated from the new legislation, contrary to what was retained in the common law of England. Likewise, the former pre-revolutionary feudal system of land and water ownership disappeared. This constituted an important departure from the original Roman system which considered three types of ownership: private, common and public.

Through the French Napoleonic Code, the civil law system was subsequently introduced in Spain, Portugal, the Netherlands, Belgium, Italy and in other continental European countries. This system has naturally strongly influenced the water laws and institutions of the countries and territories which derived their system from the European countries just mentioned, or which were under their cultural or political influence before achieving full sovereignty. Thus, the system has expanded throughout French-speaking countries and in France's former colonies, possessions and territories; it has at various times been adapted to local religious or geo-climatic requirements. In particular, the civil law system has been instrumental in establishing the water laws and administrations of North African and other African countries, Laos, Cambodia and Viet Nam (former Indochina), French speaking Canada, Louisiana (USA) and other francophone former colonies in the Americas.

It is to be noted that Spain, a civil law country, has developed an elaborated system of water law of its own and has been instrumental in shaping water legislation in most Latin American countries, in Equatorial Guinea (Africa) and in the Philippines (Asia). Spanish law itself was strongly influenced by Islamic law introduced by the Moors who occupied Spain and Sicily for centuries. The Spanish writer Joaquin da Costa states: 'The Arabs passed through Spain; their race, religion, codes, temples and graves have disappeared but their memory is undoubtedly still alive, because their irrigation has survived.'

In Central and South American countries, the native customs relating to water have influenced the water laws imported by Spain. The Pre-Columbian civilizations knew irrigation practices very well and had special regulations and institutions concerning water.

From Mexico, the original Spanish water laws were introduced to several southern states of the USA, colliding with the principles of the common law of England brought from the eastern states.

The Portuguese water law system was introduced in the former Portuguese colonies of America (Brazil), Africa (Angola, Mozambique, Guinea Bissau, Capo Verde, São Tomé and Principe), and in Asian territories.

Central European civil law countries such as Germany, Austria, Hungary and Switzerland have developed slightly modified systems of water legislation.

A variation of the doctrine, which has found its best expression in the Swiss Civil Code but which is also applied in Germany and Austria, as well as in Brazil and in China, protects the landowner against any new enterprise of his neighbours, who are not allowed to exhaust a spring by cutting the subsoil water source which feeds it.

The Dutch laws were introduced in Indonesia, Sri Lanka and South Africa, while Belgian legislation was applied in Zaire (now the Democratic Republic of Congo), Rwanda and Burundi. Italian water law principles were exported to Somalia and Libya.

For this section, as it is not possible to cover in detail all countries which have followed the civil law system, the development of the French system will be mainly followed.

5.5.2 The legal status of water resources

According to the civil law system, water may be either public or private. Public waters are those belonging to the public or national domain, and their utilization is subject to a government permit, authorization or concession. In France, until the law of 8 April, 1910, only 'navigable and floatable' waters belonged to the public domain. From that date, other types of water could be included in this category if acquired by the state for the purpose of public works. With the Water Law of 16 December, 1964, still other waters were included in the public domain, such as those necessary for domestic water supply, agricultural and industrial production.

The departure from the physical element of 'navigability and floatability' for classifying a water as public and the introduction of the new criterion of 'public interest' constitute important legal elements which have substantially limited the sector of the 'privately owned waters.' It is to be noted that the law of 1964 no longer speaks of 'private waters,' but of 'non-domanial waters,' and specifies that the government may always expropriate land, including the waters therein, in the public interest.

Under this system, private waters are those which, not being public nor declared as such, may be freely utilized on the basis of the riparian doctrine without the need for administrative intervention. Rainwater and spring waters, these latter only provided they do not become 'public flowing waters' when leaving a privately owned land, are considered private waters.

Flowing waters are considered as public or collective goods either on the basis of the civil code or of water legislation. This means that the individuals may only claim rights of use over such waters.

The régime of underground waters under the civil law system is rather ambiguous. For example, according to Article 552 of the French Civil Code, 'The ownership of land includes the ownership of anything above and below. The owner may undertake underground . . . any research which he wishes and abstract the relevant products therefrom . . .' Water, as an accessory of land, is one of those products. However the landowner only owns the water he is able to abstract; he has no rights on the aquifer itself, so that if his well dries up because of the activities of a neighbouring user, he can claim no compensation.

In France, with Law No. 92–3 of 3 January, 1992, all water resources, whether surface or underground, have become a common asset of the nation. Their protection and development must be consistent with the ecological balance, in the general interest. Thus, the distinction between private and public waters has ceased to exist, and water may only be subject to use rights.

5.5.3 The right to use water

Under the civil law system, one or more basic legal texts (constitution, land law, civil code, rural code, or other legal enactment) define the public domain, i.e., a category of goods which cannot be the object of private ownership. Generally, this category includes 'public waters.' As a consequence, the utilization of 'public domain waters' is subject to administrative authorization, permit or concession. Land laws, water laws or other legal enactments regulate the administrative régime of these waters.[23]

Conversely, privately owned waters (i.e., those not being public waters) can be freely utilized, subject to certain limitations of a statutory nature, such as servitudes or rights of way. The right to use private waters, both surface and underground, derives from land ownership, which recognizes to the owner the right to use at pleasure the water existing upon his land without limitation (*ius utendi et abutendi*). Yet the practice of the courts has limited this absolute right of use by making it subject to numerous restrictions, particularly as regards the prohibition to pollute water, the care to be exercised like the 'father of the family' and the protection of the use of water belonging to the inhabitants of a village.

In France, for instance, a law of 8 April, 1898, amended Sections 641 to 643 of the French Civil Code by allowing the landowner to change the conditions governing the outflow of natural waters, provided he compensate the owner of the lower grounds; the landowner is also forbidden to use his source to the prejudice of those owners who had erected, not less than 30 years before, works for utilizing the waters or for facilitating their passage on their own land, or in any other manner which might deprive a village of the water. In addition, the landowner cannot divert from its natural course to the prejudice of the lower riparian users, the water which, after leaving the property on which it rises, forms streams having the character of public and running waters. Court decisions also prevent the owner from using his right for the sole purpose of damaging his neighbours. Conversely, there is an obligation of the downstream landowner to receive the natural flow of the water from the higher land (drainage easement). Any damage is subject to compensation.

Law No. 92–3 of 3 January, 1992, has changed this situation by requiring all those who intend to use water to effect a declaration or to obtain an administrative authorization, depending on the type of water utilization. Those using water for domestic purposes are exempt from this requirement. It is worth noting that, while the authorizations and permits granted for the use of public waters are of a precarious nature and may be modified or revoked without the payment of an indemnity, as is the case of authorizations for the use of anything in the public domain, prior to the entry into force of Law No. 92–3 the authorizations issued for the use of non-domanial waters could not be revoked without the payment of compensation, unless otherwise provided for by law.

Finally, the right to use water by virtue of an authorization is subject to the payment of a water rate or fee. Fees may also be collected in the case of water-related activities such as, *inter alia*, the extraction of gravel from river beds and for river crossings. Water users under the permit system are required to pay water rates, fees or charges, not only in France, but also in other civil law countries.

5.5.4 Water quality and pollution control

There has been a growing concern for the maintenance of water quality. While in the past provisions for the protection of water against pollution were contained in criminal or penal

---

[23] In France, the Fluvial Public Domain and Inland Navigation Code prohibits the execution of any work in public watercourses without an administrative authorization; the expression 'work' includes both water diversions and discharges.

codes, a large number of legal enactments have been issued in civil law countries dealing specifically with water pollution control and environment protection, with a view to reconciling the public interest in economic growth and industrial development with public health and environmental requirements. Such provisions have been included either in newly enacted water laws or in environmental codes, or in separate legal enactments regulating specific types of pollution.

The need for controlling water pollution has resulted in new types of authorizations for the discharge of waste (organic, inorganic or radioactive) into water bodies. The same concern has obliged administrations to define specific standards relating to each type of water use and discharge. Also, water and environment protection legislation and the building licencing authorities now require the setting up of plants for treating effluents before they are discharged. Moreover, fees to be paid for waste and wastewater discharges have been introduced, in order to implement the 'polluter pays' principle.

The main problems that the pollution control issue raises refer to:

(i) coordination between the authorities directly or indirectly involved: water authorities, health authorities, national authorities, regional or local authorities;
(ii) the difficulty of applying the criteria and procedures elaborated in industrialized countries to developing countries at large;
(iii) the political difficulty encountered by the decision makers to implement actions relating to pollution control and environmental protection.

In France, the Water Law of 16 December, 1964, introduced important provisions concerning the control of water quality. For the purpose of implementing this law, a permit system for wastewater discharges was introduced, together with the pricing of discharges. Six financial basin agencies responsible for pollution control were instituted. Subsequently, these agencies have been replaced with water agencies (*agences de l'eau*), with a view to integrating water quantity and quality aspects into institutions operating at the river basin level.

5.5.5 Water administration

Until relatively recently in civil law countries no centralized water administrations were in existence, with the exception of those in charge of 'public waters,' i.e., the waters belonging to the public domain, which were generally administered by the ministry in charge of the public domain or registry (normally the ministry responsible for economics and/or finance). This trend has been reversed in the past decades, and water resources management functions have been vested in ministries responsible for water, or in charge of the environment. Often water resources are administered under the overall umbrella of environment protection, through specialized water departments.

Other government administrations responsible for sectorial aspects of water resources management (development and conservation) in civil law countries include the ministries of agricultural development and cooperation, irrigation, planning, public works, transport and communications, public health, through their departments of rural engineering, water and forests, fisheries, bridges and roads, cadastre, mining, urbanism and habitat, meteorology, hydraulics, hygiene and public health.

Apart from these government ministries and departments, other more or less autonomous institutions (government mixed institutions or private companies) have been set up for developing particular water utilizations such as municipal or domestic water supplies, power generation and transmission, agricultural development, ports and harbours, wastewater management.

Finally, since the river basin has been recognized as the ideal unit for water resources management, it being easier at this level to integrate all resource management aspects, there has been a trend towards creating river basin institutions. The *Confederaciones hidrográficas* of Spain offer a good example of integration of functions at the basin level. The water agencies of France play a key role in the development of river basin management plans.

Furthermore, particularly in Europe, there is a move towards allowing stakeholders to participate in decision making with regard to water resources management and planning.

5.5.6 Conclusion

In France and in many other civil law countries, the increasing intervention of the administration and the introduction of the 'water use permit system' have rendered obsolete the former subdivision between public and private water ownership, and all water utilizations, independently from the legal nature of water, have been submitted to regulatory control. The category of private waters, either surface or underground, has generally been discarded in new water legislation. This is in the case of France, Italy and Spain.

Permits for the discharge of wastewater have been introduced in parallel with those for the use of water. Also water rates and fees, both for the use of water and the discharge of effluents, have been introduced and are assessed on the basis of the prevailing economic criteria of particular areas, basins or types of use.

Coordinating institutional machineries, such as national water boards, councils or committees, or as parts of an overall environmental council, have been set up.

From a pragmatic point of view, it may be said that the differences in the legal and institutional framework for water resources management between civil law and common law countries, and particularly those relating to the ownership status of water, have slowly disappeared; this, as a consequence of the very nature of water, which calls for the implementation of certain measures, independently from the legal and institutional framework of the country where the water is located.

It is important to note that the European Union (EU), through the Water Framework Directive,[24] requires member countries to take measures in the interest of a better protection and conservation of water resources. These countries, and those which have applied for EU membership (candidate countries) must transpose the provisions of the directive into their legal frameworks. Member countries have taken steps in this direction and are now in their second river basin planning cycle.

---

24 Directive 2000/60/EC of the European Parliament and of the Council of 23 October, 2000, establishing a framework for Community action in the field of water policy. *Official Journal*, L 327, 22/12/2000, 1.

5.6 Water law principles in common law countries

5.6.1 Introduction

The ancient Roman conception of the landowner's exclusive right to use (and abuse of) all the water springing or flowing on his land without regard to what happens downstream or upstream was incorporated into the English common law system. Generally, under this system no ownership of water is possible, either public or private. Not even the Crown can own water. The ancient Roman conception of water resources being common to everybody (*res communis omnium*) has been maintained.

English water law is part of the English common law system and has given birth to the riparian doctrine. This system has influenced the water laws and institutions of the countries which have derived their legal and institutional system from England. In the nineteenth century the system of riparian rights was instituted throughout the British colonies. In several cases, however, in order to cope with the requirements of the regions to which it was exported, the original doctrine was modified, and new concepts of law for water ownership, administration and control were created. In countries with a long-established fluvial system such as India and Egypt, where the British were primarily the governors of large indigenous populations, they instituted complex and technically advanced systems of government control of water.

In Australia, the original common law principles had to be modified and adapted to the conditions of water scarcity and underpopulation of the country. Detailed legislation was enacted in order to extend the right to use water to non-riparians, thereby encouraging irrigation in the hope of achieving a closer settlement upon the land.

In the USA, 31 states, starting from the east, have adopted the 'natural flow' theory of the riparian doctrine, though limiting it in its extent through the so-called 'reasonable use' doctrine.

In countries of the British Commonwealth such as New Zealand and Canada,[25] where the climatic conditions are similar to those of England, water laws could be developed, and continued to survive, on the same lines as those whence they originated.

In many countries following the common law system, laws, ordinances, regulations or other legal enactments (principal or subsidiary) have been issued for administering or regulating specific aspects of water management. Arising from local conditions, water enactments deriving from the common law of England are not global, but pertain to a special region and often to a specific watercourse.

Until recently, this multitude of laws has seldom been unified or consolidated into one comprehensive water act. Every specific use of water has been the subject of special legislation, causing a proliferation of legal enactments. In addition, most of the water legislation issued has been concerned with domestic water supplies, sewerage or fishing, with less importance attached to other uses such as irrigation, agricultural uses and hydropower generation. Other legal enactments applied to waterworks, mining, quarrying, railways, local government, land drainage, transport, public health, town and country planning, oil in navigable waters, etc., which were incidental or relevant to water resources. This trend has now been

---

25 With the exception of Quebec, which derives its legislation from the French civil law system.

reversed, and many countries have opted for consolidated bodies of rules on water resources, covering both quantity and quality aspects. In some countries, water laws are complemented by laws on environment protection.

In countries following the common law system, customary law generally plays an important role in the legislative process. A large part of the law in England and the USA still remains attached to customary law under the appellation of common law. In these countries, the sphere of customary law in the field of water has been gradually restricted due to increasing intervention of water legislation and modern water policies and administration.

For practical purposes, rather than cover in detail all the countries which have adopted the common law system, the evolution of this system in the United Kingdom, Australia and Canada is illustrated herebelow, as these countries are more representative of recent trends.

5.6.2 The legal status of water resources

Under the common law system, of which the riparian doctrine is the expression, there can be no ownership of or right of property in the running water of streams, rivers or natural channels. This water, regarded as transient and fugitive, is *res communis, publici juris*, i.e., it is common to all who can claim a right of access to it, and may be used in a reasonable manner by a riparian landowner. Only the water which accumulates or falls on one's land and is collected in artificial or natural drains and reservoirs may be privately owned; such private ownership is limited to the time of possession. The same principle applies to underground water, which becomes the property of the person who abstracts it and retains it in his possession.

The concept of ownership is relevant only in relation to the bed or to the banks of a river and differs according to whether the river is tidal or non-tidal. According to court decisions, a river is tidal only as to such lengths as are within the regular ebb and flow of the highest tides; the bed of the river in those lengths, and of any estuaries, vests *prima facie* in the Crown as far up the river as the tide flows and up to the high water marks of ordinary tides along the shores of estuaries. The bed, however, may be the subject of statutory title, or the Crown's title may be granted to an individual. In any event, ownership is subject to the respect of public fishing and navigation rights as well as Crown's titles.[26]

As regards non-tidal rivers, there is the presumption, in the absence of evidence to the contrary, that the ownership in the bed vests in the owners of the adjacent land up to the middle of the river. The owner of land on both sides of the river is presumed to own the whole bed thereof. The public has no right to fish in such a river, except with the licence or consent of the riparian owner, but a non-tidal river which is navigable is subject to a public right of navigation.

As regards lakes and ponds, their soil and bed vest in the riparian owners up to the middle of the water body (*ad medium filum aquarum*).

The position regarding ownership and riparian rights of underground water is the same as that of surface water when it flows in a known and defined channel. It will depend on the facts of each case whether or not a particular channel is known and defined. Underground

---

26 Rees v. Miller (1882) 8 Q.B.D. 626, and Lord Fitzhardinge v. Purcell (1908) 2 Ch. 139, in: Richardson, H.J. (1975) System of Water Law and Organizations in the United Kingdom. *Proceedings of the Conference on Global Water Law Systems*. Valencia, F-9.

percolating water may be abstracted by the owner who finds it under his land without regard to the effect this may have on the supply of water to springs or other wells and boreholes.

The status of artificial watercourses is determined by their nature, the circumstances under which they are constructed and the use to which they are put. For example, an artificial watercourse may have been constructed under such circumstances, and have been so used, that the riparian owners have acquired the same rights as they would have acquired had it been a natural watercourse. Conversely, an ancient watercourse constructed and maintained solely for the use of a mill may give the purchaser of the mill no right to the use of the water in the watercourse.

Under this system, only court decisions, administrative ordinances or regulations may limit or restrict the reasonable right to use water by the riparian owner.

The legal status of water under this system is peculiar to England, where it originated. The interests involved are domestic water supply, navigation and fishing.

5.6.3 The right to use water

The common law system entitles *jure naturae* (i.e., by the law of nature) the owner or occupier of land adjacent to a natural stream to use and enjoy the water flowing past his land for the very reason that he owns or occupies the land. Such use takes place *ministerio legis*, i.e., by virtue of the law, without the administrative interference of the authorities. The right to the use of flowing water is known as the 'riparian doctrine.' Furthermore, every landowner whose property abuts the banks of a stream or body of water is entitled to have the water pass his land (or maintain a natural lake level) undiminished in quantity and not modified in quality. This is the 'natural flow' doctrine.

While the riparian owner has the right to use the water flowing through his land for purposes not inconsistent with the rights of other riparian owners upstream and downstream, an upstream landowner cannot diminish the quantity or injure the quality of water which would otherwise descend, nor can a downstream proprietor block the natural water flow without the licence or consent of the upstream proprietors. Any unreasonable and unauthorized interference with the use of the water by a person entitled to it may be subject to an action for damages and may be restrained by injunction, even if there has been no actual damage to the plaintiff.

For a riparian owner whose land abuts a navigable river, the right of access to a stream includes the right to moor vessels alongside his frontage.

Under common law, a riparian owner may abstract water for his own domestic purposes, i.e., for drinking and culinary purposes, cleansing and washing, and to satisfy the ordinary needs of livestock. If abstraction for domestic purposes exhausts the water, downstream riparian owners cannot complain. Nevertheless, the right of a riparian owner to take water is not limited to domestic purposes. He may exercise his right for extraordinary purposes, provided – according to scholars and court decisions – that he does not interfere with the right of other riparian owners, that the use is reasonable and connected with the riparian tenement, and that the water is returned to the river undiminished in quantity and unaltered in quality.

In the United Kingdom, the common law riparian system existed unchanged until 1963, when licencing for the abstraction of water was imposed by statute.[27] Nevertheless, many

---

27 England and Wales, *Water Resources Act 1963*, Chapter 38.

features of riparianism were retained. The Water Resources Act of 1963 divided England and Wales into 29 areas, each under the jurisdiction of a river authority responsible for authorizing water abstractions.

The act limited the number of recognized water users to those who either had the right to occupy land contiguous to that water, or had acquired rights of access to that water, or were given the right to acquire contiguous land compulsorily. With the exception of the last category, these users corresponded to the riparian owners at common law; for some uses they had to obtain an authorization, which could be refused at the discretion of the competent authority, and had to pay a fee. No licence was necessary for the abstraction of up to 1,000 gallons of water for any purpose, provided that this did not form part of a continuous operation, or of a series of operations, whereby in the aggregate more than 1,000 gallons of water were abstracted. In addition, the owner of land contiguous to inland water could abstract an unlimited quantity at the place where his land was contiguous to that water without a licence, not only for domestic, but also for agricultural uses (excluding spray irrigation) on a holding consisting of contiguous land 'with or without other land held therewith.'[28]

The Water Resources Act of 1973 reduced the number of river authorities to ten; these river authorities were mainly vested with water supply and sewerage functions and, to a limited extent, with the control of recreational (fishing) uses of water.

5.6.4 Water quality and pollution control

Under common law, a riparian owner has a right to receive the water flowing past his tenement in its natural state, unimpaired in its natural quality. If anyone pollutes the water, the riparian owner is entitled to bring action against him without the need to prove actual damage and may obtain an injunction to restrain from continuing to pollute, unless the polluter demonstrates that he has obtained a legal right to pollute.

The owner of a fishery has a right of action against anyone who discharges injurious or offensive matters into a river and unlawfully disturbs the exercise or enjoyment of the fishery by polluting the water and killing or driving away fish. The owner of a well and the owner of an off-shore oyster bed have similar legal rights with respect to pollution.

In an action against pollution it is not necessary to show deterioration of a stream in general, but that something has been added to the water affecting the purity and quality thereof at the point where the offending matter enters the stream. What has to be decided is whether the offending matter would appreciably pollute the stream if its waters were otherwise pure. It is no defence in an action against pollution to show that the river is already polluted from other sources; where a number of manufacturers cause a nuisance to a downstream riparian owner by discharging polluting matter into a stream, he will have a right of action against each of the polluters.

A legal right to pollute water may be acquired by statute, prescription, custom or grant. In England, for instance, an act of Parliament may empower a public body to discharge effluents purified to a stipulated standard of quality, and if that standard is complied with,

---

28 England and Wales, *Water Resources Act 1968*, Sect. 24(2).

there will be no action even if pollution occurs, because the statutory provisions supersede all common law rights.

A right to pollute a watercourse may also be claimed as an easement by a user or otherwise. In England, under the Prescription Act of 1832, a prescriptive right has been established to discharge mine water which has been impregnated with metallic substances to another person's watercourse. Prescription may only be claimed, however, for some act which can have a lawful origin under common law, but never for actions which would cause a public nuisance or be injurious to public health, or contravene a statute.

A private person whose rights have been injuriously affected may be entitled to claim damages and an injunction to prevent the continuation of the injury. An injunction will not normally be granted if the nuisance is not likely to occur and the payment of damages is considered by the court to be an adequate remedy; an injunction has also been refused in a case where the nuisance had largely abated since the action began and the defendants have taken steps to repair the injury.

Where an act of pollution causes a public nuisance, an aggrieved party may apply for the sanction of the attorney general to institute civil proceedings on behalf of the public, or the attorney general may himself institute such proceedings.

5.6.5 Recent developments of the common law system

At the end of the nineteenth century and the beginning of the twentieth, conditions were suited for a change from the riparian system throughout the British dominions. Indeed, it is possible to relate the development of irrigated agriculture to the abandonment of riparianism and the introduction of a permit system. For instance, in the period immediately preceding World War I the state of Victoria, first among the Australian states to abandon riparianism, became much more advanced in this sector than its neighbours South Australia and New South Wales, which were slow in changing their laws. At the beginning of the twentieth century the Canadian Provinces of Alberta, Saskatchewan and British Columbia vested all rights on water in the Crown, preserving only domestic and pre-existing riparian rights. For other purposes, a licence was required.

In England and Wales a permit system was introduced by statute in 1963 and was further developed over the years. A comprehensive licencing régime for large-scale water abstraction is now being implemented by virtue of a water act which received royal assent in 2003. The Australian states, also, have gradually introduced a system of licences that with time has become rather comprehensive, such as in New South Wales, Victoria and the Northern Territory. In other common law countries attempts to establish a permit system have encountered difficulties of implementation for lack of adequate legislation, shortage of qualified manpower, financial constraints, extended size of the countries and, importantly, because of the resistance of the populations to the introduction of any control over their water activities, often based on customary rights.

It is worth mentioning the efforts made in England and Wales to privatize the water industry, i.e., the provision of water supply and sewerage services to the public. Privatization has taken place through the transfer of the service provision functions of the ten water authorities to the private sector, i.e., to ten water companies operating under licence. However, since private companies usually are profit-driven, tight controls have been imposed by statute to ensure the quality of the services and that the services are provided to the advantage of the customers. In particular, a charging scheme has been introduced, by which the service charges proposed by the companies are subject to government approval.[29] The Office for Water Services (OFWAT), headed by a director general, was responsible for supervising the overall performance of the companies. The Water Act of 2003 has provided for the replacement of the director general with a Water Services Regulation Authority having the duty to protect the interests of consumers. The act has further established an independent Consumer Council for Water. The Water Act of 2014 has increased the opportunities for competition in the provision of water services.

The privatization option has been pursued more or less successfully in a number of other countries, including countries following the common law system such as Australia (Victoria) and Belize, and countries following other traditions. In this connection, in recent years policies focusing on the reform and privatization of water services have been implemented in several African, Asian and Latin American countries. These policies, which called for private investment and a commercialization of operations at all levels, have led in some instances to a substantial increase in water prices and electricity tariffs, thereby producing considerable adverse impacts on the poorest strata of the population, both in rural and in urban areas.[30] However, in the era of globalization there are trends towards the liberalization of the provision of services that until now were considered as essential and therefore public.[31]

As was mentioned earlier,[32] the differences between the legal and institutional framework for water resources management in civil law countries and that in common law countries tend to disappear because water resources management calls for the implementation of measures which apply universally. Moreover, the EU requires member states and the states applying for membership to approximate their legislation to the Water Framework Directive, among others. In England and Wales steps have been taken in this direction.

5.6.6 Water administration

Generally speaking, under the common law system no centralized organization for allocating or recognizing rights to use water has been in existence until relatively recently. Usually, there have been many water authorities or boards at the central, basin or local level, mainly concerned with domestic water supply, sewerage and fishing.

Furthermore, diverse ministries have been directly or indirectly responsible for or interested in the sectorial aspects of water resources, including those responsible for environment protection, agriculture, public works, public health, planning, finance, transport and communications, interior or local government, through departments in charge of irrigation, electricity, mines, municipal corporations, counties, lands, drainage, forests, harbours, and others.

This fragmentation is due to the fact that in the past institutions were created to deal with a particular use of water, a harmful effect or other water-related issue, as the need arose.

---

29 Nevertheless, there has been a considerable increase in prices.
30 Examples of not fully consumer-friendly privatization of water services include Cochabamba in Bolivia (concession to Bechtel, 1999), Metro Manila in the Philippines (concession to Suez-Ondeo, 1997), Jakarta (partnerships led by Suez and Thames, 1997), Buenos Aires (concession to Ondeo-Aguas Argentinas, 1993) and Dar es Salaam (lease contract with City Water, a consortium led by the British company Biwater, 2003).
31 See, for instance, Directive 2006/123/EC of the European Parliament and the Council of 12 December, 2006, on services in the internal market. *Official Journal*, L 376, 27/12/2006, 36.
32 See this chapter, Section 5.5.6.

With time, however, governments have come to realize that this proliferation of centres of power constrains water resources management, especially if a river basin approach is adopted and the integrated management goal is pursued. Thus, water resources management and water rights administration functions are now increasingly being vested in centralized institutions, such as in the case of England and Wales, where water abstraction and impoundment licences are issued and administered by the Environment Agency. If circumstances so warrant, certain water resources management responsibilities are placed in the hands of river basin agencies or boards, which are closer to water users.

5.7 Water law principles in the former Soviet system[33]

In spite of the dissolution of the Soviet Union and of the abandonment of the socialist system by the countries under its influence, we will briefly describe the Soviet system of water law, as it continues to influence the post-socialist legislation of these countries.

5.7.1 Fundamentals of Soviet water law

The Soviet system of law, which had nothing in common with other conceptions of law, as it was viewed expressly within the framework of national economic development, was deemed to be in constant evolution. From the economic viewpoint, water was considered as included in land. A passage from Karl Marx's *Capital* reads as follows: 'Land which includes from the economic standpoint water as well, as it naturally provides to men their riches, means of subsistence, appears to us, without the men's contribution, as the general object of human work.'[34]

According to Soviet theories, a water body could be used at the same time for many purposes, falling within many branches of water economics. A decree of the Central Committee of the Armenian SSR stated that 'The utilization of all water sources includes the utilization of several spots of surface land, embankments, irrigable land, canal land, etc., so that the utilization of the sources without also utilizing in one way or another the surface land is impossible.' This is the reason why water law was strictly bound to land law.

The exclusive right of state water ownership was consecrated in Article 6 of the Soviet Constitution, which provided that 'waters, as the land, the subsoil and the forests, are the property of the Socialist Soviet State, that is, a matter of all people.' The Land Law of 28 October, 1917, abolished the private ownership of water.

There were numerous sources of water law in the USSR, including decrees, resolutions, orders, regulations and instructions.

The 'Fundamentals of Water Legislation of the USSR and the Union Republics,' approved on 10 December, 1970, contained basic principles governing the utilization and protection of rivers, lakes, seas, reservoirs, and other surface and underground water bodies located within the territory of the Soviet Union. They had a legal force superior to that of

---

33 Kolbasov, O. (1975) Water Law in the USSR. *Proceedings of the Conference on Global Water Law Systems*. Valencia.
34 Marx, K., *Das Kapital*.

decrees, resolutions, orders, regulations, principles, instructions and other legal enactments regulating water utilization and protection.

On the basis of the Fundamentals, fifteen water codes were adopted in the Union Republics, most of which now form part of the Commonwealth of Independent States.

In addition, in order to regulate water relations at all levels (Union, Republic, region, district, local), subsidiary normative acts such as decrees, resolutions, orders, regulations and instructions were adopted. Provisions relating to water were also to be found in resolutions and decrees on environmental protection.[35]

5.7.2 The legal status of water

Consistent with Article 6 of the Soviet Constitution, Article 3 of the Fundamentals proclaimed the right of exclusive state ownership of water resources.

The same Fundamentals drew a distinction between water and water resources. While water was a dynamic substance, and as such could not be the subject of exclusive state property but could be appropriated by individuals and organizations as a result of lawful activities such as the provision of drinking water from municipal water supply systems, water resources were natural water reserves constituting separate material wealth. In the Soviet concept, water resources included seas, lakes, rivers, reservoirs, canals, sources or basins of underground water, mountain and polar glaciers, as well as other surface and underground bodies and sources. Due to the great diversity of water resources, the unitary notion of water body was introduced. The whole complex of water bodies constituted integrated state water resources, and their ownership was vested in the state.

Water legislation envisaged three main categories of waters: surface waters, underground waters and glaciers. Surface waters were seas, lakes, rivers, reservoirs, other water bodies and water sources. Seas could be inland or territorial, while the other water bodies could be inland or boundary. Underground water resources, which were all considered to be inland, could be fresh or saline, cold or hot (thermal). Glaciers were never classified.

5.7.3 The right to use water

Water users could be state cooperatives, public enterprises, institutions, organizations and citizens.

Water uses were classified according to their economic objectives. A first distinction was drawn between general water use and special water use.

General water uses were those carried out by individuals for simple household consumption, swimming, livestock watering, game fishing, etc.; these uses took place without the help of structures or technical installations and did not affect the condition of water bodies. Therefore, they were not subject to an administrative permit by state agencies and did not require detailed regulations.

---

35 Decree of the CPSU Central Committee and the USSR Council of Ministers *On Strengthening the Environmental Protection and Betterment of the Utilization of Natural Resources* of 29 December, 1972, *Collected Decrees of the USSR*, 1973, No. 2, 6.

On the contrary, special water uses were those which took place with the help of structures or technical installations and which could cause the deterioration of water resources and, consequently, damage to the state and to other water users. In these cases, preliminary research and project studies, together with state control, were normally required. Special water use permits were granted by the agencies responsible for regulating the utilization and protection of water resources, in consultation with the agencies responsible for state sanitary inspection, for the protection of fish resources and other agencies concerned, based on favourable advice accorded by the local authorities. The procedure for the granting of permits was established by the Council of Ministers.

A second distinction was the one between joint and exclusive water uses. Joint water uses were those whereby a water body was utilized by two or more water users having independent rights. Under an exclusive water use, a water body, or a portion thereof, was made available to one user to satisfy his needs. This was the case of water bodies made available to collective farms, fish-breeding farms and the like. Exclusive water users could allow other enterprises, institutions, organizations or citizens to use water of the water body or portion thereof (secondary or dependent water use).

A water use could be permanent or temporary. The temporary water use could be either short-term, i.e., up to a period of three years, or long-term, i.e., from three to twenty-five years. Its duration could be extended for a period not exceeding the initial one. The duration of general water uses was unlimited.

All water users were under the obligation to utilize water rationally. This was particularly important for comprehensive water uses. In this case, a water body was utilized by one or more users to satisfy various needs (irrigation, fishing, domestic needs, etc.) at the same time. Comprehensive water use did not mean equal satisfaction of all demands. In most instances some uses were given priority in conformity with local economic and natural conditions.

Water uses were subject to the payment of fees only in cases specified by the law.

5.7.4 Order of priorities

Water legislation granted priority to drinking and domestic needs over all other water uses. The latter uses could therefore be limited every time they interfered with the former.

Article 24 of the Fundamentals empowered the local authorities to prohibit or limit the utilization of drinking water for industrial purposes in favour of drinking and domestic water supplies in cases of calamity, accidents, or any other extraordinary circumstance, and in the event of excessive water consumption by an enterprise.

According to the same Fundamentals, the utilization of water of a quality suitable for drinking purposes for needs other than drinking and domestic uses was not admissible as a rule.

5.7.5 Harmful effects of water

The prevention and mitigation of the harmful effects of water were regarded by water legislation as an important task of the state. Under Article 40 of the Fundamentals, it was binding on enterprises, organizations and agencies to take action in order to prevent or eliminate floods, inundations, water encroachments, the destruction of banks, dykes and other structures, the bogging up and salinization of land, soil erosion, ravine formation, landslides, mudflows and other harmful phenomena. Such measures had to be considered by the state economic development plans.

5.7.6 Water quality and pollution control

The law regulated the discharge of wastewater into water bodies as a form of water use and made it subject to the permission of the agencies responsible for the regulation of the use and protection of natural resources, subject to the agreement of the state sanitary inspection authority, the fish protection agency and the other agencies concerned. Such permission specified the name of the author of the discharge, the water body and the place of discharge, the amount, kind, composition, temperature, degree of treatment, possible fluctuations in time of the wastewater to be discharged, and other conditions. In the case of unwarranted deviations from the specified procedure, the discharge was restricted, suspended or prohibited by the aforementioned agencies, even if this resulted in the stoppage of work of separate industrial installations, workshops, enterprises, organizations and institutions.

The ministries of health of the republics approved maximum permissible concentrations of noxious substances in waters meant for domestic water use. In addition, there were general requirements as to the composition and characteristics of water at points of drinking, domestic and recreational water uses, and fishing. Permissible concentrations of radioactive substances were rated separately.

Preventive measures played an important role in the protection of water resources. Water users were under an obligation to take appropriate measures to stop the discharge of wastewater through the modification of production technologies and of the schemes of water supply, such as the utilization of air cooling, water recycling systems and other technical means. New or restructured enterprises could be put into operation only if provided with adequate anti-pollution facilities.

In addition, special districts and zones were established to ensure the protection of water resources used for drinking and domestic water supply, therapeutic, health resort and recreational needs.

5.7.7 Enforcement

As regards the enforcement of the law, various types of responsibility were envisaged: criminal, civil, administrative and disciplinary.

The criminal codes and other special legislation of the republics provided for offences resulting from the violation of rules on water use and the pollution of water bodies. The penalties could be: imprisonment for a term of a maximum of five years, corrective labour for up to one year, or a fine in the amount of a maximum of 300 rubles.

Under the Decree of the Presidium of the Supreme Soviet of the USSR of 26 February, 1974, the pollution of inland and territorial seas as a result of unlawful discharges from vessels or other floating means of substances harmful to people and to the marine fauna (i.e., substances exceeding the maximum permissible concentrations), or as a result of the violation of international agreements in existence, was to be punished by imprisonment for a term of a maximum of two years, or by corrective labour for up to one year, or with a fine not exceeding 10,000 rubles. If a considerable damage to people or marine fauna was caused, both term and fine would be augmented.

As regards civil liabilities, the violation of the water legislation gave place to compensation for the damage caused. Moreover, the right to use water was revoked.

5.7.8 Centralized inventory and planning

A water cadastre contained basic information on all surface and underground water resources situated within the territory of the USSR. It was kept by the agencies responsible for hydro-meteorological service and water economy, which were also in charge of the preparation of schemes for complex (comprehensive) water utilization and protection, and for water balance calculations. These data are presently available in the newly independent individual states, but in many of these states the cadastres have not been brought up to date since the disbandment of the Soviet Union.

5.7.9 Water administration

In the same way as water legislation was viewed within the framework of national economic development, water administration was considered within the context of the national economy.

The administrative authorities responsible for the utilization and protection of water resources could be divided into three categories:

(i) agencies of the general state administration, i.e., the councils of ministers of the Union and of the Union republics and the local authorities at the regional (*oblast*), district (*rayon*) and local (urban and rural) levels. This category also involved the agencies responsible for planning, registration, coordination of sciences, etc. They regulated the utilization and protection of water resources as an integral part of the management of the economy.
(ii) agencies of special state administration, which included the ministries of reclamation and water economy, health, geology, the state urban technical inspectorate of the Union and its local boards. These agencies were entrusted with the regulation of the conditions of water use and protection, and with the control over the observance of water legislation. Specialized state inspectorates were present at every stage of this level of administration.
(iii) agencies of branch administration, i.e., all ministries and departments managing those sectors of the national economy which utilized water resources.

The attorney general's office, the courts, arbitration tribunals, and the people themselves were responsible for ensuring respect for the law.

As regards river, lake and sea basins, the drainage basin principle was adopted at the level of special state administration, to regulate the utilization and protection of water resources.

5.7.10 Evolution and trends[36]

As a consequence of the disbandment of the Soviet Union and of the political and economic changes which have occurred as a result of it, the newly independent states have taken steps which modify their national legal and institutional frameworks and which reflect the

---

36 Based on Nanni, M. (2003) *Legal and Institutional Aspects, Kura Basin (Southern Caucasus), Pripyat Basin, Tobol Basin*. Report prepared within the framework of the TACIS Joint Rivers Management Project, MottMac Donald/Arcadis Euroconsult.

fact that decisions relating to water resources development and management are no longer made in Moscow. Thus, new water codes or laws have been enacted in recent years, supplemented with a number of decrees, government resolutions, ministerial or departmental orders, instructions, norms and standards. These norms and standards partly coexist with former Soviet norms and standards.

In parallel with this, environment-specific legislation has been adopted in all the countries of the former Soviet Union in support of the environmental impact assessment (EIA) and state ecological expertise processes, in order to facilitate the prevention of the negative impacts of human activities on water and other natural resources. Finally, provisions relevant to water management may be found in legislation concerning other natural resources. It is to be noted that in some of the newly independent states groundwater is the subject of separate legislation on mineral resources, since it was traditionally considered as a mineral.

To a large extent these new legal frameworks reflect a compromise between the former socialist system, which saw the state as the main actor, and present world trends in water law and administration, which call for an increasing participation of the public in water resources management. Thus, the water codes and laws tend to be resource-development oriented and focused on the maintenance of the *status quo*, rather than lead to progressive changes towards integrated water resources management with full respect for the hydrologic cycle. They tend to be rich in classifications and technical details, but lack a clear-cut definition of government functions and powers and of the rights and duties of water users.

Moreover, the subsidiary legislation adopted under the new water codes and laws tends to be really abundant. In general this legislation has been formulated with reference to specific problems as they arose, without being integrated into the overall water resources management policies and strategies, and often without taking into consideration already existing legislation. The result is that there are contradictions which do not facilitate the task of water managers and do not stimulate compliance by water users.

There is a trend, however, towards amending the legislation so as to reflect the principles and requirements set for EU member states (and candidate countries) through the most recent directives in the water sector, and in particular those enshrined in the 2000 Water Framework Directive, i.e., *inter alia*,

(i) the identification of river basin districts comprising one or more river basins and connected groundwater resources;
(ii) the designation of an appropriate competent authority or authorities;
(iii) the definition of environmental objectives;
(iv) the conduct of basin analyses;
(v) the registration of protected areas;
(vi) the identification of water bodies used for the abstraction of drinking water;
(vii) the formulation of river basin management plans following procedures that require a high level of public participation;
(viii) the adoption of programmes of measures;
(ix) cost recovery for water services; and
(x) the introduction of effective, proportionate and dissuasive penalties for non-compliance.

This trend is supported by the fact that many of the countries of the former Soviet Union have entered into bilateral agreements with the EU in order to formalize their relations with it. These agreements provide a framework for bilateral cooperation in various fields and require the gradual approximation of each country's existing and future legislation, including legislation in the field of the environment, to that of the EU. It is worth mentioning that, through legislation enacted in 2004,[37] Ukraine has approved a programme to transpose the EU *acquis*[38] into its own legal system.

Armenia has also proceeded in this direction and has promulgated a new water code in 2002, which has replaced a post-Soviet code of 1992. Similarly, the Kazakh water code of 2003 abrogates the code of 1993, while in Kyrgyzstan the water code of 2004 abrogates the law on water of 1994. A water code in force in the Russian Federation since 1 January, 2007, replaces the water code of 1995. Tadjikistan adopted a water code in 2000 to replace the Soviet code of 1993, and amended it in 2012 with a view to making provision for river basin management. Finally, Belarus promulgated a new water code in 2014.

New legislation alone is not the panacea for solving water resources management issues. This legislation must be accompanied by institutional reforms; not an easy task, as the institutional set-up inherited from the Soviet Union is development-oriented.[39] It is always difficult to restructure existing institutions, let alone deal with the task of reassigning and retraining personnel.

Nevertheless, through their new water codes, Armenia, Kazakhstan and Kyrgyzstan have succeeded in their attempt to separate regulatory functions from development functions to some extent.[40] Also, these countries have divided their territories into specific river basin units placed under the jurisdiction of river basin organizations vested with resource management functions and powers. Similarly, the Russian Federation has opted for this solution.[41]

Finally, the concept of private ownership of land has been reintroduced, with the result that in some countries of the former Soviet Union, such as Estonia and Lithuania, water resources located entirely within the boundary of private land have followed this régime. The same applies in the Russian Federation as a result of the adoption of the new water code. However, in most former Soviet countries water resources continue to be considered a public good, and as such they continue to be subject to control by the public administration.

5.8 Water law principles in the Hindu subak system in Bali[42]

5.8.1 Legal-historical background

The Hindu system of water law has practically survived only in Bali and, to a certain extent and only at the users' level, in Bhutan and Nepal. The Bali *subak* system exemplifies its original philosophy, main characteristics and practical implementation.

---

37 Law of Ukraine of 18 March, 2004, No. 1629-IV, 'On the State Programme for Adaptation of Ukrainian Legislation to the Legislation of the European Union.'
38 The expression '*acquis communautaire*' refers to the body of rules that has been developed so far by the European Union.
39 In many of the newly independent states, Soviet-style ministries of water economy (sometimes translated from Russian as 'water management'), dealing with the development, operation and maintenance of irrigation schemes, have survived in spite of attempts to transform them into water management institutions. The same applies to their basin branches.
40 For a discussion of this concept, see Chapter 9.
41 The Russian basin unit is the *okrug*, which is a territorial subdivision larger than the *oblast*.
42 Taken from: Wohlwend, B.J. (1975) Hindu Water Law and Administration in Bali. *Proceedings of the Conference on Global Water Law Systems*. Valencia.

As early as the third century AD, Indian and Indo-Chinese traders brought Hinduism, and then Buddhism, to Indonesia. These were well received by the population, although in Java they were very soon given Javanese shape. By the tenth century, only Bali had maintained the Hindu culture in its purest form. From the seventh until the fourteenth century, various Hindu and Buddhist dynasties succeeded each other in the domination of the Indonesian spice market. Among them, it is worth mentioning King Dharmavamsa (985–1006), who promoted the first codification of Javanese laws, and Gadja Mada (1331–1364), Chief Minister of the Madjapahit Empire, who commissioned a new codification of the law. It is in this period that most of the Indonesian customary *adat* law became established.

During the fourteenth century, Moslem traders settled on the coast of Java, and in the sixteenth century Islam (*Shafite* school) became the dominant doctrine in Indonesia. Nonetheless, it was adopted by the Indonesians more as a religion than as a legal system, and Islamic customary water law principles were retained only where they conformed to traditional *adat* law. In this period, those who had chosen to maintain their ties with the Hindu doctrine migrated to Bali, which became the stronghold of Hinduism in Indonesia.

From the seventeenth century until the proclamation of independence in 1945, Indonesia experienced the invasion of the Portuguese, the Dutch and the British. It is during the period of Dutch rule that the land and water legislation of the Netherland Indies was promulgated. In Bali, however, Hindu traditional and customary rules of law continued to apply. Only in 1972, twelve years after Bali had become an Indonesian province, was its water legislation finally codified and patterned to a certain extent on the Dutch General Water Regulations of 1936.

5.8.2 Definition and origin

The *subak* system of Bali, present expression of the Hindu culture, is a centralized system of water management, as opposed to the administratively decentralized *desa*, or village irrigation system based on traditional *adat* law to be found elsewhere in the Indonesian archipelago.

The *subak* may be defined as a community of rice field (*sawah*) irrigators. The *subak* system covers the management and distribution of surface water within a *subak*.

The *subak* originates in the desire of farmers to put their land under irrigation, a purpose for which they organize themselves into mutual and self-help associations. All farmers possessing land within a reasonable distance from a stream jointly undertake to construct a diversion weir or intake thereon, together with a network of canals and feeders to convey water to their fields. In doing this, all participants automatically become members of the *subak*. The extension of a *subak* averages 100 ha. Larger *subaks* are subdivided into smaller units (*tempekan*).

The main characteristic of the *subak* is that, while its members may belong to several villages (*desa*) holding land within a given irrigation unit, *subak* management is totally independent from *desa* administration. As the central water management unit, the *subak* also supplies villages with the water they need for purposes other than irrigation.

As regards the *desa*, it constitutes the smallest administrative unit. It is organized into a people's legislative body, the *desa* People's Assembly, which elects one of its members as village chief (*kepala desa*).

Whereas drinking water is traditionally supplied from village wells, the flushing of village sewers is ensured by water made periodically available from the *subak* upon request of the various *kepala desa*.

5.8.3 Organization

The internal organization and life of the *subak* are governed by special rules of customary *adat* law known as *awig-awig*. These were codified between 1939 and 1940 by order of the then Raja of Bali in his capacity as provincial governor within the Dutch administration. Although differing from *subak* to *subak* especially with respect to financial matters, all *awig-awig* are based on identical principles, the purpose of which is to ensure an equitable sharing of *subak* water among irrigators in relation to crops, soil conditions, the two monsoons, the location of individual fields, *desa* requirements and other relevant factors.

The affairs of each *subak* are governed by the community of its members organized into a *subak* meeting which has sovereign jurisdiction over all water management and related matters within the *subak* area. *Subak* meeting decisions are implemented by the *kelian subak*, or chief water master, who is elected by the *subak* meeting from among its members. The charge of *kelian subak* is hereditary in practice (although not in law), provided that the designated heir shows the necessary qualifications. In the performance of his functions, the *kelian subak* is assisted by deputies (*kelian tempek*) responsible for the various sub-units (*tempekan*) of the irrigation network, by assistants (*kesinoman*), controlling the end of network diversions, and by criers (*saya*), whose function is to inform individual irrigators of their water turn.

The *subak* meeting is convened once a month under the chairmanship of the *kelian subak*, in order to discuss water management and related financial matters. The *subak* meeting also approves the various *awig-awig* which regulate, *inter alia*, the obligation of members to cultivate paddy, second and third crops on time, waterworks operation and maintenance, the settlement of disputes, penalties, and the date of performance of ritual ceremonies. In this connection, it is customary for each farmer to construct an altar on his irrigated field, and for the community to have altars or temples built at the various diversion points of the irrigation network.

5.8.4 The legal status of water resources

The Hindu doctrine symbolically equates water with life, the sustainer of plants, animals and human beings on the one hand, the support of divine influences on the other. Similarly, the *subak* may be symbolically compared with the human body, and the water of its irrigation network as the blood which necessarily irrigates all its limbs.

Because of its fluid and purifying nature, water was declared as indivisible by the *Dharma Shastras*, and has therefore escaped the definition of 'thing' or *res*. It could therefore never become an object of appropriation. The same principle is present in Indonesian traditional *adat* law, which has a series of rights to use both land and water either separately or jointly.

As regards land, the fullest and most exclusive use right is called *hak milik*.[43] Although it has all the apparent characteristics of an ownership right, it has never been considered by the

---

43 Sudargo, G. & Hornick, R.N. (1972) *An Introduction to Indonesian Law: Unity in Diversity*. Bandung, Alumni Press.

Indonesians as ownership, which virtually rests with the community, but as possession and usufruct only. The same applies to the right to use water, known as *hak guna air*.

The system of public and private land ownership was, however, introduced in Indonesia by the Dutch, along with the concept of public waters for surface springs, flowing and still waters whereon the existence of private rights could not be demonstrated or which were not exclusively intended for use by public institutions.[44] Other waters, and groundwaters in particular, were thus made to fall into the exclusive dominion of the private landowner, a concept unknown to traditional *adat* law.

With the promulgation of the Indonesian constitution, provision was expressly made for land and water resources, together with the natural riches contained therein, to become subject to state control and to be used for the benefit of the people.[45]

The agrarian law of 1960 specified that the state is empowered to regulate all matters regarding the possession and use of land, water, air space and other natural resources, in its capacity as representative of the Indonesian people to whom these resources have been entrusted by God.[46] *Adat* law was formally recognized as the agrarian law in force in the whole Indonesian archipelago.

5.8.5 The right to use water

Membership in the *subak* carries the right to use water. However, water rights are never definitive, but are allocated as established monthly by the *subak* meeting on the basis of individual needs and in proportion to the water available. In addition to *subak* members, neighbouring villages can make periodical requests for water to the *subak* meeting or to the *kelian subak*, through their *desa* people's meeting or the *kepala desa*. The *subak* is bound to supply the requested water as the circumstances permit. Within each *desa*, such water is used in the framework of a public service for domestic and sewerage purposes. It is therefore not subject to individual, but to collective water rights. The same applies to drinking water supplied from community wells, in which everyone enjoys a right of use.

Farmers who are not members of the *subak* are also entitled to make individual requests for irrigation water from the *subak* network. In this case, *subak* members may refuse to let the needed water cross their fields until the *subak* meeting has decided the case affirmatively. When a positive decision has been made, the outsider becomes automatically a member of the *subak*, and his field is incorporated into the irrigation network.

Irrigation water rights are subject to periodic modifications, in order to accommodate established cropping patterns and production objectives.

The misuse, waste and non-use of allocated water are subject to temporary reduction, suspension, re-allocation or termination of the right to use water. The final decision is always made by the *subak* meeting.

---

44 1936 General Water Regulations, Art. 1.
45 Constitution, Art. 33.
46 Agrarian Law of 1960, Art. 2. This law abrogated not only the Dutch Agrarian Law and Public Domain Declaration of 1870, but Book II of the Civil Code of 1848 governing land, water and natural resources, and the Royal Decree of 1872 which had established the right of 'agrarian ownership' (*eigendom*) and provided for the conversion of *hak milik* into *eigendom* through registration.

The right to water animals is free, but their owner is responsible for damages caused thereby to the banks of watercourses, to waterworks, to fields and crops. The same applies to fishing and fish-breeding in irrigated fields. However, the exercise of this right is unrestricted in watercourses. Artificially constructed fishponds are subject to protected collective fishing rights.

5.8.6 Order of priorities

In Bali, the general or public interest prevails over that of the individual. If normally the use of water for drinking and for animal watering has priority over other uses, in the case of fire, preference is given to fire-extinguishing needs. Irrigation has priority over orchard watering and, within an irrigated area, first crops are given priority over second and third crops, provided that this corresponds to the general interest.

5.8.7 Water quality and pollution control

In the *desa*, special rules govern the protection of drinking water quality. Moreover, due to the sacred nature of water and its constant employment for ritual purposes, there is an absolute prohibition to use surface water bodies for human or domestic waste disposal. Individual households are required to be equipped with the necessary waste disposal pits, which may not be sunk close to any surface or underground source of water. Domestic wastewater is, however, disposed of into the village drains which are flushed twice a week with the water allocated from the nearest *subak* for this purpose.

5.8.8 The water distribution system

The basic unit of the water distribution in the *subak* area is the individual field (*sawah*). Individual water rights are measured in *tek-tek* (cut into pieces), or the width of a *tembuku*, a log placed across an irrigation canal with cuts releasing water to a corresponding number of feeders. The quantity of water released to each field is in proportion to crop needs and expressed in *tek-tek* per *tenah*, or bundle of paddy seed. One *tenah* corresponds to 25–30 kg of seed. Where water is not enough, it is distributed by turns.

The basic rule of water distribution is the *desa-kala-patra*. *Desa* covers the non-agricultural needs of the villages, *kala* is time, and indicates dry season priorities of crops planted during the wet season, and *patra* refers to circumstances, i.e., to water/soil/crops relationships.

Kala has two dimensions, of which the first concerns the priority given in the dry season to crops planted in November-December, then to those planted in January-February and, lastly, to those planted in March-April. The second is a spatial dimension, and refers to the priority to be given to the fields situated closest to the intake over all other fields.

At an early stage of development there were few fields to be irrigated. When the *subak* reached its full extension and the total water supply was no longer enough to allow all fields to be irrigated simultaneously, it became necessary to establish upstream and downstream priorities.

Each field is supplied with water continuously, day and night, until the turn passes to the next category of fields. Since each field is fed from an individual feeder or ditch, there is usually no problem of water conveyance, or right of way, from one field to another. The land on which feeders and ditches are located belongs to the community and is not subject to exclusive individual use rights.

5.8.9 Financial aspects

The traditional principle on which the *subak* organization rests is that of mutual help among the members of a community. It has been constitutionally institutionalized as the second of the five pillars after that of the belief in one God.[47] Accordingly, the financial and other material resources needed for *subak* management and operations are contributed by all members.

Subak membership may be active or inactive. Active members are those farmers who cultivate between 0.3 and 0.5 ha of paddy fields. They contribute their labour for normal *subak* and waterworks construction, operation and maintenance, and in cash or kind for both normal and extraordinary works. Inactive members are usually government officials and social institutions[48] possessing paddy fields, but who are not in a position to cultivate personally, as well as farmers possessing less than 0.3 ha of paddy fields. They contribute in cash or in kind for both normal and extraordinary works. Farmers possessing more than 0.5 ha of paddy fields are considered active for 0.5 ha and inactive for the rest.

The *kelian subak* is paid by the *subak* meeting either in a land grant, a salary in cash or kind, or in an exemption from active membership contributions. In the case he is granted a land use right, his field is located at the end of the *subak* network in order to ensure an equitable distribution of water. The *kelian subak* is responsible for compensating his deputies and assistants.

5.8.10 Water law implementation

There is no register of water rights allocated for each irrigation period. There is however a register of individual fields. Water rights are allocated on the basis of a cultivation plan drawn up by the *subak* meeting for each crop season.

The *kelian subak*, together with his deputies and assistants, is responsible for the implementation of the cultivation plan, as well as for the control of the adequate operation and maintenance of the irrigation network.

5.8.11 The settlement of disputes

The disputes arising among members of a *subak* are settled internally, on the basis of a compromissory procedure. The wrong committed by a member is not considered as an offence, but as a disorder (*adharma*) which the whole community is bound to correct. The wrongdoer may be expelled from the *subak*, which means his virtual death vis-a-vis the community. The *kelian subak* and, in last resort, the *subak* meeting, have an arbitration function.

5.8.12 The statutory subak

In 1972, a Provincial Water Regulation[49] patterned on the 1936 general water regulations for Java and Madura was promulgated in order to reconcile the *subak* system of Bali administratively with the existing legal and institutional framework.

---

47 Constitution of Indonesia, Preamble; 1960 Agrarian Law, Art. 5.
48 Or religious endowments (*waqf*) in Islamic law.
49 1972 Provincial Water Regulation for Bali.

As one of the 26 autonomous Indonesian provinces, Bali is headed by a governor, assisted by a provincial people's assembly with which he formulates and approves provincial policies. The provincial administration is subdivided into separate administrative units and technical services. Administrative units are the *kabupaten*, or regency, and the municipality. The *kabupaten* is headed by a *bupati*, or regent, flanked by a regency people's assembly and subdivided into districts headed by assistant regents who control groups of *desa* or rural communities. The municipality is headed by a mayor flanked by a city people's assembly. The mayor is assisted by assistant mayors, or district officers, who control urban communities.

The technical services are subdivided on the one hand into provincial, regency and municipal services and, on the other hand, into regional and sectional services. Water resources matters are coordinated at the provincial level by a provincial irrigation commission, and at the regency level, by the Regency Irrigation Commission.

It is at the *kabupaten* level that the *subak* system has been institutionalized and integrated into the Indonesian government administration. The 1972 Provincial Water Regulation introduced government control over intakes on a main river and over the major part of the irrigation network originally constructed and operated by the *subak*. The leadership in the water distribution system has been largely transferred from the *subak* meeting to the Regency Irrigation Commission.

Subak water distribution is now based on integrated cultivation plans drawn up by the Regency Irrigation Commission under the chairmanship of the *bupati* and with the assistance of the Regency Public Works Service whose head, the chief irrigation officer, acts as chief water master of the *kabupaten*, assisted by the regional and sectional public works services within their respective areas of jurisdiction.

Operating under the chief irrigation officer are the various district irrigation officers who, as water masters, control groups of intakes on the main river and corresponding primary canals down to, and including, their respective secondary canal diversions. Tertiary canal intakes on the secondary canals are operated and controlled by assistant water masters and tax collectors, down to 50 meters on each tertiary canal. From that point, the *kelian subak* takes over water management responsibilities under the authority of the *subak* meeting. It is worth noting that, in special circumstances, a district irrigation officer may request the *subak* to take its responsibilities up to the primary canal level.

5.8.13 Conclusion

In spite of its institutionalization, the Hindu inspired *subak* water management system of Bali continues, as in the past, to operate according to its own customary rules (*awig-awig*).

Efforts have been made to export this system to other parts of the Indonesian archipelago, as it has been influencing, and continues to influence, many aspects of the administratively decentralized *desa adat* law-based irrigation system adopted, under various forms.

References

Ali ibn Muhammad, al-Mawardi (1901) *Traité de droit public musulman*. Paris, E. Leroux.
Brockelmann, C. (1939) *Geschichte der Islamischen Völker und Staaten*. Munich and Berlin, R. Oldenburg.

Bruno, H. (1913) *Le Régime de l'Eau en Droit Musulman*. Paris, A. Rousseau.

Caponera, D.A. (1973) *Water Laws in Moslem Countries*, Irrigation and Drainage Paper No.20, Rome, FAO.

Chenon, E. (1923) *Le démembrement de la proprieté foncière en France avant et après la Révolution*. Paris, Sirey.

D'Emilia (1940) *Lezioni di diritto musulmano*. Rome, Ed. Universitarie.

Ducaurroy, L. (1849) Legislation musulmane Sunnite, rite Hanéfite. *Journal asiatique*. Paris.

El-Charani, le Cheik (1898) *Balance de la loi musulmane*, translated by Perron. Algiers, Fontana.

Jansse, L. (1953) *La proprieté, le régime des biens dans les civilisations occidentales*. Paris, Ed. Ouvrières.

Kolbasov, O.S. (1975) Water Law in the URSS. *Proceedings of the Conference on Global Water Law Systems*. Valencia.

Kolbasov, O.S. (1968) *Legislation on Water Use in the USSR*, Judicial Literature, Moscow, 1965, translated by the Water Resources Center. Madison, University of Wisconsin.

Maktari, A.A. (1975) Islamic Water Law. *Proceedings of the Conference on Global Water Law Systems*, Valencia.

Richardson, H.J. (1975) System of Water Law and Organizations in the United Kingdom. *Proceedings of the Conference on Global Water Law Systems*. Valencia.

Sudargo, G. & Hornick, R.N. (1972) *An Introduction to Indonesian Law: Unity in Diversity*. Bandung, Alumni Press.

Teclaff, L.A. (1972) *Abstraction and Use of Water: A Comparison of Legal Régimes*. New York, United Nations.

Wohlwend, B.J. (1975) Hindu Water Law and Administration in Bali. *Proceedings of the Conference on Global Water Law Systems*. Valencia.

Yahya ibn Adam, Kitab al Kjaradj (1896) *Le livre de l'impôt foncier*. Leiden, E.J. Brill.

Chapter 6

Development by region

6.1 Africa

6.1.1 Introduction

African countries have derived their water laws and institutions from legal systems imported by foreign powers during the past centuries. The countries which were formerly under French, Belgian, Italian, Spanish or Portuguese administration have generally adopted the civil law system, while those which were under British rule derive their water laws and institutions from the common law system. The legal systems of other countries are the result of influences other than those just mentioned, or of a combination of two or more systems. Moslem water law also plays an important role in the countries which were once part of the Ottoman Empire and in those which were reached by Arab influence.

These legal systems have generally developed together with customary law, which varies from area to area. Customary law in Africa is particularly significant with respect to matters relating to land and water.

Since their independence, most African governments acknowledge that water resources play a key role in their social and economic development. Water is needed, *inter alia*, to shift from cash-crop production to large-scale irrigated agriculture, to develop hydropower generation and industrial schemes, to supply population centres and rural areas and, in general, to satisfy the increasing water demands resulting from rapid population growth.

These uses, if not rationally regulated and managed, can lead to undesirable results, such as floods, droughts, soil erosion, salinization, siltation and others, or have negative effects on water quality, public health and the environment. Likewise, they can produce conflicts among users.

Until recently, the water laws and institutions of most African countries were inadequate to meet the needs just mentioned. Water and water-related laws were enacted to solve specific problems arising from water use and development, the harmful effects of water, misuse and water pollution. In most cases, water laws and regulations were administered by different government ministries, departments or agencies, often without coordination among them, thus producing overlapping of functions and powers in the water sector, poor planning and a sectorial approach to water resources projects. The result was that these projects were hampered, delayed or even doomed to failure. Other consequences of this situation were the waste of natural, financial and human resources and uncertainty as to the position of water users vis-a-vis other users and the administration.

Sustainable and rational water resources management calls for adequate water laws and institutions which should be established on the basis of the particular situation and needs existing in any one country, and which should ensure the implementation of water policies and plans oriented towards the satisfaction of present and future water demands. Water quality aspects should not be ignored, as the water pollution problem, which has reached considerable proportions in industrialized countries, is also becoming a matter of concern in Africa. By the same token, the interaction of water with other natural resources should not be overlooked, since the exploitation of one of these resources may have negative repercussions on the integrity of the others. Aware of the shortcomings in their legal and institutional frameworks for water resources management, several African countries have undertaken extensive reforms and have enacted comprehensive water laws to address these challenges.

At the international level, the awareness of the need to favour larger regional and continental cooperation over small-scale economic organizations has led African governments to establish regional, sub-regional, drainage basin and continental organizations which have the purpose of pursuing common economic, financial and social goals.[1] Such arrangements may greatly facilitate integrated administration and legislation programmes aiming to harmonize national water laws and institutions and provide a useful framework for sustainable water resources management.

It is, however, the choice of the most suitable type of regional arrangement that represents the major challenge, given the different political régimes, languages, legal systems, economic and other motivations constituting unifying or dividing factors. Judging from the results they have achieved, it would seem that basin or, in the case of large basins, sub-basin organizations constitute the ideal framework for the integration of the legal and institutional aspects of water resources management at the regional level.

Also, the need to protect the environment and to manage the natural resources on a rational or sustainable basis is becoming a concern for many African governments. Environmental protection or natural resources ministries and councils have been set up in some states, and environmental or natural resources protection laws are being promulgated, together with water legislation.

In this section, water laws and institutions in African countries will be examined by groups of countries, according to the origin of their legal system. This classification is approximative, as each country has its own characteristics, depending on the prevailing local conditions.[2]

6.1.2 Customary law

Customary law in Africa is without doubt the most important of the sources of law, and of water law in particular, as it is the one which is most known and respected by the population.

---

1 An example of continental cooperation is the Organization of African Unity (OAU), created in 1963. Regional economig integration organizations with natural resources (including water) as part of their mandate include the Southern African Development Community (SADC) and the Intergovernmental Authority on Development (IGAD), created in 1992 and 1996, respectively. In the field of water resources, the most important organizations are the Organisation pour la mise en valeur du fleuve Sénégal (OMVS), established in 1972; the Lake Chad Basin Commission, established in 1964; the Niger Basin Authority, created in 1980.

2 For this classification, see Caponera, D.A. (1979) *Water Law in Selected African Countries*. Legislative Study No. 17. Rome, FAO.

With few exceptions, it is to be found throughout the continent, and has relevance in fields such as land tenure and ownership, grazing, cultivation, water administration, animal watering, fishing rights and other water rights, users' associations, land settlement and redistribution, succession, as well as in the procedures relating to the recording, 'immatriculation,' registration, adjudication and settlement of disputes on such rights.[3] Customary rules become more detailed where water is scarce.

Customary law, which is mostly unwritten, may vary from region to region, from country to country, and even from area to area within one single country. It is generally based on the principle that land and water belong to the community and, therefore, the individual has only a right to use the water, according to the communal, tribal or community customary land tenure system prevailing. Under this system, the concept of private ownership of water is usually unknown.

In some countries, those engaging in water or land use in a particular area unite to exercise community control over such use. The rights and obligations of individuals within the community are governed by local custom, and a local official is appointed for the administration of water and for the settlement of disputes and conflicting claims.

Customary law in some countries has been specifically recognized in legislation, particularly as regards land tenure, grazing rights and cultivation rights. This is the case of Benin, Burkina Faso, Cameroon, Central African Republic, Côte d'Ivoire, Democratic Republic of Congo, Congo, Gabon, Ghana, Kenya, Niger, Nigeria, Sudan, Tanzania, Togo, Uganda and Zambia. In Madagascar customary rights are recognized as *droits d'usage* (use rights). In Tunisia, a special régime is provided for the so-called *terres collectives* (collective lands), while in Libya tribal lands are treated as communal property. In Senegal, any land not statutorily forming part of the public domain or privately held under the 'immatriculation' (land registration) system belongs to the public domain but is handed back to the authorities of the community for redistribution to users.[4]

Legislation to adjust customary law to development needs represents a current policy issue in many African states. The existence of a customary legal and institutional framework facilitates the introduction of sound water resources policies, legislation and administration, and of measures for bringing all water resources under state control. In fact, modern tendencies in water resources policy require people's participation in the management of water resources, which is more easily achieved where a less individualistic and more community-oriented approach exists in respect of the ownership, use and distribution of water as well as within the organization of water users' associations, consortia or cooperatives. Such an approach, which is difficult to introduce in western societies, is congenial to the African environment, where the existence of traditional forms of community organization may facilitate its institutionalization.

6.1.3 Countries following principles of the civil law system

There are about 25 countries in Africa following the civil law system. In this group are those countries formerly within French West Africa (AOF)[5] and French Equatorial Africa

---

3 *Ibid.*, 8.
4 Mifsud, F. (1967) *Customary Land Law in Africa*. Rome, FAO.
5 Benin, Burkina Faso, Guinea, Côte d'Ivoire, Mali, Mauritania, Niger, Senegal.

(AEF),[6] Madagascar, those which after the First World War were under French trusteeship[7] and were regarded as French overseas territories or departments.[8] To this group also belong the countries formerly under Belgian,[9] Italian,[10] Spanish[11] and Portuguese[12] administration.

In the countries formerly under French administration, waters were originally classified, as in France, as public or private, public waters being those 'navigable or floatable.' Later, due to climatic circumstances, i.e., to the fact that most African streams are seasonal and therefore non-navigable during certain periods of the year, with the consequence that very little is left to the public domain, the distinction between navigable and non-navigable waters disappeared, and, generally, all waters were placed in the public domain. Under this régime, every use of public water is subject to an administrative authorization, permit or concession.

Likewise, the water administration apparatus has tended to follow that which prevailed in France,[13] but recently in many of these countries the bulk of water management functions has been transferred to ministries responsible for water resources. In addition, institutional mechanisms have been established in the form of councils, commissions or committees to allow sectorial ministries and agencies, and sometimes non-governmental organizations, to participate in decision-making processes with regard to water resources management. This is the case of Burkina Faso, Mali, Niger and Senegal, among others. The river basin is increasingly being considered as the ideal unit for integrated water resources planning and management, and institutional mechanisms are being introduced at this level to facilitate the process. Finally, specialized institutions, whether government, private or mixed, have been set up to deal with particular water development activities such as domestic and municipal water supplies, power generation and distribution, irrigation and others.

A similar type of legal régime and administration of water resources applies in the countries formerly under Belgian administration, i.e., Rwanda, Burundi and the Democratic Republic of Congo.

Libya derives its water laws and institutions from Italian law, which was superimposed over the pre-existing Islamic customary water law. Accordingly, waters were classified as public or private, and the use of public waters, which were defined as those 'having been so declared in view of their public interest,' was subject to administrative permit. The administration of water rights was centralized in the Ministry of Public Works, but more recently all waters have been declared state property and have been placed under the control of the General Water Authority. A special agency known as the Great Manmade River Water Utilization Authority is responsible for the use of water transferred from the desert to coastal regions for agricultural purposes.

The present Somalia is the result of a merger of the former Italian Somalia and the former British Somaliland; it inherited both Italian and British water legislation, but in rural

---

6 Central African Republic, Chad, Congo, Gabon.
7 Cameroon, Togo.
8 Comores, Djibouti and Réunion.
9 Rwanda, Burundi and Democratic Republic of Congo.
10 Libya, Somalia.
11 Equatorial Guinea.
12 Angola, Capo Verde, Guinea Bissau, Mozambique, São Tomé and Principe.
13 See Chapter 5, Section 5.5.

areas customary Somali law and the Islamic *Shari'a*[14] prevail. Provisions relevant to water resources are scattered in numerous legal instruments dating back to the colonial period, which have now been consolidated and updated into a new draft water law. This draft law has recently been submitted to Parliament for discussion and eventual adoption. Water resources management is the responsibility of the Ministry of Water and Energy.

Equatorial Guinea derives its water laws and institutions from Spain. A water law enacted in 2007 provides for water resources planning and management by river basin. Water resources management is centralized under the Ministry of Fisheries and Environment.

The basic water legislation of the countries formerly under Portuguese administration consists of a Decree of 1946 regulating the use of public waters in the colonies, as subsequently amended and implemented by regulations of the local governors. This legislation establishes the procedure for the granting of permits for the use of public waters, and distinguishes between free uses (for domestic purposes and the watering of animals), simple uses requiring a simple licence and large diversions affecting the public interest, which are subject to concession. This system prevails in Capo Verde, Guinea Bissau and São Tomé and Principe. Mozambique and Angola adopted water laws in 1991 and 2002, respectively, which promote the river basin management approach.

6.1.4 Countries following principles of the common law system

The African countries formerly under British administration[15] or mandate[16] have adopted the British system, according to which water is *res communis omnium* (common to all), of which the riparian landowner can make use, unless it has specifically been brought under government control through legislation or judicial decisions.

However, crown land, as defined in various ordinances during the colonial period, did not generally include water resources, with the result that every specific use of water became the subject of separate legislation. This produced a large number of legal enactments concerning specific water utilizations or regulating other activities incidental to water resources.

With a few exceptions, in the African countries following the common law system no centralized water administration existed. Nevertheless, there has recently been a tendency to set up water resources administrations, authorities, boards or commissions, which are either autonomous or attached to ministries responsible for natural resources or environment protection, with a view to coordinating activities and projects and to centralizing the administration of water rights. This is the case of Ghana, Kenya, Sierra Leone, Swaziland, Tanzania, Uganda, Zambia and Zimbabwe.

In addition to this move towards centralization and coordination, in order to facilitate water resources management and basin planning a number of countries have opted for the establishment of river basin institutions. Among them, it is worth mentioning the basin water resources committees of Kenya,[17] the river basin authorities of Nigeria,[18] the water basin

---

14 *Ibid.*, Section 5.4.2.
15 Botswana, The Gambia, Kenya, Lesotho, Malawi, Nigeria, Seychelles, Sierra Leone, South Sudan, Sudan, Swaziland, Tanzania, Uganda, Zambia, Zimbabwe.
16 Togo.
17 Water Act No. 43 of 2016.
18 River Basin Development Authorities Act of 1986.

management boards of Sierra Leone,[19] the river basin authorities of Swaziland,[20] the basin water boards of Tanzania[21] and the catchment councils of Zimbabwe.[22]

Nigeria deserves special mention because of its federal administrative setup. According to its constitution, water resources are under the jurisdiction of the component states, while the federal government retains certain powers of overall coordination and control. The Water Resources Act of 1993 vests in the federal government the right to the use and control of all surface and groundwater affecting more than one state, for the purpose of promoting the planning, development and use of the country's water resources, and coordinating their distribution, use and management. Water administration is organized on a state basis, and ministries of water resources are in place both at the federal and at the state level.

It should be noted that under the common law system the legal and institutional aspects of water resources management were mainly concerned with the provision of water for municipal water supply and, in some countries, with pollution control, sewerage, hydropower generation and recreation (fishing). Only recently, agricultural, industrial and other uses of water have been incorporated into comprehensive water laws, with the exception of countries such as Egypt and Sudan, where large-scale irrigation was practiced before the arrival of the British, and where irrigation regulations have existed since ancient times.

6.1.5 Countries following principles of other systems

This group includes countries deriving their laws and institutions from systems other than those just mentioned or from a combination of legal systems. Liberia, for instance, has derived its water legislation and administration from that of the eastern USA, while the systems existing in Ethiopia and Mauritius are a combination of systems.

In Ethiopia, customary Ethiopian law, some Italian legislation, French legislation, Swiss legislation (as regards the civil code) and, to a certain extent, Anglo-Saxon legislation, all coexist. The constitution has declared 'all resources in the water' as government property. Based on this statement, water resources management has undergone a significant evolutionary process. At the initial stage, a National Water Resources Commission equipped with an executive secretariat operated as a regulatory organ for water affairs throughout the country. The chairman of this commission had the rank of minister, but the organization enjoyed a certain degree of autonomy. Under the commission, there were three organs: (i) a water resources development agency, responsible for studies, investigations and planning; (ii) a water supply and sewerage agency, responsible for the operation and maintenance of these services throughout the country, with the exception of some important municipalities, such as Addis Ababa and Harar; and (iii) a waterworks development agency, responsible for the construction or supervision of all waterworks, with the subsequent handing over of the same to institutions or municipalities for operation and maintenance. Basin agencies under the umbrella of the Valley Development Authority (VADA) were established in 1977. The Proclamation of 1994 on the use of water resources vested water management responsibilities

---

19 National Water Resources Management Agency Act of 2017.
20 Water Act No. 7 of 2003.
21 Water Resources Management Act No. 11 of 2009.
22 Created on the basis of Water Act No. 31/98 of 1998, Chapter 20:24.

in the central government, represented by the Ministry for the Development of Natural Resources and Environment Protection, and in the regional governments for matters within their respective territorial jurisdiction. The regional governments operated through their own departments for the development of natural resources and environment protection. The 1994 Proclamation has been repealed and replaced by Proclamation 197/2000 on water resources management, and water resources management has become a function of the Ministry of Water Resources. The ministry may delegate its functions and powers to 'the appropriate body for efficient execution of its duties' (Art. 8.2), that is, to the regional governments or to river basin organizations. The latter organizations were established on the basis of the River Basin Councils and Authorities Proclamation of 2007, and are now in the process of being strengthened.

The legal framework established in Mauritius by the French was maintained until 1810, when the island passed to the British, who undertook to respect the laws, customs and religion of the Mauritians. The British system of law did not substitute, but only complemented or attempted to interpret the French legal system. Many institutions were responsible for water supply until 1971, when the Central Water Authority was created under the Ministry of Public Utilities and was vested with water resources management, planning and development functions for the whole country. One of the branches of this central water authority was the Water Rights Administration, which was responsible for the inventory, regulation and monitoring of water rights and for the issuance of permits. In 1993 the Water Rights Administration was abolished and its functions were transferred to a newly established water unit under the ministry. In the administration of water rights, both quantitative and qualitative aspects are considered. The Central Water Authority is now mainly responsible for the treatment and supply of potable water for domestic, commercial and industrial purposes.

Egypt, which in 1830 practically detached itself from the Ottoman Empire, though remaining theoretically under its sovereignty, retained ancient irrigation rules and a centralized water management administration. During the nineteenth century, water laws of French and British origin were also enacted, strengthening even more the government control over water resources use and distribution through the Ministry of Water Resources and Irrigation.

Water legislation and administration in the Republic of South Africa derive from the so-called Roman-Dutch law, a combination of the Dutch and English legal systems. Water resources management is governed by the National Water Act,[23] a comprehensive code calling for the formulation of national and river basin level water resources strategies and the creation of water reserves to ensure the satisfaction of basic human needs and the maintenance of the ecological balance of aquatic ecosystems. The act further provides for the licencing of water uses and for water charges, and for the progressive establishment of catchment management agencies within designated water management areas. Water management functions are delegated by the Ministry of Water Affairs to the catchment agencies, in which the local communities are to be involved. The act also provides the basis for the creation of local water users' associations for the performance of certain water management functions at the local level. Advisory committees may also be established, to deal with specific issues. A water tribunal handles water disputes.

---

23 No. 36 of 20 August, 1998.

Likewise, in Botswana and Zambia there are strong departments of water affairs responsible for overall water resources management. In Namibia, the Water Resources Management Act of 2004 has provided for the establishment of a water resources management agency, a water regulatory board and a water advisory council. In these countries, water courts hear appeals on decisions concerning water use rights.

6.1.6 Countries influenced by principles of the Islamic water law system

In countries north of the Sahara which once were part of the Ottoman Empire,[24] Moslem customary water law is still observed, particularly at the local level, despite the enactment of subsequent laws, such as the Ottoman Mejelle Code and other legislation. After the fall of the Ottoman Empire, provisions concerning water resources were either confirmed or amplified by newly issued laws, decrees, *dahir*, orders and regulations. In general, all waters were declared to belong to the state, the crown or the public domain as representing the Moslem community, and every use of water other than for drinking purposes previously recognized under the Ottoman Empire or under the *Shari'a* was brought under government control. In addition, special water commissions were established for the recognition or modification of existing water rights, and a cadastre was created for the registration of land and water use rights.

Significant developments have recently taken place in the countries of the Maghreb, i.e., Algeria, Morocco and Tunisia. These countries have adopted comprehensive water laws[25] that, in addition to regulating water uses and providing for the protection of water quality, consider environmental aspects. These laws follow a river basin approach by providing for river basin organizations and river basin management planning, including the participation of stakeholders. In Egypt, a new draft law on water resources and irrigation is currently being considered.

South of the Sahara, despite the introduction of written western water laws, the basic principles of Islamic water law fostered by the Arabs continue to be observed as local customary law.[26]

6.2 Asia and the Pacific

6.2.1 Introduction

The laws and institutions of most countries of Asia and the Pacific region have been influenced by systems alien to the region.

---

24 Algeria, Egypt, Libya, Morocco, Somalia, Sudan, Tunisia.
25 Algeria: Water Law No. 05–12 of 4 August, 2005; Morocco: Water Law No. 36–15 of 10 August, 2016; Tunisia: Law No. 75–16 of 1975, promulgating the Water Code.
26 Chad, Mali, Mauritania, Niger, Senegal, and by parts of the populations of Benin, Burkina Faso, Cameroon, Central African Republic, Ethiopia, The Gambia, Ghana, Guinea, Ivory Coast, Nigeria, Sierra Leone, Tanzania, and Togo.

The common law doctrine of riparian rights was introduced by the British to the countries which were under their administration during the colonial period;[27] however, the original system was often modified and adjusted to local needs.

The civil law system has provided the basis for legislation and administration in the countries of former French Indochina,[28] while Dutch water law was introduced in Indonesia. The water law system of the Philippines is a combination of the Spanish law and the USA prior appropriation doctrine. Taiwan, also, has been influenced by the USA prior appropriation doctrine.

The German water law system has influenced water administration in Japan and the Republic of Korea[29] to some extent, while China has always had its own system.

Afghanistan, Bangladesh and Pakistan have adopted the Moslem system of law.

It is to be noted that, in spite of the introduction or influence of foreign systems, indigenous legal systems, i.e., Chinese, Hindu, Shintoist, Tamil (in Sri Lanka) or other, have continued to prevail, particularly as regards the behaviour of water users at the local level.

In recent years, many Asian countries have adopted comprehensive water legislation to regulate the use and protection of surface and underground water resources, or are in the process of doing so. This legislation provides for the introduction of a system of permits for water use, and in some countries it considers the river basin as the unit for water resources management and planning. Moreover, in a number of instances attempts have been made to curb the excessive fragmentation of functions through provision for institutional mechanisms to coordinate the activities of the various government agencies at the national level. This is the case of Laos and Viet Nam. In some countries, water legislation provides for institutional arrangements for river basin management. Measures are also being introduced, through legislation, to protect water quality through licencing mechanisms coupled with effluent and quality standards and objectives. In some instances, however, pollution is not yet a matter of concern, given the low degree of industrialization. Finally, the legislation often devotes specific provisions to the management of emergency situations, such as floods and drought.

6.2.2 Countries following principles of the civil law system

Initially, the Asian countries formerly under French administration, i.e., Cambodia, Laos and Viet Nam, adopted the criterion of 'navigability and floatability,' as applied in metropolitan France to place rivers and streams in the public domain and subject their use to a permit. Due to the relative availability of water, there was no need to modify this basic criterion and to declare a water as public. Nevertheless, these countries have chosen to declare all water resources as public, or belonging to the people (Viet Nam), or to the national community (Laos). Laos, Cambodia and Viet Nam have promulgated water laws that regulate all aspects of water resources management.[30]

---

27 Bangladesh, Myanmar, India, Pakistan, Singapore, Malaysia, Australia, New Zealand and Fiji.
28 Cambodia, Laos and Viet Nam.
29 South Korea.
30 Laos, Law No. 02/96 of 11 October, 1996, on Water and Aquatic Resources, replaced with a new law in 2017; Cambodia, Law on Water Resources Management of 2007; Viet Nam, Order No. 15/2012 L-CTN of 2 July, 2012, promulgating the Law on Water Resources.

6.2.3 Countries following principles of the common law system

The common law riparian system imported by the British to the states formerly under their administration has undergone significant changes.

6.2.3.1 Australia[31]

In the arid and underpopulated states of Australia, neither human settlements nor extended irrigation would have been achieved if the use of water had been limited to riparian land. Since it was indispensable to allow non-riparians to have access to water for grazing and irrigation purposes, and there was a need to establish state control over the scarce water resources available, the states of the southern part of Australia abrogated riparian rights and introduced the permit system in the late nineteenth and early twentieth centuries. This was greatly facilitated by the fact that most of the land, contrary to the case in the United Kingdom, still belonged to the crown.

Victoria was the first Australian state to abandon riparianism. From 1861, various legal provisions had reserved frontages on rivers to the crown whenever land was sold, thereby preventing the private acquisition of riparian rights unless the law specifically designated the watercourse as a boundary. The formal abolition of riparian rights, however, took place by virtue of the Irrigation Act, 1886, which vested in the crown all rights to the use, flow and control of water in any watercourse and introduced the permit system. The act accorded riparian landowners a statutory right to withdraw water for domestic purposes and livestock watering. The right of landowners whose land had been alienated by the crown prior to the act to irrigate small gardens was recognized.[32] New South Wales followed the example of Victoria ten years later.

South Australia abolished riparian rights later and only partially. The Control of Water Act, 1919–1925, which was passed after the conclusion, in 1914, of the River Murray Waters Agreement between the federal government and the three states of Victoria, New South Wales and South Australia, vested the ownership of waters in watercourses in the crown and introduced a permit system. The right of riparian owners to withdraw water was limited to domestic uses and to the irrigation of small gardens. The state retained a strip of land along the banks of rivers, thus technically making it the riparian owner. The remaining Australian states modelled their legal framework for water resources management on that of Victoria.

Comprehensive water acts have been promulgated in Victoria, South Australia, Queensland and the arid Northern Territory in 1989, 1997, 2000 and 2001, respectively, to adjust the pre-existing system of water allocation and water rights administration to increasing water scarcity. This legislation provides for water resources planning and enables water users and stakeholders to participate in decisions of relevance to water management at the river basin or aquifer level. Generally speaking, it reflects a growing concern for the condition of the environment.

---

31 This section has been derived from Clark, S.D. (1975) The Asian Region. *Proceedings of the Conference on Global Water Law Systems*. Valencia.

32 Teclaff, L. (1972) *Abstraction and Use of Water: A Comparison of Legal Regimes*. New York, United Nations.

In New South Wales, the Water Management Act 2000, No. 92, constitutes the most important and comprehensive reform of water legislation in the state since 1912. It envisages clear, transparent and strong processes through which water resources can be shared between water users and the ecosystems, and recognizes the water rights of aboriginal title holders[33] for domestic use and livestock watering. Further, it calls for statutory water management plans to be developed by local water management committees with broad community input, improved access to water resources and clear-cut and secure water rights, with compensation claimable for changes that affect water availability,[34] thus providing confidence for business development.

At the Commonwealth level, the Water Act 2007, No. 137, provides a framework for the management of the Murray-Darling basin in the national (Commonwealth) interest. Among other things, it establishes the Murray-Darling Basin Authority to ensure the integrated and sustainable management of the basin's water resources. The authority is responsible for the development of management plans for the basin, the operation of the river system, including water delivery and the monitoring of water quantity and quality, and has enforcement powers. The act further provides for the establishment of the Commonwealth Environmental Water Holder to manage, protect and restore the environmental resources of the basin.

The existence of large aquifer systems spanning several states has called for the establishment of institutional mechanisms allowing the states to cooperate in groundwater management. This is the case of the Great Artesian Basin (GAB), which is common to Queensland, New South Wales, South Australia and the Northern Territory. Each of these states has established an advisory committee for its respective portion of the GAB, which represents all stakeholders and provides advice to the water administration on the granting of groundwater abstraction licences. At the GAB level, a consultative council comprising representatives of the Commonwealth, the states, the users and various associations of stakeholders was established in 1997 and, in 2000, adopted the first GAB Strategic Management Plan, which sets guidelines for groundwater management at the Commonwealth and state level for a 15-year period. The formulation of the plan followed a complex consultation process involving groundwater users and stakeholders at all levels. The GAB Coordinating Committee, which was set up in 2004 to replace the Consultative Council, is now in charge of monitoring the implementation of the plan at the basin level and of facilitating the exchange of information among basin states. A new plan is in the process of being developed.

In parallel with this, state legislation empowers the relevant authorities to control underground water activities, including the construction of wells, the amount of water which may be abstracted and other related activities. The licencing of groundwater use is similar to that of surface water, and provisions are in place to control the activities of professional drillers through specific licencing mechanisms.

6.2.3.2 Bangladesh

In Bangladesh the National Water Policy of 1999 calls for river basin management and planning, the implementation of a system for water allocation, decentralization, cost recovery

---

33 As defined by the Commonwealth Native Title Act, 1993.
34 This was explicitly excluded by the 1912 Water Act.

and private sector and community participation in water management. Also, under this policy a distinction is drawn between regulation and resource development and supply. A water act consistent with policy requirements was promulgated in 2013, and a national water resources plan indicating short-term, medium-term and long-term water management goals has been developed.

A number of government ministries, departments, agencies and boards are involved in water resources development and management, but a prominent role in resource development is played by the Ministry of Water Resources.[35] A national water resources council has been established under the water act and is expected to take major policy decisions and coordinate activities through its executive committee.

6.2.3.3 India

In India, the introduction by the British of public ownership and government control of surface waters was dictated by the need to satisfy the needs of large masses of people. When India became a federal state after independence, ownership of water and the right to control surface waters was attributed to the individual states.[36] The union government has only limited powers of intervention in water questions, except for those concerning interstate waters[37] and the settlement of interstate water disputes under the provisions of the Interstate Water Disputes Act 1956, but it exercises *de facto* powers of intervention through its financial policies in the allocation of funds for water resources development purposes.

The River Boards Act 1956 was enacted immediately after the Interstate Water Disputes Act as the basis for the establishment of river boards with advisory powers. The boards created under the act may execute water resources development projects, such as in the case of the Betwa River Board that was established in 1976 for the construction of the Rajghart dam between the states of Uttar Pradesh and Madhya Pradesh. An example of a river board dealing with matters of relevance to water resources management is offered by the Bansagar Control Board,[38] the powers of which, however, are only advisory. In substance, the power to make decisions on matters relating to interstate water resources rests with the state governments.

At the level of the individual states, the legal framework is not uniform. While some states develop water resources on the basis of legal provisions dating back to the nineteenth century but do not have an overall legal framework for water resources management, others have enacted comprehensive water-sector policies and legislation. This is the case of Maharashtra (2005) and Andhra Pradesh (2002). Other states, including Kerala, Tamil Nadu, Bihar, Karnataka and Assam, enacted groundwater laws in 2002, 2003, 2006, 2011 and 2012, respectively. These laws were influenced by a model groundwater bill that the Ministry of Water Resources of the Union issued in the 1970's and brings up to date from time to time. In order to overcome fragmentation in the exercise of water resources management functions and powers, water authorities and commissions have recently been established in some states. A Water Resources Regulatory Authority has been in place in Maharashtra since

---

35 Which is the successor of the Ministry of Irrigation, Water Development and Flood Control.
36 Constitution of India, List II (State list) of the Seventh Schedule.
37 *Ibid.*, Entry 56 of List I (Union list) of the Seventh Schedule.
38 Established in 1979 between Uttar Pradesh, Bihar and Madhya Pradesh.

2005, while Uttar Pradesh enacted legislation in 2008 to set up a Water Management and Regulatory Commission.

Finally, the traditional water distribution system operated through the village authorities (*panchayats*) has been maintained, although subject to state permission. A number of states have introduced laws that transfer responsibility for the management of irrigation systems – generally limited to feeder or secondary canals – to farmers' associations.

6.2.3.4 Sri Lanka (Ceylon)

In Sri Lanka water resources management is the responsibility of diverse institutions, all resource-development oriented. An example is the Mahaweli Authority, which is the successor of the Mahaweli Development Board and is primarily a project implementation institution focusing on the operation of major irrigation schemes. This authority coexists with the Irrigation Department, also an irrigation development agency, the National Water Supply and Drainage Board and other sectorial institutions, all operating in accordance with the laws regulating their functions.

In the late 1990's, in order to tackle fragmentation and promote an orderly and equitable development of water resources in line with economic progress and resource protection requirements, the government produced a draft water resources management policy and a draft water act to implement it. This undertaking was supported by the Asian Development Bank, which provided assistance in the establishment of an interim water resources council and a secretariat, and the Food and Agriculture Organization of the United Nations (FAO), which focused on the legislation. The policy calls for the introduction of water use permits (water entitlements) and for institutional mechanisms at the national and basin levels to coordinate resource management activities such as data collection, water resources planning and permit administration. Both the draft policy and the draft act are slated for discussion and further elaboration by the competent authorities.

6.2.3.5 Other countries

Other states of the region, such as New Zealand, Brunei and Singapore have adopted a system similar to that of Victoria, Australia, in that the ownership of strips of land along the banks of rivers is vested in the crown or in the state, and therefore the crown or the state has full power to control water uses.

As in Australia, in Brunei, Myanmar and New Zealand an administrative order is necessary to bring a particular area under the control of the administration. Once such an area is declared, water may only be taken with the permission of the particular authority in charge of that area. Generally, there is an express statutory provision that the authority is under no obligation to supply water, and, if it does so, it may impose such conditions and make such bylaws as it deems necessary. The authority is not liable for the failure to supply water under a contract if prevented to do so by drought or unusual circumstances.

Always in these countries, the administration may cancel rights in the case of non-use, disobedience, non-payment of fees and charges, wastage, improper use, and may resume rights for public purposes. Also, it may determine the amount of water to be supplied to each consumer in times of shortage, and the uses to which it may be put. In many countries, all lands in an irrigation district are assessed to determine the benefit they might obtain from works in the area. On the basis of such assessment the right to take water is quantified, and water rates are levied. Sometimes additional rights may be allotted on application, or, alternatively, the authority may agree to sell additional amounts of water to a particular consumer at a certain time.

As regards groundwater, in Myanmar and New Zealand, like in most states of Australia, regulations or bylaws may prescribe the depth of wells, their mode of construction and the amount of water that may be taken. The relevant authority may issue licences to pump such water and impose the necessary conditions on such licences. Applications for these licences, or to alter wells, are subject to similar rules as applications for surface concessions.

In Singapore, water legislation has the character of municipal water supply regulations. Anyone willing to pay may have the right to connect his pipes to the main reticulation system. This right was modified as a consequence of water scarcity and due to the fact that water is provided by virtue of an agreement with the Sultan of Johore (Malaysia). Before being returned to mainland China, Hong Kong experienced a similar situation, as water was provided by virtue of an agreement with China.

6.2.4 Countries following principles of other systems

6.2.4.1 People's Republic of China

In the People's Republic of China, the Constitution of 1947 (Article 143) and former water laws provided that all water resources are public property belonging to the state. This principle has been reaffirmed by Article 3 of the Water Law of 2002, which states that water resources, both surface and underground, are owned by the state. The state shall encourage water undertakings, promote the conservation of water resources and take action to protect flora, the environment and water quality. Planned allocation of water and strict water conservation measures are also contemplated in the water law. The administration of water is shared among central, basin and local administrative units. At the central level, water management functions are vested in the State Council, which performs them through the Ministry of Water Resources. Basin agencies attached to the ministry carry out water management functions at the basin level. The law calls for integrated water resources planning based on river basin and regional boundaries (Article 14). Integrated (comprehensive) plans are supplemented by sectorial (special) plans. A permit system is in place for the use of surface water and groundwater, except for household consumption, and water charges are levied, except in the case of rural collective economic organizations taking water from their own ponds or reservoirs. Procedures for the administrative settlement of disputes are also provided for, with recourse to the courts as a last resort. Finally, Article 78 stipulates that the provisions of international treaties concerning water shall prevail over those of the water law, unless express reservation has been made. The Water Pollution Prevention and Control Law of 2017 provides for the setting up of emergency and reserve water resource facilities in cities with a single water source. Governments above the county-level must provide the public with information on drinking water quality at least once a quarter. The law brings the 'river chief' system into effect, with leading government and party officials assuming responsibility for addressing water pollution issues, including prevention and control, and ecological restoration, and being rewarded if their objectives are achieved, or punished, in case of failure.

Taiwan has been influenced by the USA prior appropriation doctrine, according to which a water right belongs to the first person who claims and registers it; a water right is not valid until it has been registered with the provincial water conservancy board, which then issues a permit. Failure to possess a permit, however, does not affect the right, which is validated by registration.

6.2.4.2 Japan

In Japan, the River Law of 1964 stipulates that watercourses belong to the public domain and classifies them into two categories according to their importance. The administration of the watercourses of major importance is entrusted to the central government (now the Ministry of Land, Infrastructure, Transport and Tourism), while those of a lesser importance are administered at the prefecture and local levels. Here, also, like in the Philippines and Taiwan, water use permits are accorded priority by the date of application. The authorities may impose conditions of use, but they may not impose unreasonable duties. Unique provisions exist in order to counter problems of subsidence which have occurred because of large-scale industrial extraction of groundwater and to prevent harmful effects from polluted sources which are unsuitable for industrial or domestic use. The law provides that a cabinet order may designate special areas, and any person wishing to extract groundwater for industrial or domestic use in such areas must apply for a ministerial permit. When it is felt that there is the danger of depletion of certain underground water resources, the ministry may restrict the extraction thereof.

As regards water pollution, Japanese law declares that the objectives of pollution control should be the 'mutual harmony of industries' and 'the improvement of public health.' Industrial development must not be weighed against other water uses and should not be hindered by too strict a control of pollution. To implement this policy, the Ministry of the Environment may designate certain areas and fix standards of water quality in each area. Persons wishing to discharge industrial waste may only do so with the consent of the minister, whose task is to see that the discharge complies with the prescribed standards for that area. In addition to these provisions, the authority responsible for a river in a non-designated area is responsible for controlling pollution.

The Basic Act on the Water Cycle, promulgated in 2014, regards water as the common property of mankind and affirms the necessity to preserve the integrity of the hydrologic cycle. Since the river is at the center of the water cycle, a river is to be managed in an integrated manner and, when it flows across administrative boundaries, the central government is to seek coordination with the local governments. The act calls for the development of a water-cycle basic plan, which is subject to review every five years. The central and local governments are to take all the measures needed to maintain and improve water retention and recharge functions, and to management watersheds.

6.2.4.3 The Philippines

Water laws and institutions in the Philippines are the result of a combination of the Spanish system and the western USA prior appropriation doctrine.[39] From the laws of Spain, the Philippines derived the distinction between public and private waters, but this distinction has been abandoned, and all waters are now public. Water use is subject to administrative permit,

---

39 See this chapter, Section 6.5.

except for domestic purposes. The granting of permits follows the principles of beneficial use and priority of application or use introduced by the Irrigation Law of 1912 under the influence of the USA. These principles were confirmed by the Water Code of 1976. Administrative permits, which are granted for the duration of the beneficial use, are given priority according to the date of application. Hence, in the relationship between private persons, the right to use water in times of shortage depends on the date of the original application.[40] Water rights are granted if there is still unappropriated water in the source from which the applicant wishes to draw. Permits are issued centrally but are also registered locally. The rights acquired prior to the code of 1976 are recognized and cannot be usurped by later appropriators. For persons who were using water at the time of promulgation of the code, but had not perfected their rights under the old legislation, special provision is made to register their claims. As regards groundwater, private consumers must obtain a permit from local municipal councils; their operations are supervised by the local district engineer. Exploration licences may be granted to prospect for groundwater on public land. A permit is necessary to extract the water so found, which becomes the property of the finder.

6.2.4.4 Other countries

In Afghanistan, Islamic law is the basis of water resources management, particularly at the users' level. Accordingly, water resources are community property to be controlled by the responsible authorities. Subsequent control exercised by the government follows this rule. In addition, a community right to quench thirst from private waters in case of need exists, but it may only be exercised if there is surplus water. While recognizing the principles of Islamic law and existing customs and traditions, the Water Law of 2009 sets out the functions and powers of government institutions and makes provision for water use and wastewater discharge permits and licences.

The basis for state control over water resources in Indonesia is provided by the Constitution of 1945, which stipulates that, 'Land, water and natural resources are a gift of Almighty God, and shall be controlled by the state and utilized for the greatest welfare of the people in a just and equitable manner.' The use of water for irrigation purposes is viewed by legislation as a community matter, which in practice is translated into a water distribution system aimed at satisfying all users. The Water Resources Law of 2004[41] subjects beneficial water uses to licence requirements and adopts a river basin approach to water resources management and planning. Water resources inventories and plans, to be developed by river basin or aquifer with the participation of stakeholders and water users, must take into consideration the relationship between surface water and groundwater. The law provides for the establishment of water resources boards at various territorial levels and for the participation of farmers in the development and management of irrigation schemes.

A special mention is to be given to the countries of the Middle East which have followed the Moslem system of law from its inception. In Bahrain, Kuwait, Oman, Qatar, Saudi Arabia, United Arab Emirates (UAE) and Yemen, the basis of water law continues to be

---

40 But in emergency cases, domestic and municipal uses come first.
41 Law No. 7 of 2004, promulgated on 8 March, 2004 and published in Republic of Indonesia State Gazette No. 32/2004.

governed by the provisions of the *Shari'a*. No water legislation has been enacted in spite of efforts in some countries to do so, except in Yemen.[42]

In Oman and the UAE, a water resources council has been set up to coordinate the activities of the various ministries responsible for sectorial aspects of water resources. In the UAE this council coordinates the activities of the water departments of the component Emirates, which are autonomous.

Lebanon has a mixture of Ottoman, Islamic and modern legislation, scattered in different legal texts. A draft water code was prepared with the assistance of French advisors in 2004 and was enacted in 2018. In 2000 the country undertook reforms in the water sector by vesting water resources management responsibilities in the Ministry of Energy and Water and merging the existing 22 water boards into four regional water authorities.

In Jordan, the Jordan Water Authority Law of 1988 has declared all waters state property, and the Jordan Water Authority, which succeeded the Jordan Natural Resources Authority, has been established. This authority is responsible for the control of all the water management activities of the country: planning, surveying, groundwater abstraction, pollution control, sewerage, etc. In addition, a Jordan Valley Authority has been created as a development institution.

In Iran, the water law of 1968 has nationalized water resources and has introduced a system of water use permits. However, the *Shari'a* law of the *Shi'ite* school still prevails, particularly at the local level.

Syria has undergone many changes in its water-related legislation, and has passed a law organizing water management into seven river basin units. All water resources are the property of the people, i.e., of the state.

In Iraq, the legislation left by the British is still applied in the field of water, together with Moslem law. Water, as a natural resource, is state property, and any use is subject to licence.

Since water is scarce in all of these countries, conflicts are present at the local level and also, in some instances, at the international level.

6.2.5 Water administration

In the People's Republic of China the administration of water resources is entrusted to the Ministry of Water Resources, which is a department of the State Council. The same principle was applied in Viet Nam, but there the water administration was transferred to the Ministry of Agriculture and, later on, to the Ministry of Natural Resources and Environment. Other countries, such as Iran and Pakistan, have a single ministry of water and power which deals with all water and water related matters. In Afghanistan, water management is a responsibility of the Ministry of Irrigation, Water Resources and Environment. In Nepal, the Ministry of Energy, Water Resources and Irrigation is in charge of water resources administration, but other ministries and agencies bear responsibilities in the water sector. The Water and Energy Commission is responsible for water resources policy and plan formulation, on which it reports to the National Planning Commission. In Bangladesh, the Ministry of

---

42 Where a water law was enacted in August, 2002.

Water Resources is in charge of policy and plan formulation, but the ministry is resource-development oriented. The central body for water management in New Zealand is the Ministry of Environment. The ministry interacts with the regional, district and city councils in the preparation of resource management policies and plans,[43] the implementation of which is then devolved to the councils. In the Philippines water resources are administered by the National Water Resources Board.

In other countries of Asia, however, different responsibilities and powers are conferred to separate institutions. In Japan, the administration of water resources is shared between the Ministry of Land, Infrastructure, Transport and Tourism and the Ministry of Environment, the latter having several public service corporations under its supervision. The Japan Water Agency[44] is in charge of the construction, operation and maintenance of hydraulic works in each of the seven river systems designated by the water resources development master plan. In Thailand, water resources management functions are entrusted to the Royal Irrigation Department, the Department of Water Resources and the municipal authorities, but a number of other institutions, such as the Electricity Generation Authority, deal with sectorial aspects, all development-oriented. In Laos, the administration of water resources is vested in the Ministry of Water Resources and Environment and several other government ministries. However, an interministerial body for the coordination of water management was established in the late 1990's, and river basin committees are being established. In Indonesia, water resources administration is entrusted to the Ministry of Agriculture, the Ministry of Public Works, the Ministry of Mines and Energy and the Ministry of Environment. In the Republic of Korea, it is shared between the Ministry of Land, Infrastructure and Transport, which has a Water Resources Bureau, and the Ministry of Environment, with a central River Management Committee coordinating activities in the water sector and four river basin management offices and committees. Local river management committees coordinate water resources management and resolve water disputes at the basin level. The Water Management Committee, which was established in 2015, performs advisory functions on major policy issues. Papua New Guinea relies on the Department of Environment and Conservation, the National Water Supply and Sewerage Board, the Water Board and the Department of Health.

Federal countries such as Australia, India and Malaysia have a decentralized water administration. Australia, however, has established the Australian Water Resources Council for the purpose of coordinating water and water-related activities. India has established the Central Water Commission for the same purpose.

As regards coordination and planning, existing divisions of administrative responsibilities often render the governments unaccommodating to present needs. To give an example, in the Republic of Korea, as in Japan, rivers are classified according to their importance. The administration of rivers of major importance is entrusted to the central government, while the various provincial governments or prefectures have the task of administering and controlling less important rivers, or the tributaries of main rivers. This division of responsibility obviously neglects to consider the relationship between the water flowing in tributaries and water in the main stream of a river. The situation is further complicated by the fact that applications to develop and use water resources are filed with the local authorities, sometimes

---

43 Plan preparation takes place with the participation, at some stage, of the stakeholders.
44 Formerly known as the Water Resources Development Public Corporation.

without reference to either the provincial/prefectoral or the central government and without lodging documents notifying the nature and extent of the use. To overcome this problem, the Korean government has promoted the establishment of river basin management entities.

Even more significant is the proliferation of agencies with powers directly bearing on water management. While in the Republic of Korea the Ministry of Land, Infrastructure and Transport and the Ministry of Environment have primary importance for waters controlled by the central government, the activities of the Ministries of Interior, Agriculture, Food and Rural Affairs, Commerce and Industry all overlap. The same applies to the activities of numerous semi-independent special development authorities such as the Korea Water Resources Corporation (K-Water), the Korea Environment Corporation (KECO), the Koreal Rural Community Corporation (KRC) and the Korea Hydro- and Nuclear Power Corporation (KHNP). Each provincial government maintains its different bureaus and departments which, similarly, have responsibilities which either overlap or potentially conflict.

6.3 Central and South America

6.3.1 Introduction[45]

In general, in Central and South America water resources are abundant. Some South American rivers are among the longest of the world. However, water distribution, both in seasonal and geographical terms, is uneven. Torrential watercourses cause floods and erosion in some parts of the region, and there are wide arid areas. In many instances little attention has been paid to water location, quality and régime. Likewise, not much is known about groundwater.

The size and the navigability of the main South American rivers allowed European civilization and trade to penetrate to the heart of the continent, paving the way, however, for the international conflicts that erupted during the nineteenth century. There was no building of inland navigation systems such as those connecting the river systems of North America or Europe. Many river ports now stand idle, and public investment in this sector has increased relatively little.

Approximately half of the population of Central and South America has piped drinking water supply. This does not mean, however, that urban dwellers have been able to receive water in the quantity and of the quality necessary to satisfy their needs. The same applies to sewerage services. There is also a challenge created by the impact of the major concentrations of population and industry on water resources, which has led to a quest for private sector participation in the delivery of water and sanitation services as a way to relieve governments of the task of improving these services and of the responsibility for raising tariffs to cover operation costs. It is worth mentioning the repression of street protests – the 'water war' – following the privatization of the municipal water supply company of Cochabamba, Bolivia, in 1999, through the concession to a private consortium led by the American company Bechtel, and the enactment in the same year of a drinking water and sanitation law which withdrew subsidies. Following the protests, the concession was terminated by the government in 2000, and Bechtel was forced to leave. In 2001, Bechtel filed a lawsuit

---

45 Caponera, D.A. (1983) *Water Legislation in South American Countries*. Legislative Study No. 19. Rome, FAO.

against Bolivia in order to claim compensation for the profit lost.[46] This and other failures in privatization efforts have ignited the debate over the recognition of a human right to water in domestic legal contexts.[47]

In most of Central and South America, agricultural production is not possible without irrigation. The limiting factor is not, however, any shortage of water, but rather the vast human resources required for its development and the slow pace of such development.

The increasing demand for electricity, coupled with progressively higher oil prices, makes for the feasibility of hydroelectric schemes. Electric power produced by such means thus becomes a driving force in the development of water resources and an instrument of integration and expansion. South American countries have grasped this point, and are planning or building dams which rank among the largest in the world. The siting of large dams in international basins has given rise to diplomatic and political problems, and much effort has been spent on seeking solutions for them. Hydropower development may be constrained in the long term by less river flows being available as a result of melting glaciers in the Andean region due to the impacts of climate change.

The increasing use of water for industries is now competing with other uses such as municipal water supply and, in some cases, with agricultural water uses. The main incipient problem, however, is the impact of industrial water uses on the environment, which, together with untreated biological wastes discharged into the water, limits re-use potential. Mining uses, which in the past have been protected by the law, are now the object of attempts to bring them under control.

6.3.2 Central and South American water law principles

The issues outlined above have given rise to a widespread interest in the study and development of water law as regards both its national and international aspects. Many countries in the region have introduced new water laws and water codes in recent years, and the process of legislative and administrative change continues where water is concerned. At the international level, numerous treaties[48] have been concluded, and the views of Latin American countries are repeatedly made known in various international forums.

6.3.2.1 Latin American countries

In Latin American countries, legislation takes its origin in the Ordinances of the Spanish Crown issued for all its possessions in America (the Laws of the Indies). While these enactments emphasize the public interest and the states' decision-making powers, in the nineteenth century a movement towards codification was responsible for the introduction into many countries of the privatizing tendencies of the Civil Code of France. The doctrinal influences of Spanish water laws introduced in 1866 and 1879, the 1933 Consolidated Text

---

46 Another example of failure in the privatization of water services in South America is offered by Buenos Aires (concession to Ondeo-Aguas Argentinas, 1993).
47 See Chapter 7, Section 7.12.
48 A list of these will be found in FAO (1978) *Systematic Index of International Water Resources.* Legislative Study No.15. Rome, FAO.

of Italian water laws and, more recently, the diffusion of legislative models as promoted by international organizations and by meetings of experts in water law[49] have expanded the range of options open to South American lawmakers. This fact did not give rise to major differences in legislation, as the problems that lawmakers were called upon to deal with were similar in all countries concerned. However, if this was the situation until not too long ago, the models promoted by international organizations nowadays do not always address the needs of the countries under consideration. Thus, lawmakers are increasingly aware of the fact that they have to meet national expectations, rather than follow policies designed by outsiders.

Early legislation promoted expansion in water use, in that it offered the user certainty before the law and a freedom of action by assigning to him prerogatives that were the same or almost the same as those associated with ownership. The law assigned water rights to landowners or to a given class of users, or allowed their exercise over specified phases of the hydrological cycle, legitimating *de facto* the deprivation of pre-existing indigenous communities of their customary water rights. This system called for a very simple administrative organization for its application.

The different criteria applied for assessing the value of the different uses of water over time and geographically led to the passing of special laws for respective uses or situations, to the laying down of priorities in the water codes, and to assigning the administrative management of the various uses and problems to distinct technical bodies, some of which were centralized while others were to be found at different levels of government.

This has meant that in certain areas water rights assigned to individuals committed more of the resource than was actually available and that more water was abstracted than was compatible with reasonable resource management. Moreover, basin administration conducted from different centers of decision and under legal rules which were formally in conflict with each other resulted in unjustified delays in the implementation of plans, dispersion of the basic information necessary for planning water resources development and tension between sectors and between regions, all of which stood in the way of an integrated or rational use of water.

In order to obviate these negative situations, governments began to assume wider powers where water was concerned, this circumstance in turn rendering necessary a reform of juridical and administrative systems. The reform process went forward at a different pace and at different times in different countries, but the overall trend converged on similar objectives.[50] The work of assembling statistical data needed for efficient use, protection and conservation of water resources has been intensified, and the relevant government departments have been strengthened. Recognizing that water is a key driver of socio-economic development and that water must not be viewed in isolation from other natural resources, a number of countries have adopted an integrated water resources management and ecosystem-friendly approach, and the river basin as the unit for water resources management and planning.

---

49 See Minutes and Proceedings of the International Water Law Association, in *Annales Juris Aquarum*, Vol.1, Buenos Aires, 1968.
50 Valls, M. F. (1976) La estructura juridica y administrativa como instrumento de la politica del agua en America Latina. *Annales Juris Aquarum*, Vol.II. Caracas, 226–249.

Participatory approaches to planning and decision making are increasingly being followed, involving to varying extent the indigenous communities.

A distinction is to be drawn between large federal countries, such as Argentina and Brazil, and the remaining countries, which, to the exception of Mexico, are unitary. Both Argentina and Brazil have a national/federal constitution and a constitution for each province/state. Moreover, they have enacted federal legislation laying down the basic principles of water resources management, while provincial/state legislation takes care of more localized issues. In turn, this legislation is detailed through local/municipal legislation.

6.3.2.2 Other countries of Central and South America

In Central and South America there are countries which have inherited a different legal and institutional framework relating to water resources management. Barbados, Belize, Guyana, Jamaica, St. Lucia, Trinidad and Tobago all have adopted the common law riparianism concept imported by the British in modified versions. French Guyana, Guadalupe, Haiti and Martinique have inherited the French civil law system. Surinam and Curaçao follow the Dutch water law and administration system. As these systems have already been analyzed, attention will here be paid to the system of water law followed in Latin American countries influenced by the Spanish and Portuguese systems.

6.3.3 The legal status of water resources

In Latin America, that is, in countries influenced by the Spanish and Portuguese systems, including Argentina, Bolivia, Brazil, Chile, Colombia, Ecuador, Guatemala, Nicaragua, Mexico, Panama, Peru and Venezuela, public ownership has been extended to all water resources almost everywhere. In Panama, even meteoric waters are public. In Uruguay, all waters, except rainwater harvested on private land, are part of the public domain. Similarly, under the Water Law of Costa Rica rainwater and waters located or flowing entirely on private land are the property of the landowner.

Consistent with the ancient civil law tradition,[51] the legal régime of groundwater is rather ambiguous in some Latin American countries, due to the fact that, although no explicit mention is made of the resource by their civil codes, everything to be found above and below the land is owned by the landowner. This is the case of Costa Rica,[52] where groundwater abstracted from wells located on private land belongs to the landowner. In Argentina, El Salvador, Panama, Paraguay, Mexico and Uruguay groundwater belongs to the public domain, as in Guatemala and Honduras, but landowners are entitled to abstract the water they need for minor uses and in some cases for irrigation purposes. In 2000 Uruguay adopted a decree[53] regulating the planning and management of the portion of the Guaraní Aquifer

---

51 See Chapter 5, Section 5.5.2.
52 This was the case of Argentina before the 1968 reform of the Civil Code, and of Paraguay prior to the promulgation of the Water Law of 2007.
53 Decree No. 214/000 of 26 July 2000. The Guaraní Aquifer System underlies parts of Argentina, Brazil, Paraguay and Uruguay.

System located within its territory. The decree sets up the Guaraní Aquifer Advisory Board and provides for public participation in aquifer management through public audiences.

In Ecuador, the civil code declares that water is a good common to all, and therefore is not susceptible to appropriation. The Water Act of 2014 deems all waters to be national property for public use and explicitly prohibits all forms of privatization of water. Rights on water existing before the introduction of the act are conditional upon the efficient use of the resource. These rights are subject to review with a view to determining whether there are cases of water hoarding and, in such cases, they may be subject to modification and, if necessary, termination.

6.3.4 The right to use water

With the extension of public ownership on water resources, individual water use rights have been limited and subjected to regulation by the public authorities in order to secure better protection.

Rights on publicly owned water may be acquired by virtue of water use authorizations, permits or concessions granted by the competent administrative authorities. The law specifies the cases in which an authorization, permit or concession is not required. Generally speaking, the exemption applies to drinking and domestic water uses, the watering of livestock and bathing. It may extend to fishing and navigation, such as in the case of Bolivia, Uruguay and Venezuela, and to farming and irrigation when the amount of water used and the means employed are not significant (Brazil, Guatemala, Honduras). In Costa Rica, irrigation with rainwater flowing in public roads or riverbeds does not require a concession when permanent intakes are not employed. The same applies to the irrigation of riparian land with water drawn from navigable watercourses, when the common use is not affected. In Mexico, a concession is not required for groundwater abstraction, except when the water is located within protected or reserved areas, or in areas subject to special régime so declared by the federal government.

The Water Act of Ecuador vests an autonomous right to water conservation in nature, or Mother Earth (*Pacha Mama*), on the grounds that water supports all forms of life. This right entails further entitlements to the protection of water sources, natural water bodies, catchment areas, recharge areas, wetlands and glaciers, to the maintenance of minimum ecological flows so as to preserve ecosystems and biological diversity, to the integrity of the hydrologic cycle, to the protection of river basins and their ecosystems from pollution, and to the restoration and recovery of degraded ecosystems. The right of Mother Earth to water conservation acts as a limit to the power of the public administration to authorize water uses and potentially polluting activities.

In Chile, amendments to the 1981 water code (2005) have introduced a charge for the non-use of a water right, while the previous legal régime, which favoured water markets, allowed water right holders to hoard water and speculate on it.

6.3.5 Order of priorities

The order of priorities among different water uses varies from country to country. As a general rule, domestic and drinking needs range first, followed by the other uses in an order of preference which may be established according to the general interest or considering regional and local differences and needs. In Brazil, priorities are established in accordance with water resources plans, but like in other countries, human consumption and the watering of animals range higher in the order. In Colombia, priorities are determined on a regional basis in the light of ecological, economic and social considerations. In Peru, basic human needs are to be satisfied first as a matter of principle, since access to water is a fundamental right. Similarly, in Ecuador and Venezuela access to water is viewed as a fundamental human right. Therefore, it must be guaranteed. In Ecuador, irrigation aiming at ensuring food security, and the maintenance of ecologic flows in water bodies, also enjoy priority over water uses for production purposes. In Mexico, priorities are determined by the National Water Commission and the basin organizations in consultation with the water users. Recent constitutional amendments, however, have recognized the right to water and sanitation as a human right. Therefore, it has priority over other water uses. In Paraguay, the water use rights of indigenous communities enjoy priority over other water uses.

6.3.6 Legislation on water use, quality and pollution control

Central and South American countries are increasingly interested in the preservation of water resources, both quantitatively and qualitatively.

From a quantitative viewpoint, provisions are to be found in water statutes aimed at preventing the waste or other improper uses of water. These provisions may cover the prohibition to use water for a purpose other than that for which the relevant concession has been granted (Argentina, Brazil) or in excess of that provided for in the grant (Argentina, Mexico, Peru), the obligation to meter the water (Argentina, Chile, Mexico) and to maintain the installations needed for the efficient use of water (Argentina, Chile, Ecuador), the obligation to use water rationally, efficiently or economically (Ecuador, Honduras, Venezuela), and limitations in quantity of the right to use water (Paraguay). In Chile, the 2005 amendments to the 1981 water code require the applicant for a water right to justify the amount of water requested in the application. This requirement also aims to prevent water right holders from hoarding unused water and speculating on it.

As regards the qualitative aspect, national legislation subjects waste and wastewater discharges from industrial, domestic and other sources into bodies of water to administrative authorization and to compliance with certain conditions, such as treatment prior to discharge (Argentina, Brazil, Ecuador, Mexico, Peru), the provision for cleanup or natural drainage (Brazil), the respect of technical rules and regulations established by the authorities (Ecuador, Paraguay), and to the obligation to compensate any loss or damage caused (Brazil, Mexico). Industrial activities are placed under special control in Colombia and in Uruguay.

For the purpose of ensuring a quality of water suitable for human consumption and for other uses, in some cases national legislation provides for the classification of water resources according to the intended purpose of use, and for the determination of the conditions governing water collection, supply, distribution and quality (Brazil, Colombia, Peru). For each class of water, limit values are issued with reference to given parameters set with respect to specified substances (Brazil, Venezuela). In Venezuela, the legislation also calls for the introduction of measures to control non-point pollution, i.e., pollution from diffused sources.[54]

---

54 See Chapter 7, Section 7.14.5.

In Ecuador, the water act places on users the obligation to carry out timely analyses of the water that they use, and to provide suitable treatment for water exceeding the pollution tolerances prescribed by the water authorities. Together with the Ministry of Health, the authorities supervise all installations for the treatment of water that has become polluted. Similar obligations are imposed on providers of water services by the Mexican Law on National Waters, as amended (2004).[55]

The general prohibition to pollute watercourses is present (Argentina, in respect of navigable watercourses), also with a view to protecting aquatic life (Colombia). The penal code of El Salvador views the death of a person caused by the introduction of noxious substances into water as murder.

The water legislation of some countries (Argentina, Brazil, Chile, Colombia, Mexico, Peru) also envisages the possibility of reusing water, subject to administrative authorization and in respect of the conditions laid down by the competent authorities. In the case of wastewater or drainage water, conditions include treatment requirements in relation to the intended water use. These waters may not be used directly to irrigate crops to be eaten raw (Chile, Peru), or for raising shellfish.

6.3.7 Water administration

In order to obviate the negative situations created by the assignment of water resources management to distinct technical bodies, some of which were centralized while others were to be found at different government levels, some Central and South American governments have concentrated, or coordinated, the activities of various administrative bodies where water resources development and conservation are concerned. Thus, Argentina has set up the Under-Secretariat for Water Resources; Brazil, the Under-Secretariat for Water Resources under the Ministry of Environment, the National Water Resources Council and the National Water Agency (ANA); Colombia, the Institute for the Development of Renewable Natural Resources; Costa Rica, the National Water Council; Ecuador, the Intercultural and Multinational Water Council and the National Secretariat for Water; Cuba, the National Institute for Water Resources; Bolivia, the Ministry of Environment and Water; Mexico and Panama, the National Water Commission; Paraguay, the General Directorate for the Protection and Conservation of Water Resources under the Environment Secretariat; Peru, the National Water Authority; Uruguay, the National Water Directorate under the Ministry of Housing, Territorial Planning and Environment; and, Venezuela, the National Water Council and the Ministry of Eco-Socialism and Water.

In Mexico, the amended Law on National Waters provides for the transformation of the National Water Commission, attached to the Secretariat of the Environment and Natural Resources, into an autonomous legal entity with its own assets. Further, it provides for the transformation of the regional directorates of the commission into basin organizations supported by basin councils integrated with the participation of water users. The law explicitly recognizes the basin and the aquifer as the basic territorial unit for integrated water resources management.

---

55 Published in the Official Gazette of the Federation on 29 April, 2004.

The powers vested by their respective constitutions in the central governments, together with a greater potential for financing, installing and operating hydraulic works, have enabled them to coordinate their actions with those of local governments and, in federal countries, with those of state/provincial governments. In Argentina, for example, various agreements have been stipulated between the national and provincial governments and between the provincial governments themselves. Based on a law of 2002,[56] however, the river or lake basin, and the aquifer, are considered as the unit for water management. The law calls for the establishment of basin committees to advise the authority responsible for water resources on matters relevant to the management of water and the environment within interjurisdictional (interprovincial) basins. Thus, when a water use subject to administrative permit is likely to cause significant environmental impact across administrative boundaries, the approval of the relevant basin committee is required.

The principle by which water resources management is to take place by river basin is enshrined in the national water resources policy of Brazil,[57] and is the basis for the establishment of river basin committees and their executive secretariats, i.e., the water agencies. River basin management institutions in the form of basin committees and councils and their secretariats are also provided for by the water legislation of Ecuador, Honduras, Peru and Venezuela. The objectives of these institutions may vary from country to country. For instance, the basin councils of Honduras, to be established under legislation adopted in 2017, are expected to propose, agree on, and implement projects and actions relating to water resources management, the development of hydraulic structures and the protection and conservation of water resources, while the basin councils of Peru and Venezuela participate in the river basin planning process as a whole. In Uruguay, the Constitution, as amended in 2004, requires the participation of water users and the civil society in river basin planning and management; the national water resources policy, enacted in 2009 and implemented through a decree issued in 2013, provides for the establishment of river basin and aquifer commissions to advise the Regional Water Councils on plan formulation and implementation. A characteristic common to all these basin institutions is that they include representatives of a broad range of stakeholders, and in some instances representatives of farmer and indigenous communities. Finally, Panama created a National Directorate of Integrated Hydrologic Basin Management in 2005. Although endowed with basin management responsibilities, this is a national institution, since the small size of the country does not justify the creation of separate basin agencies.

The Water Act of Ecuador provides for the establishment of community-based organizations for supplying drinking water to rural areas not serviced by a municipal water utility. Similar organizations may be established for the management of irrigation and drainage infrastructure. The act, like, the Water Act of Peru of 2009, explicitly recognizes all traditional and community-based forms of water resources management and the traditional water rights of indigenous communities.

---

56 Law No. 25688 of 28 November, 2002, on the Environmental Management Regime for Water Resources.
57 Law No. 9433 of 8 January, 1997, establishing the National Water Resources Policy, and the National Water Resources Management System.

6.4 Europe

6.4.1 Introduction

In the various countries of Europe, water law has developed and persisted in accordance with climatic conditions and following different social and economic needs. Most continental countries have absorbed the principles elaborated originally in the French Napoleonic Code into their own legislation and are referred to as civil law countries. The legal systems of some central and north European countries have, in addition, been influenced by Germanic and customary law principles. We have seen how the United Kingdom has given birth to the so-called riparian doctrine.

The former Soviet Union and the eastern European countries had a system of water laws and institutions which differed from those of the countries just mentioned. The political events occurring after 1988 have brought about economic liberalization and the establishment of relations with the European Union (EU), so this system is undergoing significant changes.

A distinction may also be drawn between the countries belonging to the EU,[58] those which have applied for EU membership (candidate countries),[59] and those which are not members. Member countries are subject to the 'regulations' and 'directives' of the Union on water questions. The regulations are immediately binding on them, while the directives become so after obligatory approval. Candidate countries are required to transpose EU law into their legal systems as a prerequisite for acquiring EU membership. The EU countries, together with other industrialized European countries, the USA, Canada, Mexico, Australia, New Zealand, Japan and Korea, belong to the Organization for Economic Cooperation and Development (OECD), which also issues recommendations on legal and institutional aspects of water resources management.

Particular attention has to be devoted to the German situation, as the reunification of the two Germanies has had considerable effects on the legal and institutional aspects of water resources management. In the Federal Republic of Germany (FRG), most responsibilities in water management are under the exclusive jurisdiction of the component *Länder* (states), with the federal government retaining only limited powers. Instead, the former German Democratic Republic (GDR) was a unitary state, where all the legal and institutional aspects of water resources management were under the sole responsibility of the central government. While the overall management of water resources in the FRG would prove to be a difficult task to achieve at the national level in view of the institutional constraints, in the GDR this problem did not arise. After reunification, the GDR has merged with the FRG and has adopted its legal system.

---

58 Austria, Belgium, Bulgaria, Croatia, Cyprus, Czech Republic, Denmark, Estonia, Finland, France, Germany, Greece, Hungary, Ireland, Italy, Latvia, Lithuania, Luxembourg, Malta, The Netherlands, Poland, Portugal, Romania, Slovakia, Slovenia, Spain, Sweden and the United Kingdom. Further to the 'Brexit' referendum of 2016 and to the passing of the European Union (Notification of Withdrawal) Act by Parliament in 2017, the United Kingdom is due to cease to be a member of the EU in 2019.

59 The status of candidate country is acquired by a country applying for EU membership when the European Council formally accepts the application. At present, Albania, Macedonia, Montenegro, Serbia and Turkey are candidate countries.

Many European governments have felt the need for a more stringent control over surface and underground water use and more effective conservation measures, which has led to the enactment of special water legislation to control the volume of water utilized. This is in spite of favourable climatic conditions which *per se* would never have warranted such measures in view of the relative availability of water, particularly in central and north European countries. In parallel with this, because of high industrialization, urbanization and the development of agro-allied industries, stringent controls have been introduced to protect the quality of water. More recently, efforts towards water resources development, protection and conservation are better coordinated within the framework of resource management plans. After the entry into force of the EU Water Framework Directive, river basin plans have become mandatory for EU member states and for states applying for EU membership.[60]

In order to better regulate and control water resources, various instruments are being utilized through appropriate legislation, among which are the following:[61]

(i) the introduction of administrative permits for the use of water[62] and for effluent discharge;
(ii) the establishment of water quality standards, objectives and criteria;
(iii) the obligation to treat effluents in order to meet specified standards (emission limits);
(iv) the recycling or reuse of wastewater, under certain conditions;
(v) the implementation of sludge collection, treatment and disposal;
(vi) the introduction of financial instruments such as subsidies, loans and tax exemptions to facilitate water pollution control;
(vii) the tarification, on a differential basis, of water supplies and water discharges;
(viii) the introduction of heavy penalties and sanctions against offenders;
(ix) the introduction of the 'polluter pays principle;'
(x) water resources planning.

6.4.2 The legal status and the right to use water

There is a tendency in European water legislation either to abolish or to restrict the concept of private ownership of water, and to extend government control over all water uses and activities.

In France, as in other civil law countries, the traditional distinction between public and private waters has gradually been abandoned to facilitate the administrative control of water uses and the prevention of water pollution. Water resources are now owned by the nation, which protects and conserves them in the interest of present and future generations of users. The water users may obtain water rights on the basis of administrative authorizations.[63] In Italy, also, Law No. 36 of 5 January, 1994, states that all water resources, whether surface or underground, are public. The code on the environment of 2006[64] has confirmed this

---

60 See this chapter, Section 6.4.5.
61 For a discussion of resource regulation and control tools, see Chapter 7.
62 Except minor ones.
63 Water Law No. 92–3 of 3 January, 1992.
64 Legislative Decree No. 153 of 3 April, 2006, replacing the law of 1994.

approach. Permits or concessions are required for most water uses, subject to the payment of water rates and fees.

In the United Kingdom, a permit system was introduced by the Water Resources Act of 1963. Under subsequent legislation, no licence for the abstraction of surface or underground water may be granted without prior public notice being given, so that persons affected in their rights or interests are able to file their objections. A special licencing régime has been introduced for large-scale water abstractions by the Water Act of 2003, and water undertakers and suppliers may be fined up to 10% of turnover for the breach of licence conditions, standards of performance or other obligations. Water abstractions are subject to the payment of a fee which may vary according to the quantity of water withdrawn, as metered.

In the republics of the former Soviet Union and in eastern European countries, all water resources are state property. However, with the reintroduction of the concept of private land ownership, in Estonia and Lithuania the ownership of the water situated entirely within the boundaries of private land has been made to follow the ownership of this land. In the Russian Federation, also, the water code of 2006[65] provides that small water bodies which are not hydrologically connected to other water resources may be privately owned.

A special situation arises in Finland, where the water act of 2011, which is mostly based on the water act of 1961, distinguishes the ownership right on a 'water area,' i.e., an area permanently covered with water, from the ownership right on water. While a 'water area' may belong to the state, to villages or to other political subdivisions, or to individuals, no one owns water as such. The occupier of a water area has, however, priority in its use. A water abstraction permit is required when it serves the purposes of a water utility, when the intended abstraction exerts a negative impact on water quantity, water quality or other water users and, in the case of groundwater, when the abstraction exceeds 250 m^3 per day. The act protects the water rights of the Sami indigenous communities.

6.4.3 Water quality and pollution control

The need to prevent and to abate water pollution is one of the main concerns of many European countries. Over time, a substantial amount of special water pollution control or environment protection legislation has been enacted for this purpose. This legislation provides, *inter alia*, for the control of waste and wastewater discharges and other polluting activities, the establishment of water quality and effluent standards and the protection of groundwater from pollution.

In most European countries, waste and wastewater discharges are subject to administrative permits. These permits may be issued only if certain minimum effluent standards established within the framework of a basic law or regulations are complied with through appropriate treatment. In some instance, the legislation requires the use of the best available technology to treat certain effluents. This is the case of Germany, among others. Conditions for a wastewater discharge are generally contained in the relevant permit and may include treatment requirements, conditions as regards the place, time and mode of discharge and other requirements. In Spain, voluntary agreements may be entered into between given

---

65 In force as of 1 January, 2007.

industries on the one hand, and the central or regional governments on the other, to progressively reduce pollution. These agreements are usually based on a prior environmental audit and on the commitment to take corrective action if the need arises to do so.

Like water use permits, wastewater discharge permits may be modified, suspended or revoked in the public interest or if the conditions established are not complied with. Charges or fees calculated with reference to the amount of effluent released, its noxiousness and/or other criteria may be levied, in accordance to the 'polluter pays' principle, such as in France, The Netherlands, Germany, the Czech and Slovak Republics and the United Kingdom. These charges or fees serve the purpose of subsidizing the defence against pollution. In some countries provision is made for the inspection of discharges, such as in the case of Denmark.

Financial incentives in the form of loans, grants or tax exemptions are sometimes established or granted for the construction of wastewater treatment plants and/or for the undertaking of other anti-pollution activities. In Austria, the Water Management Fund is responsible, at the federal level, for providing financial support to water supply and wastewater treatment facilities. In Hungary, economic incentives have been created to encourage compliance with legal requirements.

Unlike the legislation of other European countries, the Spanish Water Law of 1985 provides for the reuse of wastewater, subject to administrative permit and under certain conditions. The Consolidated Text (*Texto refundido*) of 2001 allows the holder of a concession for wastewater reuse to acquire the relevant wastewater discharge authorization from its holder, on the basis of a contract. In Romania, the Environment Protection Act allows for the use of sewage effluents in agriculture during the summer months.

Provisions are also made for the control of other activities which may cause water pollution. The French Dangerous or Polluting Premises Act of 1917, revised in 1976, establishes a system of licences for the operation of commercial and industrial installations. In Austria, permission is required for activities which are likely to cause water pollution or injury to the interests of other water users. A similar provision applies in Norway. In Denmark, the storage of polluting substances in the ground may not take place without permission, and regulations have been issued for specific substances (oil, sewage, liquid fertilizers and manure containers).

In particular, the protection of groundwater is ensured through the prohibition of certain activities (such as discharges into wells in France) and the establishment of protected areas or zones (Belgium, Germany). In some countries, groundwater is explicitly reserved for certain purposes, such as drinking (Czech and Slovak Republics, Romania), fire-extinguishing and manufacturing processes requiring water of the best quality (Romania). The Austrian Federal Act on Water Affairs, 1990, provides for continued monitoring of water, both surface and underground.

In compliance with the first wave of EU directives[66] concerning the quality of drinking water, or, in broader terms, the quality of water intended for human consumption,[67] EU member countries enacted regulations establishing water quality standards for both surface and underground waters. However, a new approach to water pollution control has been introduced by virtue of recently adopted directives. Thus, EU countries are adjusting

---

66 See this chapter, Section 6.4.5.1.
67 The term includes household uses, bathing, cooking.

their legislation to the new requirements, or are in the process of doing so. In particular, the EU Water Framework Directive (2000) calls for the adoption of the so-called 'combined approach,' consisting of the setting of emission limit values and quality standards with the aim to achieve environmental objectives for surface water and groundwater within each basin, leading to the attainment of a good water status by a set deadline. The directive requires the progressive reduction of emissions, losses and discharges of priority substances, as well as the phasing-out or cessation of emissions, losses and discharges of priority hazardous substances. A list of these substances[68] is annexed to the directive.

Protected areas or zones may be established for various purposes, such as for the control of water quality in relation to specified uses (Bulgaria), the protection of important natural aquifers (Czech and Slovak Republics) or groundwater recharge areas (Germany), or the protection of water intakes (Denmark, France).

Finally, fish are given special protection in some Scandinavian countries (Denmark, Iceland), as they represent a source of income.

6.4.4 Institutional framework

Water resources administration in Europe varies according to the political-administrative framework of each country. In this connection, European states may be roughly subdivided into three groups: federal, unitary and states undergoing a process of decentralization.

There is a tendency, however, common in various European countries independently from their form of government and territorial organization, to entrust water management responsibilities to ministries for the environment. This is the case of Denmark, Finland, France, Germany and Poland. In the United Kingdom, the Secretary of State for the Environment has overall responsibility for the promotion of an effective national water policy. In Finland, the National Board of Waters and the Environment, founded in 1970, was placed under the Ministry of Environment in 1986. In 1988, a State Committee on Nature Conservation was established in the USSR, with the task to control the rational use of water and to protect it from pollution and exhaustion.[69] After the disbandment of the USSR, this task was vested in similar pre-existing committees in the individual independent republics, or in newly created ministries for the environment. In Hungary, water management was the responsibility of the Ministry of Water Management and Environment Protection until 1990; it was subsequently transferred to the Ministry of Transport and Telecommunications and then was vested in the National Water Authority, which operates according to the water law of 1995 and its implementing decrees. The authority has recently been disbanded and its functions have been vested in the National Directorate for Environment, Nature Conservation and Water under the Ministry of Environment and Water and in the 12 regional directorates based on river catchments.

In EU member countries and the countries which have applied for EU membership, river basin institutions are playing an increasingly important role, given that the EU Water Framework Directive of 2000 has set the requirement for the identification and delimitation

---

68 Adopted by Decision No 2455/2001/EC of the European Parliament and Council amending Directive No. 2000/60/EC.
69 WHO (1989) *Water and Sanitation Services in Europe*.

of river basin districts and the development of basin management plans. These countries have taken steps in this direction by enacting and implementing legislation in line with the EU *acquis* in the water sector.

Mention should also be made of the role assigned to ministries of public health throughout the European continent, for what concerns the monitoring of water quality, public health protection and the prevention of water pollution. These ministries are usually made responsible, either alone or in cooperation with the ministries in charge of water resources management, for the establishment of water quality standards.

Another common feature among European countries is that the services of water supply, sewerage and sanitation (wastewater management, treatment, discharges into water bodies and sludge disposal) are operated at the local (municipal, communal or intercommunal) level. The form, structure, functions, powers and procedures of the service providers vary from country to country and even from one local administration to another within a single country. The providers are numerous and coexist, in spite of their diversity. As regards their form, they may be public, private or mixed, individual or collective. An example of privatization of the water supply and sewerage services is provided by England and Wales.[70] Their functions vary, as they may be responsible for water supply or sewerage only, for water supply and sewerage, for water supply, sewerage and sanitation, including or excluding in this case wastewater treatment or wastewater management, or for any other service resulting from a combination of those just mentioned.

6.4.4.1 Federal states in Europe

Federal countries in Europe include Austria, the Federal Republic of Germany and Switzerland. In each one of these countries, the form of water management is different. Formerly, the Soviet Union, Czechoslovakia and Yugoslavia also belonged to this category.

In Austria, the water divisions of the Federal Ministry of Sustainability and Tourism[71] and of the nine provincial administrations have water management responsibilities as regards decision making. Water management planning and environment protection bodies operating at the federal and provincial level provide advice to the ministry. A water management information system containing data and information on water management for the entire country is kept at the federal level and is open to the public; similar systems exist at the provincial level. At the federal level, the Water Management Fund provides funds for the financing of water supply and wastewater facilities; the Environment Agency carries out water quality investigations, monitors transboundary water pollution and carries out investigations on flow capacities and catchment areas. The hydrographical service and senior experts in administration, hydraulic engineering and public health operate both at the federal and the provincial levels. The water quality surveillance services are responsible for monitoring wastewater discharges and river quality at the provincial level, under the guidance of the federal services.

In the Federal Republic of Germany, the *Länder* (states) are responsible for water resources allocation and water pollution control in accordance with the regulatory framework

---

70 See Chapter 5.
71 Which replaced the Federal Ministry of Agriculture, Forestry, Environment and Water in 2018.

established at the federal level. At the regional level (*Regierungsbezirk*), water supply, wastewater and pollution control are integrated within a water authority supported by technical agencies or by the *Landesamt* (state office). Also at the county level there are water authorities generally assisted by a technical department. The municipalities represent the lowest level, at which a large number of water supply undertakings operate. As a result of municipal reorganization and mergers of companies, however, their number has been reduced. In addition, due to an increasing shortage of good quality water, water supply undertakings are being induced to cooperative efforts. Such cooperation often takes place in the form of water boards which sometimes also deal with wastewater treatment and discharges, subject to the control of the *Länder*.

In Switzerland, there are three, and sometimes four, levels of water administration: the federal, the cantonal, sometimes the regional (inter-cantonal) and the municipal levels. At the federal level, water resources development and conservation responsibilities lie with the Confederation. The Confederation is also responsible for the control of hydropower generation. The Federal Office for Environmental Protection is in charge of water services and of the provision of technical and supervisory services for water protection with the advice of the Federal Commission for Water Protection, a technical and advisory body established under the Department of the Interior. The Federal Office of Public Health is responsible for maintaining water quality standards. A constitutional amendment of 1954 has brought water pollution control and hydropower generation under federal jurisdiction. The cantons are responsible for water resources protection and river conservation, and are obliged to implement the regulations and measures necessary to prevent pollution. Also, the cantons have been made responsible for the planning and financing of water supply projects, in liaison with the municipalities and, where they exist, with the regions. The municipalities, in turn, are responsible for domestic water supply. There is a wide range of water supply companies in Switzerland, generally owned by the municipalities.

Prior to the dissolution of the Federal Republic of Yugoslavia, federal agencies were responsible for the enactment, implementation and enforcement of water or water-related legislation concerning the state as a whole or two or more republics or autonomous provinces.[72] The basic principles concerning the administration and use of water resources were determined by the republics or autonomous provinces. It is at this level that the management of watercourses intersecting communal, republican or state borders took place, while other watercourses were managed at the local level. All republics and autonomous provinces adopted water legislation. Some republics issued laws on the protection of specific waters located within their territories, or special legislation concerning the protection of drinking water. Local water regulations were also enacted at the communal level. Communities were competent to decide on the protection of water sources, either separately or within the framework of town planning schemes. Likewise, they could decide on the establishment and organization of water services and on the conditions for waste disposal. In addition, the communities were responsible for water supply and sewerage, the protection and maintenance of plants and water supply networks, and for the protection of drinking water quality. In accordance with the socialist principles of self-management, communal organizations having a common interest in the field of water management were set up. They were responsible,

---

72 See *Water Law in Selected European Countries*, op. cit., 122.

inter alia, for establishing relations between water services and the consumers, and formulating and implementing water development policies. Communes, labour associations and other organizations tended to establish water protection communities for specific lakes or rivers. For this purpose they were empowered to conclude legally binding social contracts and to allocate funds for implementing protection measures. The republics and autonomous provinces of the former Federal Republic of Yugoslavia are now independent states and have made their own legal and institutional arrangements for water resources management. Most of them have become members of the EU or are candidate to become members. Therefore, they have adopted legislation in line with the EU water *acquis*, or are in the process of doing so.

Before the separation of the two republics composing Czechoslovakia (Czech Socialist Republic and Slovak Socialist Republic), there were two parallel water administrations. In each republic, the Ministry of Forestry and Water Management had overall responsibilities concerning the management of water affairs. It was responsible, *inter alia*, for preparing water and water-related legislation, drawing up water management plans and monitoring water quality. The directives issued by the ministry were implemented by the regions and the districts. At the regional and district levels, water services were dealt with by the water management department of the regional national and district national committees. These committees were responsible for granting consent for water utilizations and for the construction of water schemes, for the control of water pollution and for emergency measures in the case of accidental spillage or contamination. A number of public entities carried out various tasks under the overall supervision of the ministry. At present, water resources management functions in the Czech Republic are shared between the Ministry of Environment and that of Agriculture, with regard to quality and quantity aspects, respectively. In Slovakia, they are vested in the Ministry of Environment.

The situation in the former Soviet Union has already been analyzed.[73]

6.4.4.2 Unitary states in Europe

In states with a unitary form of government, water management activities are generally carried out or coordinated at the central level. In some countries these activities are deconcentrated at territorial levels (intermediate, local, basin or other), while in others there is a trend towards decentralization.[74]

In Denmark, the Ministry of the Environment and Food is responsible for water resources management at the national level, mainly through its National Agency for Environmental Protection. This agency is responsible, *inter alia*, for implementing legislation on environment protection, water supply and watercourses. The county councils are in charge of water resources planning, the approval of sewerage and wastewater treatment plants, planning the quality of receiving waters, water quality control and the granting of water use permits for irrigation, industrial purposes and water supply. Local councils are responsible for the planning of water supply, sewerage and wastewater treatment, the operation of public

---

73 Chapter 5, Section 5.7.
74 See Chapter 9, Section 9.3.2.

waterworks and wastewater treatment plants (through limited liability companies), industrial water pollution control and the granting of permits for minor water uses.

A National Water Council was established in Bulgaria in 1969 under the Council of Ministers. Except for tasks relating to water quality, which were vested in the Ministry of Environment, this council performed functions concerning policy-making, planning, data collection and water rights administration with regard to both surface and underground water. In addition, it was responsible for the settlement of water disputes and the formulation of legislation on the quantitative aspects of water resources management. The council also administered a water resources fund through which it financed its activities, as well as development projects. Irrigation was the responsibility of the Ministry of Agriculture and of the agencies under its supervision. In July, 1999, a Water Act was adopted in order to reflect the requirements of the European Water Framework Directive, yet to be adopted, into the national legislation.[75] A new Water Act was then promulgated in 2006 to seek full transposition of the EU *acquis*. The act has vested the bulk of water management functions in the Ministry of Environment, renamed Ministry of Environment and Water. The Supreme Water Council, which is made up of representatives of other water-related ministries, ensures the integration of sectorial water management issues into policy and planning processes. Irrigation management functions have been transferred to water users' associations established on the basis of an act adopted on 22 March, 2001.

Water resources management in Sweden is entrusted to various national institutions: the Environment Protection Agency (SEPA), the Agency for Marine and Water Management, the Swedish Geological Survey, the Meteorological and Hydrological Institute, the National Food Agency and others. The SEPA and the Agency for Marine and Water Management, both under the umbrella of the Ministry of Envronment and Energy, are policy-making bodies. The latter agency is also responsible for monitoring progress in the implementation of water quality standards and objectives, and for matters related to fisheries. Furthermore, it coordinates the activities of the country's five water district authorities. The National Food Agency is in charge of drinking water quality, while the Meteorological and Hydrological Institute and the Geological Survey are responsible for surveying surface water and groundwater respectively, and for providing data and information thereon. At the county level, special environment protection units within the county administrations provide for the regional planning of environment protection, enforce environmental regulations and take action on environmental matters.

In Finland, the Ministry of Agriculture and Forestry deals with water quantity aspects, i.e., the use of water, and the Ministry of the Environment is responsible for water quality and pollution control. Water transport is a matter for the Ministry of Transport and Communications, while the Ministry of Social Affairs and Health deals with drinking and bathing water quality. The Finnish Environment Institute (SYKE) is Finland's national centre for environmental research and development and produces data on the state of the environment, including water resources. A number of tasks have been delegated to regional and local authorities. At the regional level, fifteen Centers for Economic Development, Transport and the Environment (ELY Centers) coordinate the preparation of river basin management plans, supervise compliance with the permits issued by the six Regional State Administrative

---

75 Bulgaria joined the EU on 1 January, 2007.

Agencies (AVI)[76] and carry out field monitoring work. Both the centers and the agencies are under the supervision of the Ministry of the Environment. At the local level, municipal environment boards act as supervisory authorities and resolve minor problems regarding drainage, water supply and wastewater. Finland has three administrative water rights courts which handle legal matters concerning the allocation of water rights, applications for projects affecting water resources, wastewater discharges, waterworks and other matters. In addition, there is a Water Rights Appeal Court. The Supreme Administrative Court and the Supreme Court of Justice also act as courts of appeal in water affairs.

6.4.4.3 States undergoing a process of decentralization

To this group belong countries with a unitary form of government in which, for various reasons, powers are being, or have been, transferred or delegated from the central level to government administrations or institutions acting more or less autonomously at another territorial level of jurisdiction.

In Italy, the central water administration, which in the past was made up of the Ministry of Public Works and Hydroelectric Plants, the Ministry of Environment, the Ministry of Agriculture and Forestry, the Ministry of Industry and the Ministry of Public Health, has been entrusted to the Ministry for the Environment, Land and the Sea. Following the political decentralization based on the Constitution of 1947, previously centralized water management responsibilities have been transferred to the regions, the provinces and the municipalities. The regions now have full administrative and legislative powers with respect to numerous water resources affairs, such as inland navigation, fishing and aquaculture in inland waters, mineral and thermal water exploration and exploitation, certain categories of waterworks, regional public works including local water supply and sewerage, wells and other matters. In addition, they are responsible for the granting and cancellation of concessions for water diversions.[77] The so-called regions with special status, i.e., those which are almost completely autonomous,[78] enjoy extended and independent powers over the management of their respective water resources. The provinces are generally responsible for the issuance of fishing and aquaculture permits and for authorizing industrial waste disposal. Municipalities are responsible for the construction, operation and maintenance of water supply, sewerage and wastewater treatment installations and for water quality and pollution control. In recent years there has been a movement towards the privatization of water services. Finally, it is worth mentioning the existence of specialized water tribunals.

Always in Italy, the law of 1989 concerning the protection of the soil introduced the idea of river basin administration, not only for the purpose of protecting the soil against the harmful effects of water, but also for the management and planning of water resources by river basin. This idea is not new to the Italian system. In the past, the Po River Valley Authority was responsible for water resources management at the basin level. However, the

---

76 The delivery of permits involving water quantity considerations requires the participation of the Ministry of Agriculture and Forestry. The ELY also participate in the process.
77 Before 2001, concessions for major diversions were issued by the Ministry of Public Works, while the regions were responsible for granting concessions for minor diversions.
78 The Valle d'Aosta, Friuli-Venezia Giulia, Sardinia, Sicily and Trentino Alto Adige regions.

implementation of the law has faced a number of challenges, due to the existence of water management institutions at various territorial levels within the boundaries of each basin, the functions of which would overlap with those of the new basin agencies. The situation did not improve after the entry into force of the Code of the Environment of 2006, which transposed the provisions of the EU Water Framework Directive into domestic law. The code provides for the setting up of river basin district organizations, but fails to indicate a strategic roadmap for their establishment.

In the Netherlands, the responsibility for water management is subdivided into three levels. At the central level, the Ministry of Transport, Public Works and Water Management is generally competent to decide on all matters concerning waters at the national level, i.e., the main rivers (Rhine, Scheldt and Meuse), lakes, etc. In particular, it is responsible for defence against floods, for the construction and direction of works concerning ports and navigable waterways and of other important waterworks, for the direction of water-related affairs as regards potable water, agriculture and navigation, for water quality monitoring and the operations related thereto. Other ministries are responsible for sectorial aspects of water management. The provinces have the same competences as the state with respect to the waters under their jurisdiction, including groundwater. As regards water quality, the provinces may delegate powers to the water boards (*waterschappen*) or to special water purification boards which operate under their control. The water boards, which have customary origins, are public territorial collectivities at the local level which can be created or suppressed by provincial regulation. They carry out numerous functions, which vary from board to board. Such functions may include the construction, operation and maintenance of protection works, surface water policing, water quality control and wastewater treatment. In addition, they are responsible for issuing permits for water diversions and discharge. The provision of drinking water is undertaken by numerous companies whose area of jurisdiction does not coincide with that of the water boards.

In Belgium, the water administration has undergone a process of decentralization, which is based on the different languages spoken in the country. The basic management responsibilities for water supply, wastewater treatment and sludge disposal were transferred to the regions in 1981, and special companies were established at the municipal level to organize, plan, construct, operate and maintain, although separately, drinking water supply and wastewater facilities. Nevertheless, the central government retains a certain number of responsibilities, such as the establishment of water quality standards, effluent standards, and the construction of major infrastructural works. Customary users' associations (*wateringues*) have specific water management powers at the local level within given areas.

6.4.4.4 Basin level

In Spain, the water law of 1985 confirmed important water management functions to basin administrations which had been in existence since the beginning of the last century, under the overall supervision of the Ministry of Public Works and Urban Planning. In addition, and this constituted a novelty, the Law of 1985 made the same basin agencies (hydrographic confederations) responsible for the management of groundwater resources previously under the régime of private ownership. These basin-level organizations have remained intact in spite of the movement towards decentralization which took place in Spain, and of the legislative reforms leading to the adoption of the consolidated text (*texto refundido*) of 2001 and to its subsequent amendments in view of the transposition of the EU Water Framework Directive.

Because of its long tradition in water resources management, Spain has been able to achieve administrative decentralization successfully, while maintaining the management of water resources by river basin and at the national level. At present, the basin agencies are placed under the supervision of the Ministry of Agriculture and Fisheries, Food and Environment.

In the United Kingdom, the River Boards Act of 1948 assigned to river boards the responsibility for river pollution control which was formerly carried out by numerous county borough and county district councils. These river boards also had functions connected with land drainage, flood control and the maintenance of navigability conditions on some of the main waterways. In 1951 the boards were made responsible for authorizing wastewater discharges and establishing effluent standards. The 1963 Water Resources Act replaced the existing river boards with 29 river authorities, which were also made responsible for authorizing water withdrawals. With the Water Resources Act of 1973, the river authorities were replaced with ten water authorities, the geographical areas of which generally conformed to river basins. Within its own area of jurisdiction, each authority was responsible for water supply, sewerage and sewage treatment and disposal. These services are now provided by private companies within the framework of the Water Act of 1989, while regulatory functions were first transferred to the newly established National Rivers Authority, and then to the Ministry of the Environment and the Environment Agency.

In France, the Law on the regime and distribution of water and water pollution control of 1964 established six financial river basin agencies (*agences financières de bassin*) responsible for the control of and the prevention of water pollution. These agencies, which in the 1990's were turned into water agencies (*agences de l'eau*), collect charges from the communities for the water and sanitation services provided. This capital is invested in mechanisms for the protection of water quality and pollution control and in the improvement of the services. As a result of the legal reforms introduced after the entry into force of the EU Water Framework Directive, the agencies are now also responsible for implementing the objectives of the river basin management plans by promoting the efficient management of water resources and related ecosystems.[79] The Law of 1964 also established river basin committees composed of representatives of the water users and of the public administration in order to control the activities of the agencies and to determine the rates of the charges levied by them. The Water Law of 1992 has vested the committees with water resources planning functions, which are performed in consultation with the local authorities and the water users. The agencies and the committees operate under the supervision of the ministry in charge of the environment.[80] Water supply, sewerage and sewage management remain the responsibility of the municipalities.

In Italy, further to the promulgation of the law of 1989, basin authorities have also been created for the river basins declared of national interest by the law itself. There are interregional basins, which cross the boundaries of two or more regions, and purely regional basins. While the activities relating to interregional basins are the subject of agreements between the regions concerned by territory, which are to establish an institutional committee and a technical committee, decisions relating to the regional basins are taken by the regions. The law vests the basin authorities with planning functions and with the formulation of work

---

79 See this chapter, Section 6.4.5.2.
80 Now the Ministry for Ecological and Inclusive Transition.

plans for water and soil conservation, without prejudice to the responsibilities of the state, the regions, the provinces and the municipalities, which are provided for in other laws. Based on the provisions of the EU Water Framework Directive, the Code of the Environment of 2006 has subdivided the national territory into eight river basin districts, each placed under the jurisdiction of a river basin district authority charged with the preparation of the river basin district management plan. This reform has faced a number of challenges, since it entailed a transfer of functions from the authorities established under the 1989 law to the new authorities.

In Bulgaria, four river basin management districts have been identified under the water act, which cover 'water regions' corresponding to river basins or groups thereof. These districts are under the jurisdiction of river basin directorates attached to the Ministry of Environment and Water, which are responsible for organizing river basin planning, issuing and registering permits for water use and wastewater discharge and coordinating water resources monitoring programmes. The directorates are supported by basin councils.

In the Czech Republic, the existing basin authorities report directly to the Ministry of Agriculture. The authorities are in charge of the management of watercourses and surface waters. In the Slovak Republic, four basin authorities have been established for the Danube, the Hron, the Bodrog and the Hornag. Water management by river basin also takes place in Poland and Romania.

Finally, under the 2006 Water Code of the Russian Federation the territory of the federation is to be divided into twenty river basin *okrugs* (areas), with basin councils being established to advise the government on matters relating to the use and protection of water resources and on the preparation of 'schemes for the complex use and protection of water' (water resources development master plans).

6.4.5 The process of transposition of the European legal framework

6.4.5.1 Background[81]

Water resources protection has been on the agenda of the European Commission since the first wave of European water legislation began with standards for the rivers and lakes used for drinking water abstraction in 1975, and culminating in 1980 with the setting of binding quality targets for drinking water. EU legislation included quality objectives for fish waters, shellfish waters, bathing waters and groundwater. Its main instrument for emission control was the Dangerous Substances Directive (1976).

In 1988, a number of potential improvements to this legislation were identified, resulting in a second wave of water legislation in the 1990's, namely, the Urban Waste Water Treatment Directive, providing for secondary (biological) wastewater treatment, and even more stringent treatment where necessary, and the Nitrates Directive, addressing water pollution caused by nitrates used in agriculture.[82] Both directives were adopted in 1991. In

---

81 For an overview of the Water Framework Directive, visit http://ec.europa.eu/environment/water/water-framework/info/intro_en.htm
82 Council Directive 91/676/EEC of 12 December, 1991, concerning the protection of waters against pollution caused by nitrates, in: *Official Journal* L 135, 30/5/1991, 40.

particular, the Nitrates Directive mandates EU member states to designate as vulnerable zones all land areas draining into waters that are or are likely to become affected by pollution ('nitrate-sensitive' areas). Member states are also called upon to establish codes of good agricultural practices binding on farmers and to implement programmes of measures relevant to the prevention of non-point source pollution. Monitoring obligations are also provided for in this directive.

Other developments were a Directive for Integrated Pollution and Prevention Control (IPPC), adopted in 1996, addressing pollution from large industrial installations, and a Drinking Water Directive adopted in 1998, reviewing quality standards and provisions for tightening them where necessary. This directive, which was amended in 2003, 2009 and 2015, is in the process of being further revised to provide for better access to safe drinking water to all, as well as better public information.[83]

In the mid-1990's, a more global approach to water resources management was considered in order to address the increasing demand by citizens and environmental organizations for cleaner rivers and lakes, groundwater and coastal beaches. Thus, the commission initiated a process of consultation leading to a widespread consensus that, while considerable progress had been made in tackling individual issues, the current water policy was fragmented, both in terms of objectives and of means. All parties agreed on the need for a comprehensive framework. In response to this, the commission presented a Proposal for a Water Framework Directive with the following key aims:

(i) expanding the scope of water protection to all waters, including surface water and groundwater, considered within defined 'river basin districts';[84]
(ii) achieving 'good status' for all waters[85] by fixed deadlines, through the setting and coordination of objectives within each river basin, and the introduction and implementation of programmes of measures;
(iii) establishing a river basin management plan for each river basin district. The plan is essentially a detailed account of how the objectives set for the river basin (ecological status, quantitative status, chemical status and protected area objectives) are to be reached within set deadlines. The plan provides an analysis of the characteristics of the the river basin, a review of the impact of human activity on the status of waters in the basin, an estimation of the effects of existing legislation and of what remains to be done to meet the objectives, and an indication of the measures required to fill the gap. An economic analysis of water use within the basin must be carried out in order to facilitate an assessment of the cost-effectiveness of the various measures;
(iv) a 'combined approach' to emission limit values (effluent standards) and water quality standards requiring the adoption of the best available technology for point-sources and, on the effect side, the coordination of environmental objectives with a view to achieving the overall good status objective;

---

83 The relevant proposal was approved by the European Commission on 1 February, 2018.
84 According to the Water Framework Directive (Article 2.15), a river basin district is 'the area of land and sea, made up of one or more neighbouring river basins together with their associated groundwaters and coastal waters, which is identified under Article 3(1) as the main unit for management of river basins.'
85 That is, good ecological and chemical status for surface water and good chemical and quantitative status for groundwater.

(v) water prices reflecting the true cost and acting as an incentive for sustainable water use, thus helping to achieve the environmental objectives set under the directive;
(vi) public participation in river basin planning and management, to balance the interests of various groups and ensure plan implementation and enforceability;
(vii) streamlining legislation by repealing seven of the first wave directives.

The Water Framework Directive, which for the first time provides for the management of surface water and groundwater at the EU level in a comprehensive manner, was adopted in 2000. Since then, EU member states have taken action to approximate their respective legal systems and institutional arrangements to its precepts and to those of the second wave directives which remain in force. The same applies to those European countries which have applied for EU membership (candidate countries), because as a condition for becoming members, they are also required to transpose EU legislation into their legal frameworks. The example of these countries is followed by those which do not belong to Europe but have concluded partnership and cooperation agreements with the EU, or fall under the area of influence of the United Nations Economic Commission for Europe (UNECE), such as is the case of Kazakhstan.

Since the implementation of the Water Framework Directive has resulted to be more complex than originally envisaged, in 2001 the member states have agreed with the Commission and Norway on a common implementation strategy that aims at developing a common understanding of approaches and informal technical guidance on best practices, thereby reducing the chances of wrong application. The strategy is largely based on the sharing of information and experiences and the production of guidelines on technical issues.

To complete the European legal framework just described, the European Parliament and the Council adopted a 'daughter' directive on groundwater and a directive on flood management on 12 December, 2006,[86] and 23 October, 2007,[87] respectively. A further directive adopted in 2008[88] lays down environmental quality standards for priority substances and other pollutants.

Among other things, the daughter directive on groundwater aims at defining criteria for assessing the good chemical status and groundwater quality trends, identifying significant and sustained upward trends in the concentration of pollutants and defining starting points for trend reversal. For groundwater bodies which are considered to be at risk pursuant to the analysis of pressures and impacts to be carried out under the Water Framework Directive, member states must establish threshold values and report on them to the European Commission. The Water Framework Directive stipulates that the good quantitative and chemical status objective for groundwater is to be arrived at through the prevention or limitation of the input of pollutants and the maintenance of a balance between groundwater abstraction and recharge. EU member states have the duty to implement all the measures necessary to

---

86 Directive 2006/118/EC of the European Parliament and the Council of 12 December, 2006, on the protection of groundwater against pollution and deterioration, in: *Official Journal*, L 372, 27/12/2006, 19.
87 Directive 2007/60/EC of the European Parliament and the Council of 23 October, 2007, on the assessment and management of flood risks, in: *Official Journal*, L 288, 6/11/2007, 27.
88 Directive 2008/105/EC of the European Parliament and the Council of 16 December, 2008, on environmental quality standards in the field of water policy, in: *Official Journal*, L 348/84, 24/12/2008. This directive was amended in 2013.

'reverse any significant and sustained upward trend in the concentration of any pollutant resulting from the impact of human activity in order progressively to reduce pollution of groundwater,'[89] and to establish programmes for monitoring the status of groundwater.[90] The directive prohibits the direct discharge of pollutants into groundwater.

The flood management directive aims to reduce and manage the risks that floods pose to human health, the environment, cultural heritage and economic activity. It requires member states to carry out preliminary assessments to identify the river basins and associated coastal areas at risk of flooding and, for the areas so identified, to develop flood risk maps and establish flood risk management plans that take into consideration developments in the long term, as well as the possible effects of climate change. These plans are to be coordinated with the river basin management plans prepared under the Water Framework Directive.

A 'fitness check' of European directives is currently being conducted in order to assess their relevance, effectiveness, efficiency and coherence. By using a cost-benefit approach to weigh advantages and disadvantages, this assessment will determine whether the directives should undergo simplification. Results are expected to be available at the end of 2019.

6.4.5.2 Developments

EU member states have made significant progress towards the transposition of the Water Framework Directive and other EU directives into their legislation.[91] Among them, Belgium, France, England and Wales and Germany have amended their water legislation or adopted new legislation between 2002 and 2004. In a nutshell, this legislation provides for enhanced resource planning within formally identified river basin districts, and introduces mechanisms and procedures for public participation in this exercise. It further provides for the establishment of environmental objectives and programmes for monitoring the water status, and for the adoption and implementation of programmes of measures within the river basin district framework. The basin management plans must include all the information specified in the Water Framework Directive.

France, which has a long-standing tradition with respect to river basin planning and public participation, has amended its environment code (*code de l'environnement*) in 2004, through a law aiming at transposing the Water Framework Directive into the domestic legal system.[92] This law refines and completes the provision of the 1992 water law in respect of river basin management planning. In particular, it provides for the formal delimitation of river basin districts, the enhancement of the content of the basin plans, the expansion of public participation in the planning process and the establishment of environmental objectives. The designated 'competent authority' is the basin prefect (*préfet coordonnateur de bassin*). What is more, the new law provides for the development of programmes of measures for the achievement of the objectives. Further, the new law introduces modified provisions on cost recovery and pollution fees. The law of 2006 on water and the aquatic environments strengthens the

---

89 Water Framework Directive, Art. 4 (ii).
90 *Ibid.*, Art. 8.
91 The deadline for the transposition of the Water Framework Directive was 31 December, 2003.
92 Law No. 2004–338 of 21 April, 2004, on the transposition of Directive 2000/60/CE of the European Parliament and of the Council of 23 October, 2000, establishing a framework for Community action in the field of water policy.

legal framework for water resources management by focusing on the preservation of water resources and related ecosystems. The provisions of this law and the other laws relating to water have been incorporated into the environment code.

In England and Wales, the Water Environment (Water Framework Directive) Regulations 2003[93] set detailed procedures with regard to the participation of the public in the production of river basin management plans. In this connection, the duties of the Environment Agency are as follows:

(i) to publish a statement of the steps and consultation measures to be taken for plan preparation (and the relevant schedules) not less than three years before the beginning of the plan period;
(ii) to publish a summary of the issues that should be considered in the plan, not less than two years before the beginning of the plan period;
(iii) to publish a draft plan not less than one year before the beginning of the plan period;
(iv) to take into account the statements, representations and proposals made by the public in the period for reply to the draft plan (generally six months from the date of publication).

The regulations then enter into detail as to the manner of publication of the above information, the process of finalization, submission and approval of basin plans and the institutions and persons to be consulted throughout the process.

The German Federal Water Act of 2002 provides for the definition of river basin districts, and vests the competent authorities of the *Länder* (states) with the task of preparing basin management plans and programmes of measures to achieve the objectives indicated in the plans. Five *Länder*, however, have failed to transpose the Water Framework Directive, so that in 2005 Germany was sanctioned by the European Court of Justice.[94] With the Federal Water Act of 2009, which followed the federalism reform of 2006, broader powers, and in particular legislative powers, have been vested in the federal government, so as to guarantee uniformity in the implementation of the EU *acquis*. The Surface Water Ordinance and the Groundwater Ordinance, both adopted in 2010, transpose specific provisions of EU legislation into national law.

Another example of transposition of the EU *acquis* on water pollution control into national legislation is offered by the water code of the Belgian Walloon region, which is part of the Environment Code and was adopted by decree in 2004. The code contains provisions on the improvement of surface and underground water quality through specific measures, leading to a progressive reduction of emissions and discharges of priority substances into water, the soil and the subsoil and to put a stop to the emission and discharge of dangerous substances. However, due to the delay in the transposition of the Water Framework Directive, Belgium has incurred sanctions by the European Court of Justice.[95]

Italy has also experienced delays in the transposition process, and has justified this inaction with the presence of a conspicuous body of rules which have hindered further legal developments. At the end of 2005, it received a final warning from the European Commission

---

93 Amended in 2015.
94 europa/eu/rapid/press-release_IP-05–73.
95 http://europa.eu/rapid/press-release_IP-05-35_en.htm.

for not complying with the Water Framework Directive requirement to designate river basin districts and to report on the relevant institutional arrangements, and for failing to submit the result of the environmental studies on the status of the basins in a timely fashion.[96] In 2006, the European Court of Justice ruled against Italy for failure to transpose the Water Framework Directive into its legal system.[97] Portugal has also been sanctioned by the court, for the same reason, and Luxembourg for incorrect directive implementation.[98]

New EU member states, including Croatia, Estonia, Latvia, the Czech Republic, Bulgaria, Slovakia and Romania, have taken steps towards incorporating the EU *acquis* into their legislation, while Macedonia, which has applied for EU membership, is making efforts to reform its own water sector legislation so as to be consistent with EU requirements.

6.5 United States of America

6.5.1 Introduction[99]

The United States of America is a federal country, and the jurisdiction for the management of water resources is shared between the national (federal) government and the states. Federal water law is uniform and nationwide, with regional flexibility in the regulations of the implementing agencies. Each of the fifty states has adopted surface and groundwater laws which vary significantly from state to state. State water quality control laws are more uniform and follow a pattern set by federal legislation.

In order to understand the laws for allocating and diverting water, it is necessary to look at land settlement practices adopted when the country was founded.

Sixty percent of the total acreage of the United States is privately owned, but a large percentage of land in the western states is under federal or public ownership. The private landholdings in most states derive from the public land system developed after the Revolutionary War (1775). To form a union, the thirteen original colonies of the eastern seaboard ceded their claims west of their boundaries to the national government, which subsequently encouraged settlement and reclamation of these lands through private ownership by disposing of large tracts at nominal prices. Substantial acreages were also granted to new states for settlement under private ownership, for a revenue base through land leases and to dedicate parcels within towns and communities for a common school system. One-third of the nation's land remains in public or federal ownership for parks, forests, wildlife preserves and other uses in the public interest.

---

96 Reference: IP-05–1302, Brussels, 18 October, 2005.
97 European Court of Justice, ruling in Case C-85/05, 12 January, 2006. Reported in: https://eur-lex.europa.eu/legal-content/EN/ALL/?uri=CELEX%3A62005CJ0085.
98 http://europa.eu/rapid/press-release_IP-05-45_en.htm. See also Case C-32/05, Commission of the European Communities v. Grand-Duchy of Luxembourg, judgment of 30 November, 2006. Available from http://curia.europa.eu/juris/showPdf.jsf?text=&docid=66403&pageIndex=0&doclang=en&mode=req&dir=&occ=first&part=1&cid=320061
99 Radosevich, G.E. & Daines, D.R. (1975) Water Law and Administration in the United States of America. *Proceedings of the Conference on Global Water Law Systems*. Valencia; the content of this section has been derived from this paper.

The initial federal water policy was directed towards controlling navigable waterways for commerce and defence, regulation of power facilities and flood control. As water was abundant in the east, no involvement beyond these measures was necessary. Water quality was no problem, but if commerce was adversely affected, the reserved powers of the federal government were broad enough to intervene. The eastern states adopted the common law riparian doctrine of England.

To carry out the land settlement and development policy of the west, the federal government enacted many laws which not only made land available for private ownership, but recognized and granted water rights for these lands. The most important acts were the Mining Act of 1866, the Land Act of 1870 and the Desert Land Act of 1877. These acts acknowledged the validity of water rights created by local customs, laws and court decisions and declared all unappropriated water subject to 'appropriation and use of the public for irrigation, mining and manufacturing purposes, subject to existing rights.'[100] Based upon the federal government's recognition of local laws, the western states developed rules according to their particular needs. From this policy the doctrine of prior appropriation evolved, the basic principles of which were adopted by nearly every western state.

Direct government involvement in the planning and financing of large-scale water projects to supply water users' needs took place with the Reclamation Act of 1902, which created the Bureau of Reclamation.

The need for management of the nation's resources emerged as the multiple demands strained the limited resources and water pollution was seen as a major economic constraint. In 1965, the Water Resources Planning Act was enacted, declaring that Congress is 'to encourage the conservation, development and utilization of water and related land resources of the United States on a comprehensive and coordinated basis by the Federal Government, States, localities and private enterprise with cooperation of all Federal agencies, States, local governments, individuals, corporations, business enterprises and others concerned.'[101] The Act created the Water Resources Council to develop planning and evaluation policies, standards and procedures for preparation of comprehensive regional or river basin plans and federal water and related land projects.

Water pollution control still remains the primary responsibility of indivual states, as declared in the Water Pollution Contol Act of 1972. However, Congress has made it law that where the states fail in their duty the federal government will control and abate water pollution.

6.5.2 Federal water law principles

The basic legal authority of the federal government in the field of water resources is founded upon the Commerce, Property, General Welfare, and Treaty and Compact clauses of the Constitution of the United States.

The Commerce Clause states that Congress 'shall have power to regulate commerce with foreign nations and among the several States and with Indian Tribes.'[102] This provision

---

100 Desert Land Act, 19 Stat. 377 (1877) as amended, 43 U.S.C. 321 (1964).
101 Public Law 89–90, 79 Stat. 244, 42 USCA 1962.
102 U.S. Constitution, Article I, Section 8, clause 3.

has been interpreted as giving the federal government a pre-emptive right to regulate navigable waters and non-navigable tributaries of navigable waters and interstate waters from which fish are sold in interstate commerce or used by industries selling products in interstate commerce. The power is broad enough to include water quality control, flood protection, watershed development, aquatic life and habitat protection and land use planning. Many of the activities carried out by the Corps of Engineers and the Environmental Protection Agency are based on this clause. Through the interpretation of navigable waters, the federal government has been able to undertake projects within states, even over state objections, when it has been determined to be in the public interest.

While the Commerce Clause has been significant in the humid areas of the country, the Property Clause has had its greatest impact in the more arid western states where federal reserved lands are located. The crux of the 'Reservation Doctrine' lies in the clause which provides that 'the Congress shall have power to dispose of and make all needful rules and regulations respecting the territory or other property belonging to the United States.'[103] This enables the federal government to assert claims to waters arising on land withdrawn for specific purposes and, to the extent necessary, to carry out those purposes. In addition, the clause serves as the basis for many programmes of the Bureau of Land Management, National Park Service and Forest Service in maintaining and enhancing vast areas of grazing, recreational and wildlife lands. Irrigation projects partly or wholly constructed on or serving federal lands are authorized in part under the Property Clause. The clause enables the federal government to manage land, water and other resources to which it holds title, with ability to direct the type and manner of use, to sell, lease or otherwise operate the properties in the public interest.

The Treaty Clause enables the President, with the advice and consent of the Senate, to enter into treaties with other governments,[104] and the Compact Clause requires states to obtain congressional consent to any agreement or compact with another state or nation.[105] These two clauses have greatly contributed to the ability of the federal and the state governments to resolve problems and more equitably and effectively utilize international and interstate waters. Treaties have been entered into with both neighbours on the northern and southern borders of the USA, and over thirty compacts have been negotiated between states over transnational and interstate waters.

The federal government is also involved in the use and control of much of the nation's waters, either directly or indirectly, through various means.

The first is the holding of water rights on federal reclamation projects by the Bureau of Reclamation until the project reimbursement costs are paid and the water users take complete charge of the waterworks. The users must organize either into a water district or a water users' association, and water rights needed to meet the project requirements are acquired by the federal agency through assignment of rights held by the users to the agency and filings by the agency under state law for unappropriated water.

The second means of federal control over water quantity is the 'Reservation Doctrine,' which gives the federal government the power to reserve water on lands that have been

---

103 *Ibid.*, Article IV, Section 3 clause 2.
104 *Ibid.*, Article II, Section 2, clause 2.
105 *Ibid.*, Article I, Section 10, clause 3.

withdrawn from private purchase and which have been designated to specific federal purposes, i.e., national parks, forests, recreation areas and wildlife refuges, oil shale reserves and hydropower locations. The reservation extends to present and future uses, is not lost through non-use and has a priority as of the date the lands were withdrawn from entry. The doctrine has been interpreted to include not only surface waters but groundwater as well.[106] The doctrine has its origin in the Property Clause, which includes lands and appurtenant waters in the western states. As the states were admitted into the Union, they obtained power over waters of the state, but acquired no proprietary rights or title to lands owned by the federal government nor to the waters arising on or flowing through such land. Thus, unless the federal government disposed of these lands and waters, title remains therein. A considerable amount of conflict exists between the federal government and the states affected as to the parameters of this doctrine.

The reserved rights position of the federal government has created a great deal of concern over the jurisdiction of waters within the states, since most water rights held by individuals under state law predate the assertion and use of water by the federal government. Furthermore, approximately 61 percent of the total natural runoff in the eleven western states with large federal landholdings comes from these lands. The doctrine directly conflicts with the doctrine of prior appropriation in use in the western states on a number of points: i) no diversion and/or beneficial use is necessary to create the right; ii) the right is not lost through non-use; iii) existing water rights predated by the reservation are in jeopardy; and iv) no quantification of the federal rights has been made. Thus, the states maintain that appropriate water planning cannot take place.

The advantage of the doctrine is that it allows the government to provide for present and future generations as technology and demands change.

Water quality control in the United States was a matter of state and local concern in the early years. Federal involvement was oriented toward navigable waterways under the Rivers and Harbors Act of 1899. But as industrial development took place with large population concentrations forming around waterways, concern was expressed over the inadequacy of potable water supplies, estuarian degradation and increasing costs to bring water up to the level of usable quality. Several attempts were made to develop a water quality programme between 1912 and 1948, but the orientation was primarily directed toward disease control.

The basic water pollution act passed in 1956 provided matching funds for municipal sewage plants and for research into the extent and nature of pollution of water resources.

In 1965 the Water Quality Control Act[107] was enacted and the Federal Water Pollution Control Administration was created, making available more funds to states for treatment plants, and, more important, directing the states to develop stream quality standards by June, 1967; otherwise the federal government would take the matter in hand in the case of states not complying. The act was amended in 1970 and in 1972. The most significant law in this field, the Federal Water Pollution Control Act of 1972, also known as the Clean Water Act, declared as its objective, 'To restore and maintain the chemical, physical, and biological integrity of the Nation's waters. In order to achieve this objective it is hereby declared that, consistent with the provisions of the Act:

---

106 U.S. v. Cappaert 508 F 2d 313, (9 cir. 1974).
107 P.L. 89–234.

(i) it is a national goal that the discharge of pollutants into the navigable water be eliminated by 1985;
(ii) it is the national goal that wherever attainable, an interim goal of water quality which provides for the protection and propagation of fish, shellfish, and wildlife and provides for recreation in and on the water be achieved by July 1, 1983;
(iii) it is the national policy that the discharge of toxic pollutants in toxic amounts be prohibited;
(iv) it is the national policy that federal financial assistance be provided to construct publicly owned waste treatment works;
(v) it is the national policy that area-wide waste treatment management planning processes be developed and implemented to assure adequate control of sources of pollutants in each State; and,
(vi) it is the national policy that major research and demonstration effort be made to develop technology necessary to eliminate the discharge of pollutants into the navigable waters, waters of the contiguous zone, and the oceans.'[108]

The 1972 Act is the first federal water quality legislation that is comprehensive in setting target dates to accomplish certain pollution control practices and has a mechanism for dealing with stream standards and effluent discharge limitations to be established by the states. A National Pollutant Discharge Elimination System (NPDES) programme was created requiring discharges from point sources to obtain a permit, according to state programmes. If, however, a state fails to develop an acceptable programme, the federal government reserves the right to impose the national standards and permit requirements.

The Act prohibits flow augmentation as a substitute for adequate treatment (dilution is no solution to pollution). The Act further requires interstate cooperation in developing uniform laws and standards.

A great number of states have adopted identical or similar programmes. States are still primarily responsible for pollution control.

6.5.3 State water law principles

Surface water laws developed along two distinct philosophies and legal systems consistent with the geo-climatic condition of the state. In the humid eastern half of the country and along the west coast, the riparian doctrine was adopted. The more arid western half of the country was faced with the immediate problem of deciding how to allocate a scarce resource and thus was compelled to develop a system of law peculiar to arid lands. The result of trial, error and compromise is the doctrine of prior appropriation. Some states have varied water availability and have adopted a mixed riparian/prior appropriation system. However, there is a wide variation between states following the same doctrine as to the manner for determining water rights, exercise of the right, water use efficiency criteria, and system for obtaining water rights and administering and enforcing the law.

Both the riparian and prior appropriation doctrines have been at odds with the scientific world on the matter of classifying sources of water. From the legal point of view, it

---

108 P.L. 92–500, 86 Stat. 816 Sect. 101.

was considered necessary to distinguish between different 'types' of water so that different rights of use could be formulated. The scientific community, of course, views water in the context of the hydrologic cycle and finds much fault with the pragmatic approach adopted in the laws. In the United States, water is commonly classified as follows: surface waters are i) diffused; ii) flowing in watercourses; iii) waters in lakes and ponds; and iv) spring waters. Groundwaters are 1) diffused percolating waters; or ii) waters flowing in defined channels.

In addition, water may also be identified according to a functional classification, as wastewater (that has been 'used'), foreign waters (imported from another basin or watershed), salvaged or developed waters (the results of efforts). The general rule is that these waters belong to the developer.

6.5.3.1 Riparian water law

The water laws of the humid eastern states were patterned after the common law of England, where every landholder whose property was adjacent to a stream or body of water was entitled to have the water flow past his land (or maintain a natural lake level) undiminished in quantity and unaltered in quality. This rule is generally called the natural flow theory. This doctrine is accepted in 31 states.

In a water-abundant area with little withdrawal needs, this rule may be satisfactory. But, even in the humid parts of the USA, conflicts developed as emerging industries, municipalities and agriculture began diverting water.

To resolve the problem, the Rule of Reasonable Use developed. Under this rule, riparian landowners can divert a reasonable amount of water with respect to all other riparians on the stream, and non-riparian lands may, under certain conditions, make a reasonable use of the available waters.

Waters in states following the riparian doctrine are a public resource, held in trust for use by the people of the state. A riparian landowner does not have an ownership right, but rather has a fundamental right to a reasonable use of the water and to be free from unreasonable uses of others that might cause him harm. He is essentially a correlative co-user with all other riparians. His right to the use of water is not a right to a fixed quantity of flow or volume, but rather is dependent largely upon the extent of development that has taken place.

Fundamental to the riparian law is the location of land; water will first be allocated to the riparian landowner.

Since under this doctrine the right does not consist of a definite quantity or quality of water, the requirement of reasonableness in use acquires a certain importance in the allocation of water.

In determining what is a reasonable use, regard must be had to the subject matter of the use; the occasion and manner of its application; the object, extent, necessity, and duration of the use; the nature and size of the stream; the kind of business to which it is subservient; the importance and necessity of the use claimed by one party; the extent of the injury to the other party; the state of improvement of the country in regard to mills and machinery, and the use of water as a propelling power; the general and established usages of the country in similar cases; and all the other and ever-varying circumstances of each particular case, bearing upon the question of the fitness and propriety of the use of the water under consideration.

'. . . Each use is required to be beneficial, suitable to the watercourse and its economic and social value. If these requirements are met, reasonableness may require each riparian to put up with minor inconveniences and to adjust to quantity of water used. [If conflict occurs, a solution involves consideration] of whether the first user's investment and other values are entitled to protection and whether the new user ought to compensate the former user for the loss of that which the latter gained. In most of the cases in which the plaintiff has suffered substantial harm through his water supply for a reasonable use being taken, the decision has been that the taking is unreasonable . . .'[109]

Under either the natural flow or reasonable use theory, there is a preference for the 'natural wants' over all other uses. The 'natural wants' include household and limited livestock needs and have generally placed domestic uses in a preferred position.

Riparian water law does not require a landowner to use the water in order to maintain the right. Unless the right to use water from an adjacent water source has been sold or transferred to other lands or uses, the right will continue as long as land and water are contiguous. Abandonment of water rights is nonexistent under riparian law. There is the possibility, however, that a riparian who does not object to the open and notorious use by another, through prescription, may have his right reduced or lost. Misuse of the right may result in a restriction on use and/or judgement for damages to those adversely affected. Parties injured through the misuse must assert their claim in court.

There have been a number of significant changes in the water law of the states accepting the riparian doctrine which can be summarized into two major components: i) establishment of a permit system to allocate water among certain users, and ii) creation of administrative machinery to assess the water supplies and requirements and to allocate and manage the states' water resources through the permit system.

The permit system allows conditions of use to be stipulated in the permit, and it provides the state water agency with information on where, the duration of use, to whom, what for and what quantity of water is allocated.

Other changes include forfeiture provisions for non-use (three years is common), minimum flow requirements for fish, wildlife and recreation, and greater flexibility and certainty in acquiring the right to use the water.

6.5.3.2 The appropriation doctrine

The appropriation doctrine developed in the western United States when land was being opened to settlement and the first major users of water were the legendary gold and silver miners. These people applied the same principle to their water as they did to their mines, which was that the person who first discovered a mine was protected against all later claimants. In the water area this was translated into the appropriation doctrine or the doctrine that 'first in time is first in right,' so that the first person to use water acquires the right to its future use as against later takers.

The doctrine protected the first settlers; the second settlers had to take other parcels of land and use whatever water was left after the first user had satisfied his needs.

The transition from riparian rights to the appropriation doctrine was not completely smooth. Although the prior appropriation doctrine is applied in eighteen semiarid western states and in Alaska; in nine of these states, located generally in the plains area and the Pacific coast, the riparian doctrine is also recognized in varying degrees.

---

109 Red River Roller Mills v. Wright, 30 Minn. 29, 15 NW 167, 169 (1883).

The basic principles underlying the appropriation doctrine are the following:

(i) water must be used to the amount being put to beneficial use. Among the uses recognized as beneficial are domestic, irrigation, power production, municipal, industrial, recreational and minimum flow for aquatic life. Though the use to which the water is put must be beneficial, the manner in which it is used must be reasonable. The concept of reasonableness plays an increasingly important role in appropriation states. For example, it may no longer be reasonable to irrigate a crop by flooding when another method is readily available which will water the crop as well or better but will save some of the water to be used;
(ii) the appropriative right must exist for a definite amount. In general, the amount of water that an appropriator is entitled to divert is measured by the beneficial use involved. In some states the statutes prescribe the maximum 'duty of water' or amount which can be appropriated;
(iii) priority of right and not equality of right is the basis for dividing the water during periods of scarcity. As a consequence, the value of water rights depends greatly upon the priority date and the source of supply in terms of dependability of flow;
(iv) a final principle is that a water right under the appropriation doctrine may be brought into the economic market system. The principle is that the water right is of indefinite duration so long as it is exercised in accordance to the laws, as a property right is saleable and transferable separately from the land. This principle has both a positive and a negative economic effect. On the positive side optimum use can be gained; the negative side is that in practice, the administration cannot ensure that the rights are 'properly' maintained or that only the amount of water necessary is diverted. As a consequence, waste of water often occurs to keep the full amount of the right valid. In addition, the water administration loses control over the water users.

A water right acquired under the doctrine of appropriation is generally considered to be real property, as opposed to personal property. As such, it can be alienated, disposed of, inherited and subject to taxation. The right continues as long as the water is actually applied to some beneficial use or purpose.[110] Water-right filings or approved applications are property rights and constitute a possessory interest in the right to use water once the right has been perfected. Like other property rights, water rights may be transferred separately from the land on which it is used. It is still appurtenant, however, to the extent that it is transferred automatically with the land unless the deed transferring the land specifically reserves the water.[111]

In response to the recognition that some uses are necessarily more important than others, some jurisdictions have adopted statutes which recognize 'preferred uses.' In times of shortage a preferred use may condemn a non-preferred use in order to supply water for the higher use. Compensation must, however, be paid for the taking of a right. Another important function of establishing preferences is that it serves as criteria for the allocating agency

---

110 Kinney, C.S. (1912) *A Treatise on the Law of Irrigation and Water Rights* (2nd ed). San Francisco, Bender-Moss Company, 1313–1314.
111 Utah Code Annotated, 73-1-11 (1953).

when applicants for different uses are competing for the same unappropriated water. While orders of preference vary somewhat from state to state, all jurisdictions place domestic uses (which include municipal uses generally) as the highest.

The prior appropriation doctrine is remarkably dissimilar from the riparian doctrine in many respects, but in the allocation of water it is diametrically the opposite. According to appropriation, land and natural water source locations are immaterial. Under this doctrine, regardless of where the land or other use is located, within or without the watershed, a right to unappropriated water can be obtained if the water can be diverted and put to beneficial use.

Important in the manner of allocation are the particular procedures set down by the state law. Generally, an application must be filed with pertinent information relative to the user, use and source of supply. If the application is approved, the right will normally have the priority date of the application. If the use is one requiring construction of diversion, storage and delivery works over a period of years, the right, if the application is approved and notice to proceed given, will still retain the date of application. If, however, the applicant does not construct the works within the time period acceptable to both parties and the delay is inexcusable, the right may have a priority as of the date the water is put to use.

Several systems have been developed by the states to handle applications of water rights, but the predominant approach is the permit system. This is merely the filing of an application with the appropriate state agency, which takes the procedural steps of evaluating and determining the disposition of the application. If approved, a permit is issued, which may state conditions of use. If denied, the applicant may appeal the administrative decision to the court.

Water rights under the appropriation doctrine can be lost through non-use of the right to use for the following reasons:

(i) abandonment (non-use for a statutory period of time);
(ii) forfeiture (requiring a showing of non-use of all or a part of the right after the statutory term);
(iii) adverse possession (through the use of another person without opposition from the holder of the water right);
(iv) condemnation (by an agency or a court decision).

6.5.3.3 Groundwater management

Laws controlling the extraction and use of groundwater have become as complex as surface water doctrines. Basically, however, the states apply one of four doctrines: absolute ownership; reasonable use; correlative rights; or prior appropriation. These laws normally apply to 'percolating' waters as distinct from waters in underground streams, which follow the surface water law of the state.

(i) *Absolute ownership*: The doctrine of absolute ownership had its origin in the United Kingdom which inherited it from the Roman law. The doctrine holds that a landowner can withdraw any water from beneath his land even if liability to his neighbours results from such action. This doctrine was originally adopted in a number of eastern states where water was abundant. It is still in operation in Texas, where the landowner owns, under the 'rule of capture,' all the water he captures through pumping,¹¹² but the adverse effects of groundwater mining, land subsidence and adjacent landowner claims of water stealing are putting pressure to change the state legislation.

(ii) *Reasonable use*: Many states have modified the laws into what has become known as the 'American Rule of Reasonable Use.' This change is synonymous with the modifications in the surface riparian doctrine. The rule of this doctrine is that, since the rights of adjacent landowners are similar, and their enjoyment in the use of groundwaters is dependent upon the action of other overlying landowners, each landowner is restricted to a reasonable exercise of his own rights and a reasonable use of his own property, in view of the similar rights of others.¹¹³ This doctrine leaves much speculation as to what is a 'reasonable use,' but on the other hand affords some measure of protection for neighbouring properties.

(iii) *Correlative rights*: The doctrine of correlative rights in groundwater originated in California and is a further refinement of the reasonable use concept. The doctrine holds that among landowners overlying an underground water supply, each landowner can make a reasonable use of that supply so long as the source is sufficient, but when the supply becomes insufficient due to drought or the drawdown effect, then each landowner is entitled to water in proportion to the percent of his land in relation to all other lands overlying the underground waters. The net effect is to provide great flexibility of groundwater use in an effort to maximize the resource, but to have a remedy of equitable allocation when shortages occur.

(iv) *Appropriation*: Most of the western states have adopted groundwater statutes giving recognition and protection to prior users. This does not imply, however, that surface water law was automatically applicable to groundwater. In fact, several states enacted laws to control groundwater as late as the mid-1950's. Like surface water, groundwater is subject to appropriation for a beneficial use provided that the intended user complies with the statutory requirements to obtain a permit or licence. The administrative official must determine if unappropriated groundwater exists and what adverse effects would occur. In most states, legislation allows the state water official to designate an area as critical,¹¹⁴ and the users may be placed under direct control in order to protect the aquifer and vested rights, and to slow groundwater depletion. In some states, however, restrictions are imposed mainly on new groundwater development, it being difficult to limit existing uses.

It is the practice in the western states to create groundwater management or conservation districts among local landowners, groundwater users and other stakeholders in order, *inter alia*, to adopt groundwater management and development policies and programmes, and propose water allocation criteria and requirements. The administrative official is responsible for implementing the policies and programmes. In Kansas, groundwater management

---

112 See Burchi, S. & Nanni, M. (2003) How Groundwater Ownership and Rights influence Groundwater Intensive Use Management. *Intensive Use of Groundwater, Challenges and Opportunities*, Ramón Llamas & Emilio Custodio eds. Rotterdam, Balkema, 227–240.
113 Meeker v. East Orange 77 N.J.L. 623, 74 A. 379 (New Jersey, 1909).
114 The High Plains states of Texas, New Mexico, Kansas, Colorado and Nebraska, and Arizona, all have legislation on critical areas.

districts have established well spacing regulations and groundwater depletion guidelines that are used by the chief engineer to assess applications for appropriation. In Texas, groundwater conservation districts may be formed at the request of the landowners concerned or at the initiative of the administrative authority. Without defying the rule of capture, they tend to focus on education and the promotion of conservation technologies, in addition to well spacing and the regulation of withdrawals in order to minimize interferences and prevent conflicts among neighbours. Finally, under the Arizona Groundwater Management Code (1980) groundwater management plans are to be produced for designated active management areas (AMAs) and irrigation non-expansion areas, to reduce abstraction over a given period of time and prohibit new groundwater development. AMAs are critical overdraft areas corresponding to defined groundwater basins or sub-basins.

6.5.3.4 The conjunctive use of surface and underground water

In many areas throughout the country, underground and surface waters are hydrologically interconnected, so that withdrawals from one source affect another source. Usually surface water users are senior in time with a considerable investment in a diversion and delivery system for their water supply. Groundwater use began to increase at a rapid rate during the droughts of the 1930's with gas motors driving the pumps. Then in the late 1940's and 1950's, the west witnessed an enormous movement toward groundwater pumping, as the Rural Electrification Administration (REA) brought electricity to the rural areas and pumps could now be driven more economically by electric motors.

Conflicts soon arose in Colorado, New Mexico, California and Texas between surface and groundwater users. If the courts had strictly applied the rules of the appropriation doctrine, all wells would have been shut down, and a vast amount of water resources would not have been utilized. However, if the wells were to be permitted to pump, people who relied upon the security of their senior surface water rights would have been grieviously affected.

Several solutions emerged.[115] In New Mexico the state engineer used his authority to declare underground water basins as critical areas, thus gaining complete control over water management in the area. Then, for any groundwater user or applicant whose withdrawals adversely affected stream flows, continued withdrawals are to be offset by retiring surface water rights. Under this 'retirement of surface rights' approach, both surface and groundwater users' rights and economic interests are protected.

Colorado has taken a different approach and adopted the 'augmentation plan,' a system by which groundwater users in a common area can develop schemes that guarantee that any surface user calling for water will be supplied. The schemes may include the purchase of reservoirs, surface rights, locating wells at the surface user's headgate, etc. Once drafted, the augmentation plan must be acceptable to the state water officials and surface water users before it can be put into operation.

---

115 Radosevich, G.E. & Sutton, W. (1972) *Legal Problems and Solutions to Surface-Groundwater Management. The High Plains: Problem of Semiarid Environments.* MacPhail, D.P. ed. American Association for Advancement of Science.

References

Bachelet, M. (1968) *Systèmes fonciers et réformes agraires en Afrique Noire*. Paris, Libr. Gén. Droit et Jurisprudence.
Baye, K. M. (1970) Droit Africain: ses voies et ses vertus. *Revue Sénégalaise de Droit*, Vol. 4, 5.
Burchi, S. (1991) Current Developments and Trends in Law and Administration of Water Resources, A Comparative State-of-the-Art Appraisal. *Journal of Environmental Law*, Vol. 3, No. 1, Oxford University Press.
Burchi, S. (2012) A Comparative Review of Contemporary Water Resources Legislation: Trends, Developments and an Agenda for Reform. *Water International*, Vol. 37, No. 6, 613–627. UK, Routledge.
Burchi, S., & Nanni, M. (2003) How Groundwater Ownership and Rights influence Groundwater Intensive Use Management, in *Intensive Use of Groundwater, Challenges and Opportunities*. Ramón Llamas & Emilio Custodio Eds. Rotterdam, Balkema, 227–240.
Cano, G.J. & Vargas Galindes, F.F. (1956) *Las Leyes de aguas en Sudamerica*. Development Paper No. 56. Rome, FAO.
Caponera, D.A. (1953) *Water Laws in Italy*. Development Paper No. 22, Rome, FAO.
Caponera, D.A. (1957 & 1959) Water Legislation in Asia and the Far East. Water Resources Series No.s 31 and 35. New York, United Nations Publications.
Caponera, D.A. (1973) *Water Laws in Moslem Countries*, Irrigation and Drainage Paper No. 20/1. Rome, FAO.
Caponera, D.A. (1979) *Water Law in Selected African Countries*, Legislative Study No. 17. Rome, FAO.
Clark, R.E. (1968) Water Rights in the USA. *Annales Juris Aquarum*, Vol. I. Mendoza, Argentina, AIDA.
Cunha, L.V., Figueiredo, V.A., Correira, M.L & Gonçalves, A. S. (1977) *Management and Law for Water Resources*. Fort Collins, Colorado, Water Resources Publications.
Daines, D.R., & Falconi, G. (1974) *Water Legislaton in the Andean Pact Countries*, Utah, Logan.
Davis, P.N. (1968) Soviet and American Water Law: Two Approaches to a Common Problem. *Annales Juris Aquarum*, Vol. I. Mendoza, Argentina, AIDA.
Descroix (1942) *Le régime juridique des eaux souterraines en France et à l'étranger*, Thèse de Doctorat d'Etat en Droit, Faculté de Droit de Paris.
FAO (1964) *La législation des eaux souterraines en Europe*. Série législative No. 5. Rome, FAO.
FAO (1983) *Water Law in Selected European Countries*, Vol. 2. Legislative Study No. 30. Rome, FAO.
FAO (2002) *Law and Sustainable Development since Rio, Legal Trends in Agriculture and Natural Resources Management*, Legislative Study No. 73. Rome, FAO.
Harris, T.L. (1986) *Groundwater Resources Control and Management*. New York, Vantage Press.
Hutchinson & Co. (1968) *Africa and the Law*, the University of Wisconsin Press.
Mifsud, F.M. (1967) *Le droit foncier coutumier en Afrique*. Rome, FAO.
Opoku, K. (1974) Traditional Law under French Colonial Rule. *Verfassung und Recht im Ubersee*, Vol. 7, 139–153.
Salman M.A.S. & Bradlow, D.D. (2006) *Regulatory Frameworks for Water Resources Management – A Comparative Study*. Justice and Development Series. Washington, D.C., The World Bank.
Sandoval, M.T. (1975) *Legislación de Aguas en America Central y Mexico*. Legislative Study No. 8. Rome, FAO.
Teclaff, L.A. (1972) *Abstraction and Use of Water: A Comparison of Legal Régimes*. New York, United Nations.
United Nations (1966) *Proceedings of the Sixth Regional Conference on Water Resources Development in Asia and the Far East*. New York, United Nations Publications.
United Nations (1983) *Water Resources Legislation and Administration in Selected Caribbean Countries*. Natural Resources, Water Series No. 10. New York, United Nations.
Valls, M.F. (1983) *Water Legislation in South American Countries*. Legislative Study No. 19. Rome, FAO.

Chapter 7

Possible contents of and reasons for water law

7.1 Introduction

Because of the increase in water consumption and pollution, and the need to consider water not in isolation from other natural resources, diverse countries, whether in the industrialized or in the developing stage, in tropical, desert or temperate zones, with arid or humid climates, are re-examining their water legislation or studying the possibility of enacting new legal provisions. The aim of this amended or new legislation is to regulate water resources development and use, and provide a legal basis for the introduction of measures for the control of water pollution and the management of risk situations deriving from accidents, natural calamities, climate change and the like.

A government seeking to modify existing legislation or to enact new water legislation may have to make major changes in its water law system. Rights and privileges based on private ownership of waters, riparian rights, the landowner's right to underlying groundwater, the prior appropriation doctrine of the USA and rights arising *ministerio legis* may have to undergo alterations which render them almost unrecognizable. If water rights, whether on private or public waters, must be fitted into government policies and objectives requiring a control system, proprietors and users may no longer be able to act at their own discretion. Systems based on sporadic or not clearly defined grants and concessions must be replaced with continuing and unified administrative action.

Nations are rightfully reluctant to accept foreign interference in anything connected with their own legislative powers and may hesitate to follow the example of other countries, one reason being that they take the view that what happens abroad may not have much relevance to their own requirements. What is more, often, by modern standards the laws of so-called 'developed countries' cannot be taken as the best examples of adequate water legislation and administration. Finally, states may consider that, due to varying existing conditions, it is difficult to integrate new laws and institutions into their own physical, political, economic, religious, institutional and social framework.

The main requirements of modern water legislation are:

(i) to integrate existing traditional, customary systems or original common law principles into written rules which will facilitate the most rational use of available water through appropriate administrative action, as well as water pollution control and environment protection;
(ii) to provide for the establishment of regulatory institutions and define their functions and powers, and their relationship with development institutions;

(iii) to delimitate the functions and powers of the various branches of government in the sectorial aspects of water resources management; and
(iv) to promote and facilitate the participation of water users, stakeholders and civil society in water resources planning and decision making.

A basic water law, act or code, should not be too detailed, but should contain fundamental principles and create instrumentalities leading to the attainment of its objectives. The details for actual implementation of specific provisions of a law are best handled in subsidiary or implementing regulations. This is because lengthy parliamentary procedures are usually required for the enactment or amendment of a basic law, whereas implementing regulations may well be issued by one or more ministries empowered to do so under the law.

7.2 The contribution of the lawyer

The contribution of the water lawyer to the process of drafting or amending a water law is important, but he should know something of the historical, social, economic, governmental, legal, physical, climatological and hydrological conditions of the country before drafting the law. If he is an adviser from abroad, he will benefit from field trips to existing waterworks, dams, projects, mining districts, irrigation and urban areas, taken in the company of local counterparts, in order to understand the physical conditions in which the law is to operate.

The water lawyer or drafter should also consult with ministers, administrators and water users in order to ascertain government policies and obtain the views of those most directly concerned with sectorial aspects of water resources management. In this way he will be able to discern the problems to be solved and the constraints that the law should overcome.

A knowledge of water laws enacted in other countries of similar background is useful, as it enables the lawyer to ascertain whether or not a law which has worked well elsewhere will fit the local situation. By way of simulation, the lawyer should apply the provisions under consideration to existing and foreseeable problems in order to see whether law implementation will produce the desired result. If particular schemes or solutions are suggested, he should help develop legislation that facilitates their implementation.

When drafting innovative water legislation, the water lawyer must also be be prepared to negotiate with dominant or competing pressure groups or lobbies, powerful landowners, industrialists or environmental groups seeking special treatment. The positions of the various actors should be either reconciled or overruled, but in some cases a compromise which will placate a group without sacrificing major objectives of the legislation may be opportune. Powerful landowners or their representatives in the government may have to be convinced that, in view of future water uses, the abolition of riparian rights is a necessity, and that government control over such uses would benefit them. Part of the job of a water lawyer should be to demonstrate how the law will work and how it has to be implemented in specific instances by future administrators of the law.

7.3 General considerations

When planning to draft or amend water legislation, and after studying similar instances and general patterns, it will be possible to point out some basic issues which may be encountered and devise possible options and methods of approach. The problems to be addressed through water legislation differ from country to country and may be greater in some countries than in others. Nevertheless, in spite of differences and of the need to tailor the law to fit the particular requirements of a country, some universal principles will find a place in every water law, while some widespread problems may have a common solution.

It is obvious that an adequate water administration is essential for effective implementation and enforcement of the provisions contained in a water act or code.

The basic requirements for modern water legislation are indicated herebelow without regard for the order of their importance, bearing in mind that some of them may be suited to a particular country but that not all of them may necessarily apply.

The philosophy of water legislation must consider:

(i) the availability and quality of water in a country, basin or region, and climatic conditions;
(ii) the level of economic and industrial development;
(iii) the existing water uses and the amount of water utilized, by whom, at what location and for what purpose;
(iv) the existence of land uses influencing the régime and quality of water resources, and their intensity;
(v) the cost of development of different sources of water;
(vi) present and future water requirements of the country, basin or region;
(vii) existing custom and social conditions.

While it is not possible to formulate specific definitions with respect to the content of sound water legislation, nor to the kind of legal enactment required (code, regulations, ordinance, etc.) as these depend on the legal framework of the individual state concerned, it may be useful to illustrate some aspects that should be considered or addressed by water legislation.

Throughout the present chapter the terms 'code,' 'law,' 'act,' 'regulation,' 'ordinance,' etc., are employed to designate a legal enactment or instrument purporting to regulate different aspects of water resources management without regard for their legal form and value within the legal system of a country.

The first requirement of a water law is that it should encourage rather than hinder the activities connected with water resources development, protection and conservation. As far as possible, its general tone should be permissive rather than prohibitive, while at the same time safeguarding both public and private interests.

A water law should encourage water use and also provide a climate conducive to private investment in the water sector. A potential water user will put his capital and labour into an activity such as the construction of wells, dams, water distribution systems and treatment works when he has sufficient assurance that he will eventually recover the cost of his investment and that he will receive a fair return for a period long enough to make the venture worthwhile. As one author has said, 'The use of water by people and firms can be guided and controlled, but it cannot be forced.'[1]

The law should be as simple as possible and should require the minimum machinery for its implementation. In order to avoid legal arguments it should not be ambiguous or circumlocutory in its wording.

---

1 Trelease, F.J. (1975) New Water Legislation, Drafting for Development, Efficient Allocation and Environment Protection. *Proceedings of the Conference on Global Water Law Systems*. Valencia.

A modern water law should be as comprehensive as possible; all the water available in usable form should be subject to the same rules. The law should be consistent with the hydrological cycle, and no water, whether public, private or artificially classified in whatever other manner, should be placed beyond the law's reach.

The major objectives of a modern water law system should be:

(i) to achieve or at least to promote sustainable and rational management of the available water resources, taking into account both economic efficiency and social requirements;
(ii) to promote the welfare of the water users;
(iii) to accomplish the state's social and economic objectives;
(iv) to coordinate private activities among themselves and with state projects;
(v) to protect the interests of the public in common uses and environmental values;
(vi) to facilitate adaptation to climate change, and
(vii) to integrate the activities of individual and collective users into comprehensive national and basin plans for water resources development and management.

As of now, in many countries the various aspects of water resources management are governed by different legal enactments. Legal provisions concerning water may be found in constitutions, as well as in civil, administrative, penal, health and other types of legislation. It is advisable that all provisions on water resources be contained in a single water code or law.

By the same token, different aspects of water resources management are usually administered by separate government departments. The institutional aspects of water resources management at every level should be dealt with either through the water code itself or through another special legal enactment. In this way, the security of water rights and duties, the permanence of established authorities and procedures, and certainty with respect to existing situations for all water users, who are, in fact, all the people of a country, can be achieved.

A water law should provide the widest possible framework for the implementation of water policies within flexible directives.

Innovative water legislation must be socially acceptable and administratively enforceable. Before deciding upon the authority which shall be responsible for the allocation or reallocation of water and the recognition, modification or abolishment of traditional or pre-existing water rights, attention should be paid to the political, social, economic, administrative or religious implications. A new system must be reconciled with the pre-existing system in the interest of fairness and equity and should be based not only on what is desirable, but also on what is feasible.

In the following sections, principles, issues and possible solutions will be indicated. Most of them have universal application, although specific local problems may require variations in form, procedures, principles, implementation and enforcement.

7.4 Water policy

Legislation by itself does not constitute the panacea for problems connected to water resources development, protection and conservation. To be effective, water legislation must be the result of policy decisions which should precede its enactment, based on prevailing political, religious, technical, economic, social, legal and institutional factors. Water legislation, in turn, is strongly influenced by the existing legal system of a particular country or region.

As a consequence, one of the first tasks of water law is to define a water policy, either in its preamble or by reference to other legal enactments. In view of the importance of water resources and of its incidence on all economic and social aspects of a country's activities, the water policy, as adopted by the country's legislative or other institutional organ empowered to do so, should be the subject of a formal legal declaration in a basic legal enactment, such as the constitution, the water law, a natural resources or environmental law or the like.

Policy decisions should be made in broad terms, and may include determinations concerning water resources ownership, water use rights, priorities (by region or uses), limitations, restrictions or obligations governing uses, the payment of water rates or charges, rights and obligations of individuals and the administration, a multi-purpose or integrated approach, public participation in river basin planning, the relationship between water and other natural resources, criteria relating to economic and financial justifications, and so forth.

7.5 Collection and use of data and information

The collection, evaluation and analysis of hydrologic and other water-related data are instrumental to the achievement of the objective of sound planning and management of water resources, to the protection and enhancement of water quality and to the planning, construction and operation of water resources projects. Furthermore, they serve to detect the effects of climate change. Data and information cover hydrology, hydrogeology, hydrography, hydrometry, meteorology, water quality and any other relevant physical, economic, social, legal, institutional or related information.

In most countries data are collected by a number of institutions which operate either at different levels (national, regional, basin or local) or are limited to specific water utilizations (domestic, agricultural, industrial, hydropower generation, navigation and the like). Each of these institutions follows its own procedures and standards as required by the particular objectives of the institution itself or by those of the user for whom the institution performs its work. The data collected may be published or exchanged, but in most cases they are not. Generally speaking, little or no coordination exists among the institutions responsible for the collection of data.

As a consequence, existing data are often unattainable for the users, either public or private, or for the general public. It may also happen that, since they are not aware of the availability of the data they need, individual or public planners will undertake yet other data collections, with consequent waste of manpower, time and money. This is particularly true in countries where it is felt that information is power and therefore it is jealously harboured, to be utilized only when the particular agency sees fit.

It is now widely recognized that harmonized or coordinated monitoring procedures or programmes concerning both water quantity and quality are essential for the enhancement of the water resources knowledge base. A good – and perhaps the best – example of legislation calling for the acquisition of data and information within the framework of monitoring programmes is offered by the EU Water Framework Directive (art. 8), which requires EU member states to establish such programmes with a view to obtaining 'a coherent and comprehensive overview of the water status within each river basin district.' As part of the planning cycle, monitoring programmes are subject to periodical review and updating.

It follows that it is indispensable to provide for legal mechanisms which:

(i) promote the coordination of all water-data collection and resource monitoring activities;

(ii) promote the standardization of data collection methods and procedures, i.e., the different types of details to be included in the data collection;
(iii) ensure the maintenance of up-to-date records of the data received, possibly filed by river basin and aquifer;
(iv) promote the creation of a centralized hydrologic data unit or data bank at the national level or at the basin level, to receive and store all data collected;
(v) introduce data exchange mechanisms and protocols;
(vi) ensure that data are published periodically, following a standardized methodology and format;
(vii) make data and information available to the users.

Therefore, a water law should contain provisions to the effect that the institutions responsible for the collection of data be required to communicate the data to a centralized data bank on a regular basis; and that those collecting or possessing hydrological data be obliged to furnish such data to the centralized data bank upon request.

The water data bank should constitute a centralized water resources inventory for the whole nation or basin. In view of its importance, this administrative unit should be part of a regulatory water resources management authority at the national level, under the overall control of an interministerial body or at the basin level.

7.6 Water resources planning

On the basis of water availability, water quality and existing uses, it is possible to plan future uses and to implement measures to prevent, control or reduce water pollution. The planning of water resources and the formulation of specific plans by uses (for water supply, irrigation, hydropower generation), or at different levels or areas (national, regional, basin, local), are questions which the drafters of water legislation should contemplate. However, it is to be noted that, at present, emphasis is placed on basin plans.

Extensive use and the need to conserve water resources for future utilization and for environment protection have made planning the basic instrument for ensuring the sustainable and rational management of the water resources available.

The instruments of water resources planning may take various forms and have different objectives and characteristics. These instruments are relatively recent and are gradually substituting the traditional public interest concept for the allocation or reallocation of water use rights in the past.

A water resources plan, *inter alia*,

(i) facilitates the implementation of the water legislation, particularly as regards the criteria for allocating or reallocating water resources, water use rights, wastewater discharge rights, etc.;
(ii) facilitates the determination of priorities among different uses, among different areas, and between older and more recent rights to use water;
(iii) facilitates the preservation of water quality and the control of pollution;
(iv) facilitates the allocation of financial resources to water resources development projects;
(v) provides for basic criteria to reserve the best available water for human consumption;
(vi) sets out criteria for the determination of minimum ecological flow requirements for water bodies;

(vii) facilitates adaptation to climate change;
(viii) constitutes the basis for elaborating and assessing water rates and charges,
(ix) provides a basis for the implementation of wastewater management policies; and
(x) indicates the areas in which institutional coordination is necessary.

For these reasons, the water law should make it obligatory to formulate a water resources plan, and should give the plan the legal status of a binding instrument for the management of water resources.[2]

7.7 Ownership or other juridical status of water

The concept of the juridical status of water, including ownership, may be relevant for establishing successful control over the rights connected with the use of water resources.

The concept of ownership should not be confused with that of right of use. Ownership includes the right to use, but the contrary is not true. Ownership gives the right to alienate a property through sale, donation, transfer, inheritance, or to constitute different rights on the same, whatever their nature, at the discretion of the owner. It derives from the Roman concept according to which one had the right to use (and abuse of) his property (*ius utendi et abutendi*).

The juridical status of water varies from country to country, depending on the legal system prevailing. Its concept may be found in legal enactments such as constitutions, civil codes, land laws, water laws or laws establishing the public domain. It may happen that different definitions concerning the legal status of water are found in different legal texts within the same country. Furthermore, the legal status of water may vary with the type of water resource under consideration: surface, underground (phreatic, artesian, etc.), atmospheric or reclaimed water (wastewater). In many countries the legal status of water, or of water ownership, is not defined.

Generally speaking, water resources may be public, private, common (*res communis omnium*), of nobody (*res nullius*), or community (tribal).

7.7.1 Public waters

Public waters are those which are considered to be either the property of the state or to be held in trust by the state or the crown or the public authority. Normally, these waters are inalienable, i.e., they cannot be given away permanently, but can only be the object of an administrative grant. Public waters are also imprescriptible, i.e., in spite of long use they do not confer upon the user any right whatsoever. This legal status is to be found in civil law countries, where public water is generally considered as belonging to the public domain, i.e., a category of things which, in the public interest, are said to be the property of the state. As a consequence, only the state has the right to grant to others the right to use them. Public waters may be considered as such either because they have been so defined by the law establishing the public domain or because they have been so declared subsequently. Generally,

---

2 For a more detailed analysis of water resources plans, see Chapter 8.

the criteria to identify public waters vary from country to country and may include their navigability or floatability, their being of public interest or their likelyhood to become so.

7.7.2 Private waters

Private waters are those which the legislation considers to be the possible object of private ownership. This notion stems from the concept of land ownership to which it is closely attached. It is particularly relevant in the case of groundwater, on the basis of the legal maxim that the owner of the land owns everything located above and below his land, including groundwater. Generally, private surface water ownership is limited to rainwater and to springs or waters located within one's land. Private ownership of water is either so defined in the legislation, or, most generally, derives from the legal framework prevailing in any one country. The concept of private ownership of water, whether surface or underground, can create problems where the achievement of rational water resources management is concerned.

7.7.3 Res nullius

Waters not belonging to anyone (*res nullius*) should be considered as nobody's waters which, however, are still subject to the possibility of appropriation. In some countries the legal framework allows for the existence of this legal status of water.

7.7.4 Common waters: community or tribal waters

Common waters (*res communis omnium*) are those waters considered as the common entitlement of the whole community. This is the doctrine followed by common law countries; the doctrine includes waters which, because of their transitory, elusive nature, defy ordinary concepts of ownership. In such countries the expression 'water ownership' is avoided in legislative texts, which instead declare generally that the state has the power to control water utilizations.

Community or tribal waters are those belonging to a given community or tribe at the local level. This legal status exists under customary law in many countries and sometimes is recognized by modern water legislation. It also applies at the 'municipal' level.

7.7.5 Conclusion

The consideration of the legal status of water entails that attention be paid to the definition of the legal status and/or ownership status of several appurtenances to water such as river beds, embankments, riparian land, etc. This definition is necessary in order to protect the watercourse and to delimitate the respective powers of the individual riparian owners and of the administration.

For achieving successful administrative control over water resources in countries where private water ownership exists, there is an increasing tendency to consider water as community, public, crown, or state property. In other countries, particularly in those following the common law system, the notion of water ownership is irrelevant, and all waters are deemed to be the object of state control under powers specified in the water legislation. In the countries where customary rules exist regarding the ownership of water, such ownership, generally deemed to be community ownership, should be recognized in the legislation.

Public ownership of water resources or state control over water is inevitable and beneficial, and requires that a distinction be drawn between water ownership and ownership of land underlying or overlying the water. Such a distinction seems logical for the following reasons:

(i) water, being mobile, does not remain attached to the land but is distributed by natural or artificial means and leaves autonomously its original site;
(ii) while ownership of land and artificial structures or wells may be determined, in the case of underground waters evidence as to the amount, size, extent and location of such waters cannot be determined by the owner of the land; ownership of any property including water requires an exact delimitation as to these physical characteristics;
(iii) to safeguard the public interest; this concept exists in every country;
(iv) for religious, social, or political exigencies.

In consideration of the above, all waters, whether surface or underground, wherever and however they may occur, should be declared state, crown, domanial or public property or, in any case, subjected to state control.

Where a dual sector of ownership is retained, public and private, the law should authorize the water administration to declare by public announcement the waters which are to be considered as public. The compilation of a special register indexing public waters, kept up to date by the water administration, is also advisable. Waters so indexed should not be utilized for any purpose without a government permit, authorization or concession.

7.8 The right to use water

7.8.1 Basic concepts

The right to use water is a concept distinct from water ownership, although, often, such distinction does not appear in the water legislation. Also, individuals may confuse the right of use with an ownership right. For instance, the owner of a well certainly owns the structure of the well, but this does not entail the right of ownership of the water in the well if, according to legislation, groundwaters cannot be the object of private ownership.

In the case where waters are public property, any use may be subject to a government authorization, permit, licence or concession. In the case of privately owned waters, when they exist, their use is free and not subject to government interference; however, their use is limited by the general principles of law governing the use of private property.

The right to use water may be acquired in the following ways:

(i) *ministerio legis*, by the mere effect of the law, through inheritance, gift, sale, acquisition or donation of the land on or under which water is located; this is based on the principle according to which the landowner owns everything on or below his land;
(ii) by appropriation, i.e., through the prior use and, in some cases, registration of the water right with the competent water administration;
(iii) through the granting of an administrative authorization, permit, licence or concession.

7.8.2 Legal régimes governing the right to use water freely or by virtue of a simple declaration/registration

As to the right to use water, legal régimes may differ according to the system prevailing in any particular country. In general, these régimes may be subdivided into three groups: (i) free use of water; (ii) declaration/registration; (iii) permit and concession.

7.8.2.1 The free use of water

A right to use water without need for administrative action should be clearly specified in the water legislation. The extent of such right will depend on the conditions prevailing in any one country. The right could include the use of rainwater or that of the water flowing or springing within a private land, provided that such uses do not cause damage to or conflict with the rights of neighbouring landowners. The right to the free use of water may be applied to both public and private waters. It is universally recognized that individuals may take public water to quench their thirst and for limited domestic or household purposes; some legal systems also acknowledge the right to water domestic animals and to irrigate private gardens.

Water legislation should include provisions concerning the definition of the right to the free use of water as regards the quantity of water to be taken, the purposes for which it may be used and the conditions under which the right may be exercised.

The water administration should be granted general powers to qualify even these basic rights of use in times of emergency and to prevent harmful effects which might be caused through the improper exercise of such rights.

7.8.2.2 Declaration/registration

Under this régime, the responsible water administration may request the users of surface water or groundwater to declare their utilizations and register them. The water law should define the amount of water which may be utilized under this category and the manner of utilization, which generally should exclude the diversion of surface water or the abstraction of groundwater by mechanical means. This régime may apply to shallow wells or small surface water diversions or utilizations for irrigating limited areas or certain types of crops. Under this régime, the water administration does not play an active role, but merely acts as a registrar of water rights.

The declaration/registration procedure should be implemented by region, basin or sub-basin, and progressively, according to time schedules to be determined in accordance with procedures established under the water law, as the water administration, especially in countries of a large size, might not be in a position to handle all declarations within a deadline which is usually short.

The declaration/registration régime allows the water resources administration to set up, gradually and by areas or river basins, a centralized inventory of water users; in so doing, the administration is in a better position to know the existing situation with regard to water utilizations, even those of minor importance.

7.8.3 The permit system

In most countries, the legislation requires an administrative permission to use public waters. Some countries have established different categories of permits for different types of use;

for convenience the term 'permit' is used here indiscriminately to designate 'authorizations,' 'licences,' 'leases,' 'water rights' or 'water use rights.' In addition, for the sake of simplicity, the expression 'permit system' includes the régime of 'concessions,' the special characteristics of which are outlined in Section 7.8.4.1 below.

7.8.3.1 Applicability

The permit system should apply to all those surface and underground water utilizations or effluent discharge processes requiring the use of mechanical means. In such cases, the granting of a permit should be preceded by technical investigations. The permit system may apply to water diversions more important than those indicated under Section 7.8.2.2, the details of which should be specified by the law. It may also apply in the case of wells deeper than x meters, or equipped with a pump, or to irrigate land with an extension larger than x hectares, or to irrigate certain types of crops, or in any other case as it may be specified by the law.

A permit is granted only after the water administration has made an inquiry on the application and has given publicity to it, in order to allow anyone having conflicting interests to raise objections.

Generally, the use of public waters, whether owned by or under the control of the government, the state, the community or the crown, should be subject to a government permit or concession to be granted by the water administration. This is the case both in countries where all water resources are state property and in countries where some waters are declared or proclaimed as such. This basic rule should be specified in the water law. However, in consideration of the administrative difficulties which may occur, some adjustments and adaptations as to the type of use, volume of water, timing of implementation, areas, basins or regions may be required.

7.8.3.2 Different types of permit

Different types of water use may be specified in the water law so as to define clearly those uses which are considered lawful as opposed to others which the law generally does not recognize, such as, in some countries, the sale of water. A distinction may be made between the use of water for:

  (i) drinking, household and domestic purposes, including the watering of domestic animals;
 (ii) municipal, rural or community water supply and sewerage purposes;
(iii) irrigation, drainage, reclamation and other agricultural purposes;
 (iv) hydropower production;
  (v) mining, cooling and other industrial purposes;
 (vi) transportation, navigation, floating and other similar purposes;
(vii) commercial or similar purposes;
(viii) other public purposes, such as recharging of aquifers;
 (ix) reuse, recycling and discharge of effluents and wastewater.

A further distinction may be established either within each one of the above purposes or separately with respect to:

 (i) large and small diversions, utilizations or discharges;
(ii) types and means of utilization, treatment or discharge;

(iii) geographic location;
(iv) timing and periods of the activities.

The water law should specify the limits of each type of utilization, discharge or category of use. Some of these definitions and further distinctions may be left to the discretion of the water administration or of the authorities which frame implementation regulations. If the law considers some waters as susceptible to private ownership, the rights and obligations of the owners should be clearly specified, always keeping in mind the public interest.

According to the legal system prevailing, a permit should be required and established in the water law for the use of waters in any one of the above-described ways. The governmental permit system may be limited to cover only the most important water utilizations, leaving the administration of minor uses to local authorities or to future regulation.

Whenever different types of permits are established for different types of uses, the greater the volume of water to be used, the more detailed should be the procedures involved for the granting of a permit. The preliminary inspection or investigation made by a water administration should become more meticulous in accordance with the magnitude or importance of the intended water use or discharge.

7.8.3.3 The procedure for granting permits

The water legislation should specify the authority or authorities competent to grant water use permits or to recognize existing water rights for different purposes, and confer on them the necessary powers to do so, bearing in mind:

(i) the need for considering water rights by basin, sub-basin or by aquifer, if possible;
(ii) the need to allow maximum flexibility in the administration of water rights;
(iii) the need to consider hydrological requirements.

The procedures for the granting of water use rights should require:

(i) an *application* stating such matters as the land to be supplied, the purpose or purposes for which water is sought, the source from which it is to be drawn, the proposed point of diversion, the volume to be diverted, the detailed nature of existing or proposed hydraulic works (drains, dams, canals, wells, pumps, pipes, etc.), and, when applicable, provisions to be made for land drainage and wastewater treatment.[3] In some countries, the application might also require that the applicant undertake to abide by the conditions imposed in the permit;
(ii) an *inspection* by the water administration of the proposed sites of diversion and use, to determine whether the hydraulic structures proposed are adequate, whether the landowner can beneficially use all the water sought and whether the use might injuriously affect other users, the source of supply or the ecological balance of the water body;

---

3 In many countries, the format of the applications to be filed by prospective users under water regulations is standardized in forms to be completed by the applicant and to be submitted to the water administration together with supporting documents (drawings, maps, proof of ownership or possession of the land, etc.).

(iii) a *public announcement* to be made by the water administration or by the applicant. The purpose of such an announcement is to inform all existing water users who might be affected by the proposed use. The frequency of the announcement, as well as its media, may vary greatly according to the circumstances, but the most practical means for ensuring wide publicity should be chosen;
(iv) the *filing of objections* to the proposed use within a specified deadline;
(v) *consideration by the granting authority*; the complete dossier, including administrative recommendations for alterations in the proposal from other government departments, should be considered by the granting authority;
(vi) *approval or denial*; the application can then be approved with or without qualifications or special conditions. In the case of refusal, written justification should be provided to the applicant. In the case of competing applications for the use of the same water source, the administration should be empowered to give preference to those complying with the requirements of the relevant water resources plan, or showing the best water utilization from an hydraulic viewpoint, or satisfying a larger public interest. In the latter case, the concept of 'public interest' should be defined in the legislation so as to prevent abuse.

7.8.4 Characteristics of permits and concessions

7.8.4.1 The difference between permits and concessions

A distinction has to be drawn between the 'permit' (authorization, licence, etc.) and the 'concession.' The régime of the concession is somewhat similar to that of the permit, but in this case the relationship between the water administration and the concessionnaire is based on a detailed contract which details the respective rights and obligations of the two parties. The duration can extend to thirty, fifty, or even seventy, years, depending on the type of water utilization or the amount of the investment. Typical concessions are those granted to hydropower generating companies, large land development projects undertaken by public or private bodies or institutions, water supply and sewerage entities, etc. In general, concessions are granted to individuals or institutions (public or private) which invest large amounts of capital for the provision of public utilities, and which require a legal guarantee for the protection and reimbursement of the original investment within a certain number of years.

7.8.4.2 Common characteristics of permits and concessions

Permits (authorizations, licences) and concessions have some common characteristics which should be reflected in the water legislation. These characteristics may be summarized as follows:

(i) they are subject to a number of obligations, limitations or other restrictions placed upon the water user according to their relative importance. These may refer to technical, financial or other conditions and may include: modalities of use, protection of the rights of third parties, volume of water to be taken and timing of use, the quality to be maintained, specification and standards for construction works, payment of water rates and fees, financial costs, obligation to recycle effluents or reuse wastewater, drainage and waste treatment or disposal requirements, the circumstances under which permits and concessions may be suspended, forfeited or cancelled by the granting authority, the penalties for improper water use and the powers of the water administration to intervene. They may include provisions in the interest of embankment protection, prevention of soil erosion, pollution control, sedimentation, the maintenance of minimum flows and environment protection. The provisions established in the water law and to be expressed in the permit or concession will vary in their details according to the type of permit or concession. Sometimes it may be advisable to specify the land or the type of crop for which the water is granted, the kind and number of animals and other relevant elements.

(ii) They are personal and generally cannot be transferred without the consent of the water administration.

(iii) They are temporary, their duration depending on the kind of utilization, on the amount of water granted, on the type and size of financial investment and on other factors. In the case of permits, their duration may be limited to one year, automatically renewable.

(iv) They may be liable to suspension in the case of catastrophe, failure to comply with the conditions contained in the permit or concession, including waste and misuse of water; lack of proper upkeep and maintenance of hydraulic structures.

(v) They are subject to forfeiture to the state in cases specified in the water law, which could include: failure to comply with the conditions, non-payment of water rates, misuse, waste or pollution of water, change in type of water utilization, illegal transfer, change in conditions, non-use for a certain period of time. Particular attention should be given in the permit or concession to the situation of waterworks and structures after each period of utilization. Permits and concessions may also be terminated because the provisions of a water resources plan so require. In this case the water law should entitle the permit or concession holder to compensation, because termination is not due to his fault and he should be guaranteed against any possible loss of the capital invested.

The procedure in the case of renunciation or abandonment of a water permit or concession at the initiative of the permit or concession holder should also be indicated in the water law.

The legal titles of permits and concessions should be recorded by the water administration in a special register of water rights, specifying, by entry, the source of water abstraction or diversion, the basin, sub-basin or aquifer, the volume of water granted, the place, purpose and duration of use and all other elements needed in order to have a clear picture of the 'users' universe.'[4] The water law would thus establish a cadastre of water users, which is an important element for achieving rational water management and enabling the water administration to develop water resources management plans.

7.8.5 Recognition or reallocation of pre-existing water rights

When a new water law is enacted, it must deal with existing water rights, some of which may have been enjoyed by water users since time immemorial. This may be the case of customary water rights, riparian rights or rights obtained under the appropriation doctrine.

---

4 The expression 'users' universe' is frequently used in Mexico, particularly by officials of the water rights administration of the National Water Commission. For a comparative analysis of various water rights administration systems, see Garduño Velasco, H. (2003) *Administración de derechos de aguas – Experiencias, asuntos relevantes y lineamientos*. Legislative Study No. 81. Rome, FAO.

7.8.5.1 Customary and riparian rights

As regards customary water rights, these should be recognized, but the water law should provide for their registration in such a way as to make possible the obtainment of the data and information needed for planning purposes, i.e., quantity of water employed, way of use, place of use, purpose of use, and the like.

Water rights acquired on the basis of the riparian doctrine should also be recognized in the same amount utilized before the enactment of the water law, on certain conditions. As far as private individuals are concerned, and particularly with regard to domestic use and animal watering, these rights should be recognized to the extent that beneficial use is made of the water, provided that the user does not waste or misuse the water, and that conflicts do not occur. If conflicts occur, riparian rights may become subject to the control of the water administration. Although these rights – but not the water itself – could be considered as private, the water law should regard them as rights of use by allowing them to remain in the hands of their current holders, but should provide for their gradual absorbtion into the permit system.

7.8.5.2 Rights under the appropriation doctrine

Water rights acquired under the appropriation doctrine before the enactment of new water legislation should receive the same treatment as those acquired under the riparian doctrine. In this case, however, policy decisions as to their legal nature should be made by the government authorities and be reflected in the law. If these rights are considered as a real property right, they may be regarded as duly registered private rights to use water, having priority over other rights. In this case, however, a new water law should establish procedures and vest the water administration with the power to review them and reallocate them as 'rights of use' under the permit system to be established under the new legislation. If they are considered or interpreted as 'prior rights of use' of public waters, they fall automatically within the permit system to be established. Under the new water law, earlier permittees would, in times of shortage, continue to enjoy the full amount of water allocated to them, while later permittees might have to do without water. While in some countries the appropriation system of individual priorities was the most satisfactory way of protecting the investment of water users, today it is considered inadvisable to establish such a system by legislation. In rapidly developing countries, one undesirable effect might be the restriction of future development.

7.8.5.3 Powers of the administration

The state, government or national community, or whatever formula has been adopted to designate the (public) water owner or trustee, should not alienate its property or overall right to regulate water uses, but grant only 'rights of use.' Ancient and pre-existing water rights should not be recognized *ipso facto* by the water administration, but only after appropriate review and inquiry. To such end, a commission might be established, composed of representatives of all the departments concerned (public works, agriculture, hydrological and meteorological services, judiciary, etc.). When granting new permits or concessions, preference should be given to ancient users.

In the case of private waters, the abolition of an existing water right should be accompanied by suitable compensation, but care should be exercised not to include the real value of the water as private property in the procedures when assessing the compensation. No compensation should be due if the holder of an existing water right is expropriated of his water right but supplied with an equal quantity of water derived from another source.

The sale or transfer of existing private water rights should not be permissible under the new water law without the consent of, or at least notification to, the water administration.

The proper time to consider the vested rights of existing water users is when deciding whether to grant new permits. Generally, the water administration should grant new permits to use water from a particular source only if it is satisfied that, in normal seasons, sufficient water will be available to all existing users.

7.8.5.4 The reallocation of water

The reallocation of water resources is becoming an important issue because new uses might compete with other water uses vested in earlier allocations or with customary water rights. For example, early water allocations have not always taken into consideration the need to reserve water of the best quality as the prime source for drinking, or to reserve water to maintain minimum ecological flows in water bodies. An overriding concern is that of the legal implications of planning for water resources allocation, as it affects existing rights of use. Provisions in water legislation are required to set criteria which will reconcile new water demands with pre-existing rights of use.

7.8.6 Water markets

Based on the assumption that water markets facilitate water reallocation from low-value to more productive uses and efficient water use because users will be stimulated to save water within their entitlement in order to sell it for profit, the water legislation of some countries allows the transfer of water rights from one user to another through market mechanisms. Water markets are also seen as a response to growing water scarcity, especially in arid regions. However, not all countries allow water to be transferred separately from land,[5] and in some countries water trading is explicitly prohibited.[6]

Generally speaking, water transactions are subject to rules aiming at preventing speculation, social imbalances and adverse environmental impacts, and at protecting the rights of third parties. Thus, they undergo screening and approval by the water administration, and can include a number of limitations. In some cases, they are permitted only among users of the same kind, i.e., among irrigators, industrial users, and the like. Prior to allowing water transactions, a water law should introduce a system of well-defined and secure water rights.

In Asia, notably India and Pakistan, and in the Middle East, water markets have developed informally at the local level and function successfully.

---

5 Land without water rights acquires less value.
6 Ecuador, Peru and Namibia are cases in point.

7.9 Limitations to the right to use

All water use rights, whether of public or private waters, are subject to a number of limitations and the imposition of servitudes or rights of way, or easements. Sometimes these rights may be expropriated in the public interest, generally subject to compensation.

Legislation, and water legislation in particular, usually indicates the mode of creating, exercising, and losing limitative rights, obligations and appurtenances. The term employed to identify such limitations and servitudes varies from country to country, depending on the prevailing legal system. For simplicity of exposition, by 'limitation' is designated a provision which the law or the permit imposes on the user to limit his right to use water.

The expression 'servitude' generally indicates a right/duty relationship between two or more water users, landowners or tenants; it is sometimes referred to as an 'easement' or a 'right of way.' The concept of dominant and servient land is central to the existence of a servitude, the servient land being the one burdened by such servitude. A servitude may apply to rainwater, river water, groundwater, protected areas, or the like.

As regards their nature, there are the following servitudes:

(i) obligation to receive naturally occurring surface water;
(ii) obligation to let naturally occurring water flow from upper land;
(iii) obligation concerning rainwater;
(iv) right to construct works to facilitate or prevent obstructions to the natural flow;
(v) rights and obligations concerning artificial or surplus drainage;
(vi) rights of access to water occurring on the land of another person.

A servitude may be acquired *ministerio legis*, that is, by virtue of the law which creates and regulates the servitude, by contract between the parties concerned, or by virtue of a judicial decision, i.e., a decision of a court of law or an arbitral tribunal.

Servitudes may be lost or extinguished by:

(i) non-use for a given period of time, through prescription;
(ii) confusion, i.e., combination into one person of two opposing interests (right/duty);
(iii) renounciation by the beneficiary;
(iv) death of one or both parties, under certain conditions.

In the case of water uses granted through government permit or concession, the limitations, restrictions and obligations concerning servitudes, rights of way, or easements should be included in the respective permits or concessions, unless clear provisions regulating these issues are contained in general legislation or in the water law.

The water administration should be empowered to take compulsory easements across private land, subject to the payment of compensation. Likewise, under circumstances specified by the law a private landowner should be able to obtain a compulsory easement over neighbouring land if the owner of that land refuses to enter into agreement. Such right should be granted only after a competent authority has assessed the fact and equities, and has awarded compensation. This provision might be of importance in countries wishing to encourage private investment in water resources development projects.

7.10 Priorities

The question of priorities is very important. Priorities may be looked at in the following ways:

(i) between different types of use, such as drinking or domestic uses having priority over irrigation, navigation over hydropower generation, and so on. It is worth noting that even within the same type of use, there may be priorities, as in the case of water supply for domestic uses, a distinction between household domestic use, water supply for municipal uses and rural or community water supply;

(ii) between different existing rights of use, such as senior and junior rights priorities in the prior appropriation system. Generally speaking, this system is undergoing modifications through legislation or court decisions, and therefore priorities between different rights are fading away;

(iii) between different areas, whenever the government wishes to promote economic development through the enactment of legislation on zoning, land use planning, natural reserves, forest reserves, natural parks, underdeveloped areas or other special areas. The priorities accorded to each one of these areas for the allocation of water are also subject to change, and usually are reflected in land use plans or water resources plans;

(iv) between different quantities of water to be utilized, when the government policy favours the development of large projects utilizing substantial quantities of water rather than minor undertakings, or vice versa.

Priorities vary in time and depend upon the stage of a country's economic and social development. They may also vary within each state, from region to region, again depending on the government economic and social policies and on the stage of development of the region. If priorities are defined in water legislation, it becomes difficult to adjust them to social, economic and technological changes and needs, or to the vagaries of climate change. Therefore it does not seem advisable to crystallize them in water legislation.

The different types of priorities should be established on the basis of national, basin or other water resources plans. Thus, water use permits or concessions and wastewater discharge permits should be granted or refused by the water administration in accordance with these plans.

The institution playing a lead role with regard to water resources planning should have the power to follow and adjust priorities from time to time, as circumstances and policies require. This power should be conferred in broad terms, thus allowing the institution to identify and take into account the various factors which may determine the specific needs of a river basin or aquifer, a region or a whole country. In order to facilitate this determination, water legislation should require interaction between the institution and the various stakeholders and water users.

The only priority which might be established in a water law is that for household and domestic purposes. This priority needs to be clearly defined. Another priority which has gained attention in recent decades as a response to a growing water demand and the consequent degradation of water quality and of the aquatic ecosystems refers to the need to reserve, allocate or reallocate water of the best quality for drinking purposes and for preserving the ecological balance of water bodies.

7.11 Beneficial uses

7.11.1 Water and other natural resources

The beneficial uses of water are often dealt with by special laws. Water legislation should contain provisions to regulate these uses through the issuance of government permits or concessions. Under the permit system, a well-defined amount of water is guaranteed to each user, and at the same time the water administration is in a position to know how much water is used, by whom, for what purpose and where. This enables it to plan further resource development and to introduce and implement measures where they are most needed, in the interest of water quality preservation, environment protection, or for other purposes.

The administration of the different water uses is closely related to the management of other natural resources such as land (urban, agricultural, industrial uses), air (cloud control, meteorology), flora, fauna and other living resources, energy resources (hydropower, geothermic resources), ocean resources (in the water, on the bottom and below the sea bed), scenic and panoramic, and recreational resources.

Unless a natural resources code or law is in existence, water legislation should contain provisions governing the interrelationship between the water administration and the management of other natural resources.

7.11.2 Domestic and municipal uses

Minor and individual domestic uses are normally regulated either by provisions of the water law or by other general legislation governing the right to use water without government permit or concession, such as a civil code or a land law.

If possible, water uses for municipal, community and larger rural water supply purposes should be governed by the water law. Municipalities and other water supply institutions, organizations or authorities should be considered as important primary users to which a concession is granted by the water administration at the appropriate territorial level. Thus, the concession would confer to its holder – the concessionnaire – a clear-cut right to distribute and sell the water so granted to secondary individual users. In this way, the national water administration has a clear picture of the situation and is sure that the organization, authority or company responsible for tapping, processing and distributing the water is not deprived of it by other applications or projects.

In the relevant legislation, domestic water uses usually include drinking, bathing, cooking and watering of livestock. The watering of a small garden or orchard may also be included. A modern trend is to supply domestic water for purposes which are not exclusively domestic, especially in urban areas where mains water is also supplied for industrial and commercial purposes. Another trend is to combine in one institution the function of water supply for domestic and municipal uses with those of sewerage management, water treatment and recycling.

A water law should consider these developments when setting up a permit system and should grant the right to use water for the purpose of supplying settlements together with those relating to wastewater treatment and disposal, under a multiple-purpose permit or concession. What is more, the extent, location, amount and means of abstraction and diversion of water and the discharge and treatment of wastewater should be specified in the permit or concession.

In order to ensure the potability of the water, no permit or concession should be granted by the water administration unless, after previous consultation, approval by the health authorities is received and monitoring by the public health authorities is secured.

7.11.3 Agricultural uses

Agricultural uses of water should be defined in the water legislation. They may include irrigation and the watering of livestock. A recent trend refers to the use of water for agro-allied industries. These uses should follow the same régime as that of all water uses.

According to the legal framework of a country, agricultural uses may fall within two different categories:

(i) uses requiring an administrative permit or concession;
(ii) uses for which no permit is required.

The uses requiring a permit or concession will be governed by the provisions of the water law, and will be subject to all requirements and procedures established within the permit system, as provided for by the water law. Those for which no permit is required may be governed either by a section of the water law governing the free uses of water or by general legislation governing land ownership, servitudes and other land rights, such as a civil code or a land law, or by a law on irrigation.

Permits and concessions for the use of water for agricultural purposes should contain all those limitations and obligations required in the public interest with respect to technical, financial, economic, geophysical and related aspects, including provisions concerning drainage and the use of fertilizers, pesticides or other chemical substances which might damage, besides human health through the food chain, the soil, the sub-soil and the waters, both surface and underground. Through the issuance or refusal of permits and concessions by the water rights administration, the implementation of government policies for sustained agricultural development becomes easier. Coordination between the water rights administration and other relevant institutions should be established.

7.11.4 Industrial uses

The water legislation should bring all industrial uses of water under the control of the water administration, through the permit system. Moreover, the legislation should contain provisions dealing with water pollution control, the reuse and recycling of water and other related matters. Possibly, it should contemplate the eventuality that if the industrial water user is also a polluter, a single permit or concession be issued, containing conditions relating to both the use of water and the discharge of effluents. Should this not be feasible, then the legislation should require coordination between the water administration and the institution responsible for water quality.

In order to avoid uneconomical investment, permits and concessions for the use of water for industrial purposes should be secured by the prospective user prior to obtaining any other permit issued by a separate government authority.

7.11.5 Hydropower production

Hydropower production is undertaken by public, mixed or private corporations or institutions. It is normally the subject of special laws that regulate the construction of plants and the technical specifications for the production and distribution of the electricity produced, and that determine the right of the state to control and/or stabilize and revise the maximum water tariffs payable by the consumers to the concessionnaire. Given the sizeable economic interests involved, hydropower production usually falls under the supervision either of ministries carrying responsibilities in the field of energy, or, in some cases, of entities attached to the council of ministers, or the prime minister's office.

Since hydropower corporations and institutions use public waters, they should be treated as users and be subject to the provisions of the water law. It follows that the water law should provide for an integration into the permit system of rules concerning hydroelectric power production and distribution, and for coordination between the use of water for this purpose and other water uses, such as navigation, agricultural uses and municipal water supply. This coordination may be achieved both through water resources plans and through appropriate institutional arrangements for interaction between the water administration and the institution in charge of energy. Special attention should be paid to the relationship between irrigation, which is a consumptive use, and hydropower production, because in some arid regions they may become incompatible during certain periods of the year. Finally, a new water law should provide for the recognition of pre-existing water use rights for hydropower production.

In the last decades, mini-hydraulics has come to the attention of some governments. This is the use of water for hydroelectric production below a certain number of Kw (generally 3000 Kw). This use, also, should be subject to plan determinations and to permit requirements in accordance with the water law.

Every electric power plant or similar installation should be indexed in a special register of hydroelectric and/or thermoelectric production, transmission, distribution and transformation plants.

The water law could provide for the subdivision of permits and concessions into more than one category, according to the amount of water utilized or power produced, and for the granting of financial contributions by the state for concessions of public interest.

Provisions should be made in the water law, and reproduced in the contracts of concession, to deterrminine which of those works constructed by the concessionnaire are to become state property at the end of the concession, or before, if required in the public interest. The state should retain this faculty in every concession for hydropower production.

7.11.6 Setting of minimum flow requirements

A recent legislative trend is that of setting minimum flow requirements, or to provide for the reservation of flows, to protect ecosystems, to preserve fish life, to protect scenic values, or to satisfy basic human needs. This is the case of South Africa, among others. These requirements function as a limit to the power of the water administration to issue permits or concessions for other purposes of water use.

7.11.7 Other public uses

Often, non-consumptive uses of public waters, such as navigation, transportation, floating, site and wildlife preservation and recreation, are governed by special legislation and administered by departments other than the water administration. Therefore, the water law should establish the coordination and prior consultation procedures necessary before a permit or concession is issued by another administration.

Water use permits or concessions granted by the water administration should be referred to the other administrations concerned for advice and consultation, and vice-versa, so that all aspects of the water uses are taken into consideration, and interference between these uses, whether existing or potential, is avoided.

With respect to water utilizations for medicinal and thermal purposes, permits may be granted by the water administration if the waters are within the scope of the water law. In this case, however, the grant should take place after consultation with the public health authorities. The water law should contain provisions in order to create protected areas around medicinal and thermal water sources.

7.11.8 Conclusion

It would be desirable to have one water rights administration, as part of a central water resources management administration, to retain primary power to control all uses except, perhaps, non-consumptive uses, which are usually controlled by other government authorities.

The precise division of responsibilities among administrative bodies to be established by the water law should be determined by each country in the light of its own institutional framework. The water law should contain provisions for resolving potential interjurisdictional conflicts between responsible ministries, departments or authorities, and for ensuring an institutionalized and obligatory coordination among them at the appropriate level.

7.12 The right to water

The concept of right to water differs from that of water rights, or the right to use water.[7] While water rights refer to the entitlement to use water for one or more purposes by virtue of a law or under a permit issued by the water administration, the right to water is the right of everybody to sufficient, safe, acceptable, physically accessible and affordable water for personal and domestic uses.

The existence of a self-standing human right to water in international customary law has been the subject of extensive debate during the past decades, which is still open.[8] The global community has recognized that, although significant progress has been made in the achievement of the UN Millennium Development Goals, one of which was to halve, by the year 2015, the proportion of people who lack access to safe drinking water, much remains to be done.[9] Renewed efforts have thus led, on 25 September, 2015, to the adoption, by the

---

7 See this chapter, Section 7.8.
8 See Chapter 11, Section 11.8.
9 WHO/UNICEF (2017) Progress on Drinking Water, Sanitation and Hygiene: 2017 Update and SDG Baselines. Geneva, WHO and UNICEF. According to this report, about 2.1 billion people lacked access to safe, readily available water at home in 2015, and 4.5 billion lacked access to suitable sanitation.

UN General Assembly,[10] of Sustainable Development Goal No. 6, among other goals, calling for clean water and sanitation for all by the year 2030.

The International Covenant on Economic, Social and Cultural Rights (1966), as interpreted through General Comment No. 15 on the right to water,[11] requires states to take steps towards the progressive realization of the right to water through appropriate means, including the adoption of legislative measures. Progressivity stems from the consideration that realization may not necessarily take place overnight, due to financial and other constraints. General Comment No. 15 places emphasis in particular on the obligations of states to *respect, protect* and *fulfill* the right to water.

The obligation to *respect* entails the duty of states not to interfere with the enjoyment of the right to water. The obligation to *protect* requires states to prevent any interference by third parties with the enjoyment of the right to water, through the adoption of measures to control water abstractions, assess the impact of proposed developments and prevent water pollution compromising the satisfaction of drinking and other basic water needs. The obligation to *fulfill* refers to the recognition of this right in national policies and legislation, to the establishment of water resources management plans prioritizing its realization and setting clear objectives, and to the adoption of tariff structures ensuring that water is affordable to all, among other things.

The human right to water is now acknowledged in a number of national constitutions, whether directly or indirectly. The constitutions of Bolivia, the Democratic Republic of Congo, Ecuador, Kenya, Mexico, Morocco, Slovenia (through amendments of 2016), South Africa, Uganda, Uruguay and Zimbabwe are cases in point. In India, a Supreme Court pronouncement protects the right to water and sanitation as part of the right to life guaranteed under the constitution. The constitutions of Bolivia, Ecuador and Uruguay go well beyond the recognition of the right to water (and sanitation) as a human right, as they prohibit the privatization of the relevant services.

The recognition of the right to water is also a feature of water laws recently enacted in a number of countries, including Algeria, Angola, Argentina, Belgium (Brussels-Capital Region and Walloon Region), Burkina Faso, Central African Republic, Costa Rica, Ecuador, France, Guatemala, Guinea, Mauritania, Mexico, Morocco, Namibia, Nicaragua, Paraguay, Peru, South Africa, Tanzania and Venezuela.[12] These enactments place emphasis on the following:

- the establishment of priorities among water uses in national and basin plans, top priority being afforded to the satisfaction of basic water needs;
- a water allocation (and reallocation) system that respects this priority;
- the need to protect water sources intended for meeting drinking and household needs from depletion and quality degradation;

---

10 UN General Assembly, Resolution 70/1 of 25 September, 2015. *Transforming our World: The 2030 Agenda for Sustainable Development.* UN Doc. A/RES/70/1.
11 Adopted at the 29 session of the Committee on Economic, Social and Cultural Rights. Document E/C.12/2002/11 (2003). Available from www.refworld.org/pdfid/4538838d11.pdf. See Chapter 11, Section 8.
12 For a survey of legislation recognizing the right to water, see Various Authors (2012) *The Human Right to Safe Drinking Water and Sanitation in Law and Policy. A Source-Book.* WASH United, Freshwater Action Network and WaterLex.

- the reservation of sufficient amounts of good quality water to the satisfaction of these needs;
- the principle that social considerations must prevail over economic considerations in the provision of water services, and that tariff structures must be based on equity and solidarity criteria;
- the participation of users and civil society in the planning and management of water resources;
- the provision of access to information;
- non-discrimination and protection of vulnerable and marginalized groups, including their customary water rights.

The right to water has also been recognized through decisions of domestic courts in cases of disconnection from water supply networks and pollution of drinking water. In *City of Cape Town v. Strumpher* (2012), the Supreme Court of Appeal of South Africa ordered the restoration of the water supply by the service provider, on the grounds that the right to water is not a contractual matter, but stems from the constitution, which protects the right to water. Always in South Africa, the Constitutional Court reached a different conclusion in *Mazibuko and others v. City of Johannesburg and others* (2009), by holding that the modifications to water services introduced by the service provider were reasonable, and thus consistent with the constitutional right to water.[13] In *Subhash Kumar v. State of Bihar and others* (1991), the Supreme Court of India acknowledged that the right to life under Article 21 of the Indian Constitution includes the right of enjoyment of pollution-free water. In dealing with similar cases, courts in Argentina have based their decisions on other rights, such as that to a healthy environment, from which the right of water implicitly derives.

7.13 Harmful effects of water

Harmful effects of water include those damages caused by water, either naturally or through the action of man. The expression may cover floods, interference with drainage, siltation, waterlogging, salinization, overflow of rivers and lakes, droughts, embankment degradation, soil erosion, intrusion of sea water in deltaic areas, land subsidence due to overpumping of groundwater, wastewater, sewerage and the poor management of floodplains.

Until now, harmful effects have generally been dealt with in separate legal enactments and have been administered by different government institutions.

If water legislation establishes a permit system for water utilizations, many of the harmful effects of water may be handled through appropriate preventative provisions, obligations and limitations inserted in the permits or concessions. However, this is not possible in all cases, and other provisions of a general nature should be included in a special section of the water law.

These provisions could cover, *inter alia*,

(i) the undertaking of special works intended to prevent or abate specific harmful effects of water;
(ii) the authorities responsible for undertaking these works;

---

13 These cases are illustrated in Matchaya, G., Kaaba, O. & Nhemachena, C. (2018) *Justiciability of the Right to Water in the SADC Region: A Critical Appraisal.* Basel, MDPI.

(iii) the financial implications of the works;
(iv) the declaration and delimitation (and mapping) of 'control' or 'management' areas subject to special régime;
(v) risk management planning, which is preventative in nature;
(vi) the obligations deriving from the construction, operation and maintenance of sewerage systems;
(vii) the disposal of domestic, municipal and industrial wastes by means of sewers;
(viii) the régime of waste disposal;
(ix) the legal régime of wastewater reuse;
(x) any other aspect, as may be required.

As provisions governing the control of harmful effects of water are often found both in the water legislation and in other legal enactments, the water law should establish a machinery for coordination between the interested administrations or other licencing authorities, thus achieving concentration of efforts in the prevention and mitigation of the harmful effects of water, and in the provision of emergency response.

The water administration should always consult with the other responsible authorities before granting permits or concessions for the use of water in cases in which harmful effects are likely to occur.

7.14 Water quality and pollution control

The issues connected with the control of water quality are among the most vexing of modern water resources management and have become more acute during the last decades. These issues relate, *inter alia*, to the following:

(i) waste and misuse of water;
(ii) recycling and reuse of water;
(iii) recharging of aquifers;
(iv) wastewater and effluent control;
(v) health preservation; and
(vi) pollution control.

7.14.1 Waste and misuse of water

Water waste and misuse occur both in urban and rural areas. In urban or municipal areas, it should be the responsibility of the water supply organization to prevent or abate the phenomenon. However, lack of funds, manpower or spare parts, or scanty education of the users, sometimes make it difficult to remedy the situation. The legislation governing the water supply and sanitation sector may be either insufficient or, for one reason or another, not adequately enforced. Therefore, if necessary, the water legislation should be strengthened in this respect.

In rural areas, agglomerated or dispersed, the problem also exists, particularly around water points utilized by rural populations. It is often associated with the spread of water-borne or water-related diseases. In such areas, the legislation by itself is not sufficient to abate the occurrence. It is necessary to educate the users of water points in proper operation and maintenance practices. This often requires a minimum personnel training in public hygiene, in operation and maintenance of the facilities, and in other basic aspects relating to water management. Water legislation should contain the necessary provisions, as well as provisions tending to promote the organization of water users' associations.

7.14.2 Recycling, reuse of water and recharging of aquifers

The recycling of water, its reuse, and the recharging of aquifers are modern tools which provide new sources of water. These activities, if undertaken freely and without control by the water administration, may negatively affect the quality of water and, possibly, the land or the body receiving such waters. It follows that they should be subject to a régime of permits or concessions in which the details of the operations are specified.

On 28 May, 2018, the EU issued a new proposal for a regulation on minimum requirements for water reuse. The regulation sets out minimum treatment requirements, as well as the duty of treatment plant operators to monitor water quality at the plant outlet and to produce water reuse risk management plans.

7.14.3 Wastewater and effluent control

The decreasing availability of water for large consumptive uses has made it feasible to consider certain wastewaters and effluents, after proper treatment, as 'new' sources of water. Their utilization is possible for purposes such as irrigation.

However, the legal régime of these 'new' sources of water is uncertain, and their management (inventory, planning, utilization) is not as yet the object of specific regulation, even though the economic potential of wastewater use is obvious. The management of these sources requires special provisions in the water legislation in order to safeguard public health and the environment from the potential adverse effects of wastewater use, as regards, *inter alia*, the intended uses, treatment requirements in relation to such uses, and the siting of wastewater treatment facilities.

7.14.4 Health preservation

This aspect has a long legal and institutional tradition; it refers to all aspects of water resources management (use, protection, conservation, reuse, disposal, etc.). It is particularly relevant when groundwater resources are developed and used for drinking or other domestic purposes, or in food processing industries.

Responsibilities for health protection and preservation usually belong to the public health authority and are governed by health laws and regulations. In order to achieve the most efficient water quality and pollution control, the water law should provide for close institutionalized cooperation between the water resources management administration and the public health authority.

7.14.5 Pollution control

While a certain amount of water pollution has always existed, and many legal provisions concerning its control have been enacted in the past,[14] the problem has become acute, par-

---

14 The UK Act of 1870 was the first act of Parliament governing discharges into streams in England and Wales. See also the French River Fishing Act of 15 April, 1829.

ticularly in industrialized countries. The aim of water pollution control is to preserve, as far as possible, the natural quality of surface and underground water, to protect water-dependent ecosystems and to decrease the existing levels of water pollution in order to protect public health and allow the satisfaction of water needs.

One difficulty is to define legally the term 'pollution,' as it is a relative one. A water body may be polluted for one purpose such as drinking, but not for another (industrial or irrigation). This implies that the term 'pollution' can be defined only in relation to a particular water use; hence the practice, supported by the water legislation of several countries, to classify bodies of water according to their existing and/or intended use and to set water quality standards and objectives for each water body. Water (or ambient) quality standards refer to the values not to be exceeded for given substances if the desired quality of water in a water body is to be maintained, while water quality objectives designate the quality targets to be achieved. As regards groundwater, the definition of pollution should cover every form of contamination and deterioration of the physical, biological and chemical composition of water, including not only the discharge of solid, liquid or gaseous substances into it, but any other agents affecting its characteristics, such as those carried by polluted runoff from agricultural land (non-point sources).

The classification of water bodies, which applies to both surface and underground waters, is based on the consideration that not all waters are or should be of the same quality, nor should they be managed in the same way. Initially, this measure was not accepted, and in some cases it was explicitly rejected[15] as being inapplicable in the case of large residential or industrial areas, or when the watercourse or the aquifer, owing to its particular length or size, is subject to several and changing polluting factors. Water bodies can be classified according to:

(i) the existing water quality. In this case, legislation or regulations include a statement as to which uses are permitted or forbidden, with or without treatment, for each category of water;[16]
(ii) the existing uses. This system consists of drawing up a schedule with various water uses grouped together in categories. Water quality requirements are devised and set forth for each category in the schedule. Periodic tests are then made on water samples to determine water quality. When this system of classification is adopted, legislation provides for minimum quality standards (maximum allowable concentrations of pollutants) for each category of uses. This method was applied to groundwater and to water resources in general in the former Soviet Union and in the countries of eastern Europe under Soviet influence, such as Bulgaria;
(iii) their usability, or potential use. This system is similar to that of the existing uses, but differs as far as the application of the schedule is concerned. It was adopted by Poland[17] and Finland;
(iv) a combination of factors, such as the extent of existing pollution, the causes of pollution, the number of outfalls from which pollutants are discharged, the entities controlling

---

15 See the case of the French Parliament, which decades ago refused to adopt this system when discussing a draft law on the administration and classification of water and the control of water pollution.
16 See, for instance, Czechoslovakia, Directive No. 74, of 27 March, 1957, issued by the Directorate of Water Economy. See also the practice in some states of the USA, such as Rhode Island.
17 Ordinance of the Council of Ministers of 28 February, 1962.

such outfalls, the type and amount of pollutants discharged from each outfall, existing or potential water uses, the extent to which the existing pollution should be controlled or mitigated, and any other relevant information.

Pursuant to the EU Water Framework Directive, EU member states must identify, within each river basin district, all water bodies used for water abstraction for human consumption in an amount exceeding 10 m^3/day or serving more than 50 people, or intended to be used for this purpose in the future.[18] Hence, the directive considers both existing and potential uses.

The control of water pollution is but one aspect of overall water resources management and, from a legal point of view, it should not be dealt with separately. Provisions on water quality and pollution control should be included in general water resources and basin management plans, together with provisions relating to resource development and use and the control of the harmful effects of water. In some countries, however, this aspect is regulated by special laws on health, pollution, or factories; more recently, environment protection laws and regulations have been enacted to address water quality issues, in addition to air, soil and noise pollution, and the assessment of environmental impacts. It goes without saying that in these cases water pollution control is often handled by an environmental, health or other institution, but not by the designated water administration.

When water quality and pollution control is the subject of environmental or health legislation and is dealt with by an institution other than the water administration, it is essential for the water legislation to establish mechanisms for the coordination of planning and licencing procedures, so as to avoid duplication of efforts and the consequent waste of human and financial resources, and to achieve effectiveness in the prevention and abatement of water pollution.

Pollution control may take place either on a preventive basis or after pollution has occurred (cleanup). In the first instance, permits or concessions for the use of water should contain provisions relating to the subsequent disposal or discharge of effluents. These provisions should cover treatment requirements before discharge, conditions as to the effluent standards (emission limit values) to be met, the construction and operation of discharge outlets, the obligation to monitor discharges, the payment of polluter fees[19] and other measures. Compliance with the wastewater discharge conditions attached to a permit or concession is measured against the quality standards set for the receiving medium (water body). The cleanup of water after pollution has occurred may be costly and difficult to achieve, particularly in the case of groundwater. Therefore, it should be avoided. Water legislation should also provide for the control of pollution caused by a water user without a permit or concession.

Finally, water legislation may require that water resources quality be regularly monitored and assessed in order to determine whether deterioration warrants the limitation of wastewater discharges through the setting of more stringent effluent standards, or through other measures.

---

18 Water Framework Directive, Art. 7.
19 See this chapter, Section 7.19.

Special legal measures are necessary to control so-called 'non-point pollution,' i.e., that pollution originating in diffused and not clearly identifiable sources such as that of groundwater where the improper use of pesticides, fertilizers or other chemical substances, the siting of landfills and waste dumps, or other highly polluting activities such as piggeries and tanneries, result in the pollution of the water far from the site of the polluting activity. These measures should be recommended by technical reports.

Some countries have deemed that basic water pollution control legislation should be kept to a minimum, in consideration of the fact that the technical methods employed for water protection and purification are constantly changing and that, as a result, detailed regulations become obsolete in a short time. Thus, these countries are more concerned with establishing a uniform policy and a framework law, leaving the regulation of special or detailed questions to subsidiary legislation (orders, decrees, regulations and the like). The opposite approach is to enact detailed laws which attempt to cover water pollution control in the most precise manner. Even these laws, however, leave some of the details to further subsidiary regulations adopted pursuant to enabling provisions.

Financial and technical considerations are very relevant for the smooth implementation of pollution control legislation, also considering that the social and economic level of development in a country may constitute a major constraint to the enforcement of a water pollution control policy.

7.15 Environment protection: the 'greening' of water law

The concern for the protection of the environment has prompted many countries to enact specific environment or natural resources protection legislation and to set up environment protection institutions. Based on this legislation, the water law often provides for cooperative arrangements between the water resources administration and the administration responsible for the environment, unless the environment agency or ministry is at the same time the water administration. A more recent trend is to include provisions relating to environment protection in water legislation. This process has been defined as a 'greening' of water law,[20] the objective of which is to find a balance between water demands for human and productive activities on the one hand, and, on the other, the preservation of ecosystems and environmental sustainability.

In countries with well-established water rights allocation systems, rights to water for the environment are emerging as a category of entitlements enjoying the same legal protection as the water rights held by other users. This marks an evolution of minimum ecological flow requirements from a simple limit to the power of the water administration to issue permits and concessions towards a legal entitlement of the environment to such water volumes and instream flows as are required to maintain the health and ecological balance of water bodies and their dependent ecosystems. In some countries this evolution has gone as far as

---

20 United Nations Environment Programme (2010) *The Greening of Water Law: Managing Freshwater Resources for People and for the Environment*. Nairobi, UNEP.

recognizing juridical personality and autonomous rights to rivers,[21] and even to nature as such,[22] or to Mother Earth (*Pacha Mama*).[23] According to the legislation of these countries, this entitlement of the environment enjoys priority over water rights for other purposes of use, except for human consumption.

In an increasing number of countries, environmental impact assessment (EIA) requirements are included in the procedures for the issuance of water use (or wastewater discharge) permits contained in the water legislation, this meaning that an EIA must be filed together with the application for a permit. The EIA requirement applies to the types of activities specified in the legislation, such as water resources development projects of a certain size, the artificial recharge of aquifers and other activities which may have adverse effects on the environment.

A recent trend is that of subjecting policies, plans, programmes and laws that may adversely affect the environment and different social groups to EIA. The UK Water Act 2003, for instance, underwent a full regulatory impact assessment, setting out the government's consideration of the costs and benefits that would derive from the implementation of the act, and an environmental and equal treatment appraisal. A recent EU directive subjects to environmental assessment the plans and programmes relating to water management, agriculture, energy, industry and other sectors of relevance to water resources, which are required under legislative, regulatory or administrative provisions and that might significantly affect the environment.[24]

Finally, through their water legislation or legislation relating to other natural resources, some countries, including Colombia, Costa Rica and Peru, have introduced the concept of 'payment for ecosystem services' (PES). Ecosystem services may be defined as the benefits that people obtain from ecosystems.[25] Water-related ecosystem services include flood regulation, soil erosion control, the regulation of water supply and runoff, water quality preservation, the improvement of water infiltration and storage and the enhancement of groundwater recharge. They also include the preservation of spiritual, aesthetic and recreational benefits. These services largely depend on the behaviour of the communities living in the proximity of surface or underground water bodies which use the water and related ecosystem resources. PES consist of the incentives to be paid to communities or individuals to guarantee the services through sustainable agricultural practices, the preservation of

---

21 See the The Urewera Act 2014 (Public Act No. 51 of 27 July, 2014) and the Te Awa Tupua (Whanganui River Claims Settlement) Act 2017 (Public Act No. 7 of 20 March, 2017) of New Zealand. In India, by order of 20 March, 2017, the High Court of the State of Uttarakhand declared the Ganga and Yamuna rivers, and their tributaries, as legal entities. Through a further order dated 30 March, 2017, the court conferred the same status to the 'rivers, streams, rivulets, lakes, air, meadows, dales, jungles forests wetlands, grasslands, springs and waterfalls' in the state, and to glaciers, including the Gangotri and Yamunotri Glaciers, from which the Ganga and Yamuna Rivers take their sources. The 20 March decision was stayed by India's Supreme Court in July, 2017.
22 Bolivia, Colombia.
23 This is the case in Ecuador.
24 Directive 2001/42/EC of the European Parliament and of the Council on the assessment of the effects of certain plans and programmes on the environment, 27 June, 2001. Text in: *Official Journal*, L 197, 21/07/2001, 30.
25 UNECE (2007) *Recommendations on Payment for Ecosystem Services in Integrated Water Resources Management*. Adopted at the fourth meeting of the Parties to the UNECE Convention on the Protection and Use of Transboundary Watercourses and International Lakes (20–22 November, 2006). New York and Geneva, United Nations.

the forest cover, wetland restoration and other activities. PES schemes may be public when a municipality or a national or local government, or other public institution, purchases the services; private (self-organized), when both the seller and the buyer are private individuals or entities; or, based on trading, when established rights (permits) or quota are exchanged through market mechanisms.

7.16 Underground waters

7.16.1 General considerations

Groundwater management has frequently been the subject of special legislation implemented and enforced by different government departments and agencies at the national and local levels. The tendency to regulate surface and underground waters separately should not be pursued since groundwater, like surface water, forms part of the hydrologic cycle. From a technical, economic, social and environmental point of view, the management of water resources, including groundwater and atmospheric water, requires an integrated approach, i.e., an approach which includes both surface and underground water and, possibly, both quantity and quality aspects. Thus, the same water law should provide for the integrated regulation and administration of surface water and groundwater. By the same token, both surface water and groundwater should be managed by one and the same water resources administration. Efforts have been made in recent years by European Union (EU) member states and states applying for EU membership to consolidate water legislation hitherto fragmented, and to adopt an integrated water resources management and river basin approach, in line with the requirements of the Water Framework Directive.[26]

Due to the special nature of groundwater, to its invisibility and the limited knowledge as to its nature and behavior, and to the factor of relative risk involved in its survey, extraction and use, state interference has become a necessity, and specific legal measures have been devised to safeguard the resource. Based on this consideration, the water laws of several countries devote a special chapter to underground water. Groundwater-specific provisions in the water law and in subsidiary legislation are indispensable to facilitate the protection and conservation of aquifers.

Underground water may be privately or publicly owned.[27] The countries following the civil law system accept the basic Roman law principle reformulated in the French Napoleonic Code, according to which the landowner owns everything located above and below his land. As a consequence, the status of ownership of groundwater follows that of the overlying land. In common law countries, a distinction is drawn between underground streams, which cannot be privately owned, and other forms of groundwater which are susceptible to exclusive ownership rights to the benefit of the owner of the land. Islamic countries view groundwater as a public good. Only wells may be privately owned.

Given the problems created by increasing water scarcity and pollution, the idea has developed that restrictions be placed upon the owners of the land, of wells or other abstraction works, as to the kind and rates of abstraction and the amount of groundwater to be

---

26 See Chapter 6.
27 See Chapter 5, Sections 5.4.4, 5.5.2 and 5.6.2.

abstracted, depending on the circumstances.[28] Legislation has thus been enacted to vest all water resources in the state, or to recognize the state's superior right to the management of water resources. The declaration of groundwater as a 'public good' turns the former owner into a user, who must apply to the state administration for a water abstraction and use right. Once the state is the guardian or trustee of groundwater resources, it may, in addition to granting water rights, introduce measures to prevent aquifer depletion and groundwater pollution. Moreover, legislation now tends to require water resources planning at the level of an entire aquifer or river basin.

In some instances, the 'new' legislation has been challenged in courts of law, because of alleged inconsistencies with constitutional provisions protecting private ownership and requiring payment of compensation when rights are compulsorily negated. But such challenges have usually been rejected on grounds that regulating groundwater abstraction arises from the need to safeguard the public interest.[29]

The present trend to declare all water resources, whether surface or underground, as public should be pursued, but whenever private groundwater becomes public property water legislation should recognize the pre-existing right of the landowner to use underground water freely on his land for minor domestic purposes (drinking, household, watering of domestic animals and small gardens), provided that he observes the conditions imposed by the legislation and that the rights of other users are not affected. Publicly owned underground water should undergo the same régime applying to public surface water, i.e., its utilization should be subject to the issuance of a permit, authorization or concession and to overall government control.

Pursuant to water legislation, all underground waters should be subject to specific legal measures aimed at ensuring their efficient control and administration. These measures are briefly described in the following paragraphs.

7.16.2 Exploration or prospecting permits

Before undertaking drilling operations, those who intend to search for underground water should be obliged by the water law to secure an exploration or prospecting permit, to be issued by the water administration only after all necessary precautions or requirements have been accepted or complied with by the prospective user. These permits are the instrument through which the water administration keeps groundwater drilling activities under control. They should be temporary, their duration being limited to a short period of time (six months, for example), eventually renewable. An exploration or prospecting permit is different from a water use permit. After groundwater has been tapped through drilling, a water use permit should follow under the same régime as those issued for the use of surface water.[30]

---

28 Hayton, R. (1976) The Groundwater Legal Régime as Instrument of Policy Object and Management Requirement. *Annales Juris Aquarum*, Vol. II. Caracas, AIDA, 271.
29 For a description of recent trends in domestic water legislation and administration, see Burchi, S., & Nanni, M. (2003). How Groundwater Ownership and Rights Influence Groundwater Intensive Use Management. *Intensive Use of Groundwater, Challenges and Opportunities*, Ramón Llamas & Emilio Custodio eds. Rotterdam, Balkema, 227–240.
30 See this chapter, Section 7.8.

7.16.3 Drillers' licences or permits

The introduction of professional drillers' licences is also advisable. These licences should be issued only to qualified drillers, who should be obliged to notify the water administration of the intention to drill and of the drilling location, to furnish the same with cores or carrots, i.e., samples of the perforated strata, and in any case with a drilling report and, finally, to comply with any further obligation that the water administration may deem fit to impose, including, if so desired, the payment of drilling fees. Before starting a drilling operation, or the enlargement, deepening or modification of existing wells, a licenced driller should also make sure that his employer is in possession of the required exploration or prospecting permit.

Professional drillers' licences facilitate the control of drilling activities, as the licenced driller, in the case of non-compliance with that part of the water law concerning drilling operations, may be subject to the payment of fines, the impoundment of his equipment and, in the case of recidivity, to the suspension or loss of his licence to operate. This licence, on the one hand, enhances the drillers' professional status, and on the other hand makes them become auxiliaries of the water administration. Finally, it generates additional income to the government. In most countries professional drillers are not numerous.

7.16.4 Groundwater found incidentally

In addition to the above, the water law should contain provisions relating to procedures and requirements applicable in the case of groundwater found incidentally, for instance in the course of mineral prospection activities. In this case the major requirement should be the obligation to report the finding to the water administration. It follows that coordination between the mining licencing authority and the water administration should be established. Often, the mining licencing authority authorizes the use of water 'as necessary for the mining operations.' It would be desirable, in the interest of water management, that the water utilized or found incidentally be specified in quantitative terms.

7.16.5 Metering

The installation of meters by the water administration to control water abstraction is an important measure for ensuring successful management of groundwater resources. It is not always possible to introduce this measure in view of particular economic, technical and social constraints or circumstances. However, the metering of abstractions makes it possible to control the actual amount of water utilized, thus ensuring, through the limitation of abstractions, respect of the 'safe yield' principle and the water balance. The measure also allows the quantification of water rates and charges, which may serve as an instrument for limiting the excessive use of water.

7.16.6 Groundwater pollution control

The problems connected with groundwater pollution control require an adequate policy which is part of a global water resources management policy. The objectives of a groundwater pollution control policy should be to prevent this pollution, to abate already existing pollution and to promote the treatment, recycling and reuse of wastewater. However, when designing a policy, the particular characteristics of groundwater should be kept in mind.

Due to the complexity and slow rates of groundwater flows, there may be a long time-lapse between the occurrence of a pollution event and when its impact on groundwater and on groundwater-dependent receptors[31] is felt. By the time this impact is felt a large portion of an aquifer, or the ecosystems depending therefrom, may be affected. Thus, difficulties may arise while defining minimum protection requirements.

Legal provisions on groundwater pollution control should attempt to achieve the following objectives:

(i) to regulate preventive measures and the control of pollution in harmony with economic and social development policies;
(ii) to designate the groundwater basin (or groundwater body)[32] as the proper unit for the management of groundwater resources;
(iii) to protect the quality of groundwater in relation to the various water needs;
(iv) to ensure that either state ownership or state control be established over groundwater resources in order to allow monitoring and an easier enforcement of measures against polluters;
(v) to combat, as far as possible, groundwater pollution at the source.

As far as EU member states and the states which have applied for EU membership are concerned, the 'good status' objective,[33] environmental objectives and the objective of long-term planning of protection measures called for by the EU Water Framework Directive are to be added to the above.

The achievement of these objectives may be highly facilitated by a consolidation of the legislative provisions relating to water pollution, thus avoiding lack of coordination between measures of control, possible inadequacy in scope of control and lack of power and means to ensure effective action where and when required. In addition, an adequate (coordinated or centralized) institutional framework is required.

7.16.7 Reservation of good quality groundwater for drinking purposes

The legislation of some countries and the actions of the World Health Organization (WHO) have promoted the enactment of provisions reserving groundwater of the best quality for domestic uses and urban water supply. This preference of use should be sanctioned in the law.

---

31 A lake, an internal sea, a wetland, for instance. An example is that of the Caspian Sea, which receives polluted groundwater via the rivers flowing into it.
32 This is the expression employed in the EU Water Framework Directive, which defines a body of groundwater as 'a distinct body of water within an aquifer or aquifers.'
33 The definition of 'good quantitative and chemical status' of groundwater is provided in Annex V to the EU Water Framework Directive.

7.16.8 Zoning mechanisms

The adoption of special zoning mechanisms to facilitate groundwater conservation and protection has become a major feature of modern legal régimes. Under the legislation of a number of countries, special 'control' or 'management' zones or areas may be declared and delimited by the designated authority or authorities when it is necessary to protect aquifer recharge areas, discharge areas (to protect wetlands) or given 'hot spots', i.e., highly vulnerable areas, from pollution of depletion. Within these zones or areas the authorities may restrict or prohibit certain activities, such as new well drilling, groundwater extraction, waste disposal, the storage and handling of chemicals, the use of agricultural fertilizers and pesticides, the grazing of animals, landfills, quarrying and the construction of buildings for regular occupancy.[34]

Protected areas (perimeters) may also be established around wells or boreholes to prevent the pollution of a source of public or municipal groundwater supply. These areas should be under the responsibility of the municipal authority or local institution providing the service, or of a water utility operating under a concession régime.[35]

7.17 Control and protection of waterworks and structures

A water law should include provisions relating to the control of hydraulic structures with respect to their construction, operation, maintenance, safety and inspection. This control should take place through the implementation of a permit system and in accordance with the procedures established under the law. Minimum standard specifications for various types of hydraulic works and for equipment, to be detailed in the relevant permit or concession, should be provided for by implementing regulations.

In addition to the above, the water law should designate the institutions exercising responsibilities in this field, namely:

(i) the state, for waterworks and hydraulic structures constructed, operated and maintained by the state;
(ii) water users' associations, for those built, operated and maintained by them; and
(iii) private individuals, companies or special autonomous agencies, whenever waters are used and works constructed, operated and maintained for their exclusive benefit.

The water administration should reserve for itself overall control of waterworks and hydraulic structures, so that they are designed, constructed, operated and maintained according to standard specifications and do not constitute a danger for public safety.

The water legislation may also provide for the establishment of protection perimeters or areas around or in connection with waterworks and structures, both for minor waterworks such as wells, pumps and water supply reservoirs, and for large waterworks such as dams and barrages.

---

34 This is the practice followed in the United Kingdom, Finland, France, Germany, Israel, Italy, Poland and Algeria.
35 See Nanni, M. & Foster, S. (2005) Groundwater Resources – Shaping Legislation in Harmony with Real Issues and Sound Concepts. *Water Policy* 7:5. Also the Water Framework Directive, at Art. 7, empowers EU member states to establish safeguard zones for bodies of water used for the abstraction of drinking water.

In many countries, access to hydraulic structures such as dams, hydropower stations or reservoirs is unrestricted. Legislation should provide for the protection of such facilities in order to avoid hazards to public safety or health. Furthermore, it should designate the authorities responsible for an emergency response in the case of failure of hydraulic structures.

7.18 Protected zones/areas

In the water resources sector, it is possible that emergency circumstances arise in connection with:

(i) conflicts over water rights in a particular area;
(ii) problems arising in the case of beneficial uses of water, such as lack of water, generalized waste/misuse of water, overpumping, failure of hydropower generation, water shortage in water supply reservoirs and any other possible occurrence;
(iii) problems arising in connection with the harmful effects of water, such as floods, siltation, waterlogging, droughts, soil erosion;
(iv) problems arising in connection with widespread water quality degradation, widespread waterborne diseases, generalized pollution or generalized environmental degradation in a particular area;
(v) the need to proceed with the implementation of government policies concerning specific zoning or land or water use planning in certain areas.

Depending on the circumstances, affected areas may need to be declared as protected, restricted, rationed or to be submitted to a special régime. Some of the provisions relating to protected areas may be found in general legislation, others in special legislation providing for temporary or permanent measures to apply under exceptional circumstances.

The water resources administration should be empowered to act through specific emergency provisions of the water law. Actual implementation could take place on the basis of administrative regulations, orders and the like.

The content of emergency provisions may be, *inter alia*, the following:

(i) a delimitation of the area or areas to be brought under the special régime;
(ii) the classification of the area as protected, rationed, restricted, controlled or other, according to its special legal régime;
(iii) a general limitation of the existing individual rights to use water, whether granted under a permit or not, in the form of restriction, suspension, or other type of limitation;
(iv) the imposition of other limitations or obligations as regards water withdrawals or diversions;
(v) the prohibition of certain uses;
(vi) any other limitation, obligation or activity which may be dictated in the public interest, in order to safeguard public security, health and the environment.

A distinction should be drawn between exceptional provisions conferring extended powers for a limited period of time and regular administrative controls over water resources. While the latter are of a permanent nature, as they belong to the realm of overall water resources management and administration, the former are indispensable, as in emergencies they allow the water administration to intervene without delay. They are part of a sort of martial law with regard to water management.

When the establishment of protected zones or areas requires the expropriation of land, compensation of the affected landowners is normally provided for in the legislation. It is possible to provide for a system of several zones or areas, with restrictions being less stringent in those which are not adjacent to the water source being protected.

7.19 Legislation on financial aspects

It must be recognized that water, as a natural resource, has a price, and that its development, conservation and protection entail costs which should be reimbursed as far as possible by the users, taking into account market forces, social needs, political requirements, the public interest, the availability of water and, last but not least, the ability of the users to pay.

Therefore the water law should include provisions relating to the financial aspects of water resources development, use, conservation and protection.

These provisions may refer to:

(i) the possibility of and procedures for government financial participation and reimbursement policies with respect to water development projects, as these are a powerful vehicle for overall economic and social development;
(ii) the financing of the construction of municipal sewage and industrial waste treatment plants, through grants, subsidies, low interest loans, special bond issues, government guaranteed loans, tax exemptions and the like;
(iii) the possible forms of government contribution to be granted to individuals and associations under the form of grants, loans, exemptions from water charges, or other;
(iv) facilities and procedures as regards credit services, including the organization of *ad-hoc* centralized institutions, marketing arrangements and collection of revenues, which may be needed for water development activities, particularly for irrigation where the necessary capital is often inadequate at the users' level;
(v) the criteria for the establishment, collection and distribution of water rates, charges and fees, in view of their direct importance for preventing the waste and misuse of water, and for securing revenues to the state for the use of public waters.

As regards all types of water utilization, legal provisions and procedures should consider, (i) the kind of water rates, fees and charges for water use permits or concessions, including the filing fees payable by the applicants to cover the costs of processing applications, and (ii) registration fees payable by water rights holders or water users (individual, collective, municipal, etc.). Furthermore, in consideration of the fact that a certain degree of water pollution is to be tolerated if development is to be pursued, legislation should cover the fees to be paid by those intending to discharge wastewater, after treatment, into water bodies or the subsoil, in compliance with the 'polluter pays' principle.

The assessment and collection of water rates, fees and charges provide the water administration with the means to further water development and to facilitate the financing of the construction and maintenance of hydraulic works and projects. They also constitute a means for the practical implementation of a water policy.

In the determination of water rates, fees and charges, consideration should be given to the basis for the assessment by the assessing and collecting agency or administration:

volume, extension of land, kind of crop or utilization, nominal hydropower produced, time of use, increased value or improvement of land and, as far as polluter fees are concerned, the type and amount of pollutants to be discharged or disposed of.

From an economic point of view, the pricing of water based on the classic benefit-cost analysis cannot always be applied, as other social or not economically quantifiable factors come to fore. In some countries, for instance, paying for water *per se* goes against religious, customary or other legal beliefs. In these cases, charging for the 'services of water supply' rather than for the water is a possible alternative.

A basic water law should contain the principle that water, as a national economic asset, is subject to payment for its use, discharge or other water-related services. The amount of water rates, fees and charges, however, may vary in time according to factors such as the ability to pay. In this case, the determination of the amount should take into consideration the affordability element. The implementation of charging schemes may be imposed progressively by areas, uses or other criteria.

While the basic water law should establish the principle of water pricing, actual implementation as regards the tarification, the place and time of implementation may be taken care of by subsidiary regulation. The state must reserve for itself the right to approve water tariffs.

7.20 Implementation of water legislation

7.20.1 General remarks

The implementation and enforcement of water legislation can be difficult. New water legislation, in attempting to bring water uses under control, must contemplate traditional and accepted practices. Stiff opposition is to be expected from property holders and water users, as well as from existing and sometimes obsolete administrations and institutions. Thus, the first requirement for successful implementation and enforcement of a water law is its acceptance by the population and by all those institutions and organizations which may be affected by the novelties introduced by it.

Problems may arise as to the following issues:

(i) the size of a country and a variety of situations and traditions;
(ii) judicial and administrative control of water ownership, distribution and use;
(iii) juridical and administrative protection of water rights, whether granted by an act of the water administration or not;
(iv) claims against decisions of the water administration and relevant procedures;
(v) water courts, tribunals, and other judiciary or administrative law-enforcement bodies;
(vi) penalties and sanctions against offenders.

Water legislation should contain provisions with respect to these matters. These provisions should be adapted to local circumstances and conditions, as it is better not to have any law than a water law which, according to local custom, traditions and other requirements, proves to be inappropriate and difficult to enforce.

In order to clarify the existing situation and to avoid legal queries with respect to the interpretation of the water law, a new water law should provide for the amendment or repeal of any existing legal enactment or part thereof, the provisions of which have been superseded or consolidated by the new law. Such amendments or repeals should be provided for specifically in the new water law.

7.20.2 Judicial and administrative control over water ownership, use and distribution

A new water law must consider the existing legal situation with respect to ownership, use and distribution of water resources and, in setting forth new principles and procedures, should provide legal instruments for coping with the problems which may arise in this connection, including those relating to the construction, operation and maintenance of hydraulic works and structures.

The granting of new water use rights, the creation of rights of way or easements and the expropriation or temporary occupation of required land are sometimes necessary for the successful development and utilization of water resources. Such measures, however, entail restrictions, limitations or deprivation of property rights. Therefore they should be delimited and defined by the water law. In addition, the water law should consider the payment of adequate compensation to those who have suffered a loss or reduction as a consequence of these measures, and also the need for promptness in the payment of compensation, so as not to hamper the execution thereof through lengthy discussions on indemnity.

The water law should empower the water administration, under appropriate procedures, to inspect and install water measuring devices or other appropriate control apparatus.

7.20.3 Protection of individual rights and water rights

Guarantees should be established by the water law for the juridical and administrative protection of existing individual rights in general and of water rights in particular, whether granted by an act of the administration or not. The cases in which a water right undergoes prescription or forfeiture should be clearly defined. The same applies to cases in which a water permit or concession can be modified or suspended.

The water administration may limit or withdraw an existing right to use water in favour of a larger utilization or of a new use more important for the public interest. This power will normally be based on the provisions of a water resources plan. In these cases, the older right holder should be compensated as far as possible in kind, i.e., with the same amount of water from another source, a monetary compensation only being provided in cases where compensation in kind is not possible.

The water law should also provide for the protection of the rights of third parties, by establishing that compensation be paid to anyone who suffers a damage as a consequence of water use. Provision should be also made for cases of illicit use or misuse which deprive the legitimate users of their water or cause harm.

7.20.4 Administrative procedures for claims against the water administration

Anyone who feels aggrieved by a decision of the water administration on applications for the use of water or on water rights should be given the possibility to claim against such decision.

The first appeal should be addressed to the chief of the water administration, who should take a decision on an administrative basis or on technical, economic and public interest grounds. A further appeal against the negative decision of the chief of the administration could be referred to the minister responsible for water management. Finally, in the case of alleged violation of the law the claimant could resort to a decision of either a regular court or of a water court.

The water administration should also be empowered to adjudicate individual conflicts on water rights.

7.20.5 Water tribunals or courts

Due to the specialized aspects of the control over water resources, in some countries water laws have established special courts dealing exclusively with water matters. These courts can also be established as sections of existing second degree or district courts, with a further section for appeals attached to the supreme court. Competence, procedures, jurisdiction and composition have to be established by the water law.

The competence of water tribunals or courts may include:

(i) review and adjudication, in law and not in fact, of the administrative decisions rendered by the water administration acting as quasi-judiciary body, relating to the public domain, to the recognition of existing water rights, water uses and/or compensations to be paid;
(ii) conflicts between private water right holders, as well as between the water administration and autonomous or semi-autonomous agencies or associations with respect to water rights and the use of water; and
(iii) penalties and sanctions against violations of the water law or other relevant enactments.

Their jurisdiction should, if possible, encompass the basin or the sub-basin, depending on local conditions. The procedures should be tailored to those existing in the legal system of a country.

Ideally, a water court or tribunal should be composed of a judge acting as president and two specialists in technical and economic aspects of water resources, of which one could be appointed by the water administration. Judgments should be based on the provisions of the water law, on the public interest and on local customary law and requirements. Water tribunals have been established in South Africa, Kenya, Namibia and Peru, among others.

7.20.6 Penalties and sanctions

The water law should establish and/or coordinate penalties and sanctions imposed or to be imposed for an offence committed against the provisions of the water law or against those of any other legal enactment containing obligations on water matters.

Offences may include: damage to or destruction of waterworks or structures; interference with officials of the water administration in the exercise of their functions; use, reuse and discharge of water without a permit or concession when a permit or concession is required; violation of the terms of permits and concessions, or of water regulations.

Penalties should be graduated to suit the degree of offence, and may range from simple administrative fines to imprisonment. The water law should also provide for the demolition of illegally built works or structures at the expense of the offender, and for the procedures to be followed in the case the administration has to undertake this work when an offender fails to comply with a notification to do so. Procedures for the seizure or impoundment of equipment utilized for illegal construction works or drilling should also be provided for in the case of repeated violations.

7.21 The interconnection between water law and other legal enactments relevant to water law

Water legislation encompasses a variety of situations which include not only water, but also other sectors of the national economy.

In almost all countries, many enactments govern different aspects of a country's economic life and contain provisions which may directly or indirectly affect water resources, or vice versa. This may be the case of laws regulating housing, land settlement, paddy land, railways, electricity, crown land, mining and other natural resources.

These legal enactments should not be isolated from one another in water-tight compartments. While it may be necessary that different activities continue to be governed by special legislation and administered by separate departments, a water law should establish proper and adequate legal and administrative coordination among the various agencies and departments responsible for particular activities which have or may have an influence on successful administrative control over water resources as such.

Coordination established by a water law makes it mandatory that no government body shall grant a permit or concession which may interfere with the flow, régime or quality of water resources without previous consultation with the water administration and, vice versa, that no permit or concession be granted by the water administration without previous consultation with the other administrations concerned.

In order to establish adequate coordination, apart from the provisions to be included in the water law, amendments or revisions may be required in legal enactments dealing with other subjects. As an example, a building permit should not be issued before the housing authority has had an assurance from the water administration that, after construction, water will be available. The same applies to land settlement schemes and electric power plants, among others.

7.22 Customary water law and institutions

There are countries where only traditional – customary or religious – water regulations govern water resources management. This is the case of, *inter alia*, Saudi Arabia. In other countries, customary law coexists with more recent water legislation introduced by former colonial powers, particularly in rural areas.

Even in industrialized countries, local uses and customs concerning water ownership, distribution and utilization are often to be found, stemming from ancient grants or other legal provisions dating from time immemorial. As examples, perpetual water rights exist for the princely palaces in Rome; in Valencia a water tribunal decides on ancient customary rights; and ancient floating practices are to be found in Finland. Local uses and customs apply to particular groups, or uses, and prevail in given areas.

As far as possible, water legislation should contain provisions to recognize and regulate customary water management practices and institutions. These provisions should cover the legal régime of water and water rights, traditional water management institutions and their relationship with modern laws and the government water administration, financial aspects, including procedures for the assessment of water rates and fees or water service charges, procedures for the settlement of water disputes, the relationship between the national courts and traditional water courts and, last but not least, the procedures for recognizing, institutionalizing and modernizing customary water rights and institutions.

7.23 Water users' associations

The water law should provide for the establishment and functioning of water users' associations. This is a practical means to bring the customary administration of water under control, facilitate the control of water uses at the local level and provide a framework for the operation and maintenance of hydraulic infrastructure by the users themselves, including cost-recovery.[36]

7.24 National water resources administration

Finally, it is essential that the water law sets up the administrative machinery for the management of national water resources. The law should indicate the functions, powers and responsibilities of the national administration in the water field. These will depend on many factors, which have been analyzed in another part of this book.[37]

References

Bradlow, D.D., Palmieri, A. & Salman, M.A.S. (2002) *Regulatory Frameworks for Dam Safety – A Comparative Study*. Law, Justice and Development Series. Washington, D.C., The World Bank.
Burchi, S. (1994) *Preparing National Regulations for Water Resources Management – Principles and Practice*. Legislative Study No. 52. Rome, FAO.
Burchi, S. & D'Andrea, A. (2003) *Preparing National Regulations for Water Resources Management – Principles and Practice*, Legislative Study No. 80. Rome, FAO.
Burchi, S., & Nanni, M. (2003) How Groundwater Ownership and Rights Influence Groundwater Intensive Use Management. *Intensive Use of Groundwater, Challenges and Opportunities*, Ramón Llamas & Emilio Custodio eds. Rotterdam, Balkema, 227–240.
Canter, L.W. (1985) *River Quality Monitoring*, Lewis Publishers Inc.
Caponera, D.A. (1969) Essential Considerations for Drafting Water Laws, in *Water Legislation in Asia and the Far East*, Part II. Water Resources Series No. 35. New York, United Nations.
Caponera, D.A. (1975) *Outline for the Preparation of a National Water Resources Law Inventory*. Background Paper No. 7. Rome, FAO.
Clark, R.E. et.al. (1967) *Waters and Water Rights*, 1967.
Clark, S.D. (1973) *Guidelines for the Drafting of Water Codes*, Water Resources Series No. 43. New York, United Nations.
Despax, M. (1980) *Droit de l'environnement*. Paris, ITEC.
FAO Legal Office (1999) *Issues in Water Law Reform*. Legislative Study No. 67. Rome, FAO.
Garduño Velasco, H. (2003) *Administración de derechos de aguas – Experiencias, asuntos relevantes y lineamientos*. Legislative Study No. 81. Rome, FAO.

---

36 See Chapter 9, Section 9.2.7.
37 See Chapter 9.

Hayton, R. (1976) The Groundwater Legal Régime as Instrument of Policy Objectives and Management Requirements. *Annales Juris Aquarum*, Vol. II. Caracas, AIDA, 271.

Hodgson, S. (2004) *Land and Water – The Rights Interface*. Legislative Study No. 84. Rome, FAO.

Malakoff, E.R. (1966) *Water Pollution Control: National Legislation and Policy*. Rome, FAO.

Nanni, M. & Foster, S. (2005) Groundwater Resources – Shaping Legislation in Harmony with Real Issues and Sound Concepts. *Water Policy* 7:5.

Prieur, M. (1984) *Droit de l'environnement*. Paris, Dalloz.

Solanes, M. & Gonzales-Villareal, F. (1999) *The Dublin Principles for Water as Reflected in a Comparative Assessment of Institutional and Legal Arrangements for Integrated Water Resources Management*. Global Water Partnership TAC Background Paper No. 3, Stockholm.

Teclaff, L.A. (1977) *Legal and Institutional Responses to Growing Water Demand*. Legislative Study No. 14. Rome, FAO.

Trelease, F.J. (1975) New Water Legislation. Drafting for Development, Efficient Allocation and Environment Protection. *Proceedings of the Conference on Global Water Law Systems*, W-12. Valencia.

United Nations (1969) *Proceedings of the ECAFE Working Group of Experts on Water Codes*. Water Resources Series No. 32. New York, United Nations.

United Nations Environment Programme (2010) *The Greening of Water Law: Managing Freshwater Resources for People and for the Environment*. Nairobi, UNEP.

Chapter 8

Water resources planning and water law

8.1 The rationale of water resources planning

Water resources planning and management must be kept in line with current and foreseeable future needs. In particular, it is necessary to adopt measures to predict the impact of man's action on water resources, and to control resource development and minimize its negative effects.

The policies by which decisions for land use planning were made first and steps to correct their impact on water resources were taken afterwards have become obsolete. Instead, water resources and land use planning must be integrated processes pursued simultaneously at all levels from the start, especially now that the degradation of natural resources in general, and water resources in particular, occurs at an alarming rate and that these resources are under additional threat because of climate variability and change. In addition, due to the intimate relationship between surface water and groundwater, inland and coastal water, water quantity and quality issues, water resources planning must be in itself an integrated process. Given the importance of the environmental and public health aspects, water resources planning should also be fully integrated with environment protection and public health planning. Finally, water resources plans should be conceived as flexible tools facilitating mitigation of, and adaptation to, the effects of climate change, and the management of risks and uncertainties.

To address present water management issues, it is necessary to take adequate measures of a technical, economic, legal and institutional nature. The implementation of measures of the same nature oriented towards the achievement of a certain objective corresponds to the formulation of what can be defined as a water resources management *strategy*. A coherent combination of strategies can be referred to as water resources management *policy*.[1]

Technical measures may be of various kinds and are usually associated with hydraulic works. They have been used in the past more or less successfully.

Economic measures refer to water pricing for various purposes of utilization, charges imposed for the discharge of effluents, and economic incentives such as grants, loans and tax exemptions.

---

1 Vega da Cunha, L. (1983) *Water Resources Planning and Usage,* paper presented at the European Institute for Water Seminar 'Water in Western Europe,' 4–8 July, 1983.

Legal and institutional measures for water resources management relate to the enactment of water legislation and to the definition of institutions responsible for the management of water resources.

As was mentioned earlier,[2] water legislation generally takes two forms: a basic law that defines the principles of the water resources management policy, and regulations dealing with the details of implementation.

The definition of an appropriate institutional framework for water resources management is indispensable in order to ensure the coordination of the roles of the different agencies concerned with water resources, with particular regard to three types of agencies:

(i) agencies responsible for water-using sectors such as domestic supply, agriculture, hydropower generation and distribution, industry, fishing, navigation and tourism, constituting 'water development agencies;'
(ii) agencies not having any responsibilities as regards water use, but having responsibilities in related domains, such as environment, health, public works and finance;
(iii) agencies responsible for the administration of water-related data and information, the planning of water resources and the administration of water rights. These are the functions of a 'regulatory institution.'[3]

A water resources management policy is a coherent combination of water resources management strategies. Policy selection should be based on alternative scenarios that correspond to the specification of certain consistent combinations of the more significant factors, relating to future conditions of economic and social development, technological progress, standards of living, behaviour of water users and managers, the effects of climate change and public policies.

The main aspects of a water resources policy are the following:

(i) identification of existing and potential water users and their requirements, i.e., the water demand in terms of quantity and quality;
(ii) identification of ecological and minimum flow requirements, where appropriate;
(iii) identification of alternative water sources, both within and outside the river basin under consideration, and definition of their characteristics;
(iv) definition and specification of alternative water distribution systems connecting water users to the water source;
(v) water quality control through the establishment of quality standards and objectives, effluent treatment and other requirements;
(vi) specification of regulations and operating rules applicable to the different water uses and water users;
(vii) price fixing for the different water uses, including the use of water bodies for receiving polluted effluents, in order to optimize the use of the water and to control the environmental impact.

---

2 See Chapter 4.
3 See Chapter 9, Section 9.3.4.

Experience shows that the institutional setup is often inadequate, having been developed in respect of economic and social requirements foreign to the global hydrological context. This inadequacy gives rise to the need for water resources planning at different levels, to which water law provides the foundations, besides defining state responsibilities with regard to the formulation of general water resources development and management policies. With an adequate law, a water resources plan becomes the framework within which the water administration grants concessions and permits for water use and wastewater discharge, among other things.

Every water resources plan must depend on institutions facilitating the implementation of the strategies and directives that it establishes. This means that it must be possible to rely on a water resources administration system capable of generating a plan which ensures an effective control of the various water utilizations and facilitates water protection, and not on a system which presents obstacles to the implementation thereof.

8.2 Objectives of a water resources plan

Given the importance of water in all sectors of a country's economy, water resources management and administration must not be oriented towards one particular water development or use, as has been the case in the past, but should consider water resources as a whole, i.e., have regard to resource development, protection and conservation, at the same time paying attention to the interdependence of surface water and groundwater, and to the relationship between water and other natural resources. In this connection, national and basin water resources plans play a key role, as they allow governments to reconcile these aspects and bring them all under one umbrella, thereby rendering problems more visible and facilitating the introduction of management measures where they are most needed. National and basin water resources management plans do not overrule the existence of sectorial plans based on the various uses of water or on the levels at which such plans are to operate. Rather, they are expected to bring these plans together and ensure their reciprocal compatibility.

As their main objective, water resources management plans should have an effective and rational organization for satisfying present and foreseeable water demands with the resources available and for ensuring that these resources meet good quality standards. In order to attain this general objective, attention should be paid to the following specific objectives:

(i) as regards the relationship between water demand and water availability, to respect the hydrological equilibrium as much as possible;
(ii) as water is directly or indirectly present in all activities, to ensure that its relative scarcity does not delay the optimal development of other natural resources;
(iii) to establish rational, equitable and justified water distribution procedures aimed at satisfying the water demands of different users in time and space;
(iv) to control polluting activities and protect water quality;
(v) to achieve long-term water resources management, which takes into account future demands and resource protection requirements in relation with population growth;
(vi) to ensure the defence against the harmful effects of water, taking into consideration floods, soil erosion, salinization and drought forecasts;
(vii) to establish flexible priorities among water resources development programmes, in time and space, for all utilizations (drinking water supply, agricultural uses, industrial uses, production of hydroelectric power, etc.).

In order to attain these objectives, it is necessary to assess the quantity and quality of the water resources available in a country at given locations, as well as to draw up an inventory of existing uses and pollution sources. In addition, it is necessary to evaluate and ensure the compatibility of all sectorial water development programmes, projects or plans with the general objectives of the plan, taking into account the effects of climate change.

In the case of non-recharging groundwater aquifers,[4] the objective of a management plan may be orderly depletion, i.e., since the water resources therein are not renewable, aquifer mining over a calculated period of time, until the resources are exhausted. In this event, the plan should contemplate an 'exit strategy,' that is, it should provide an indication of the action to be taken once depletion occurs. This strategy could focus on alternative water sources, or on the relocation of water uses (and users), if feasible.[5]

It is the law which establishes the institutions and legal rules necessary for the preparation, adoption and implementation of water resources plans.

8.3 Types and characteristics of plans

Water resources plans may be of different types.

(i) *national*: a national water resources plan will cover the whole national territory and will have to consider territorial seas, as these are influenced by internal waters. It will cover the formulation, implementation and control of the strategies and criteria intended to ensure a correct assessment of water availability and demand, and to prevent conflicts among water uses. For instance, an inter-basin water diversion at the national level will be provided for through a national plan.[6] There will be one national water resources plan for each nation, for all waters available and all water uses within its boundaries. The national plan must coordinate all internal basin plans;

(ii) *regional*: a regional plan will cover part of the national territory and will be formulated on the basis of studies backed up by the overall planning;

(iii) *basin*: a basin plan is formulated within the framework of a river basin, on the basis of data relating to water availability and quality in the basin and to existing and projected water uses. When the circumstances so warrant, it also covers related groundwater resources. A basin plan normally makes provision for programmes and projects, the implementation of which takes place at the local level;

(iv) *aquifer*: within an aquifer or aquifer system, an aquifer plan aims at balancing groundwater demands and land use pressures with the needs of aquatic and terrestrial

---

4 See Chapter 12, Section 12.4.
5 Nanni, M., Burchi, S., Mechlem, K. & Stephan, R. M. (2006) Legal and Institutional Considerations. *Non-Renewable Groundwater Resources, A Guidebook on Socially-Sustainable Management for Water Policy-Makers* (Foster, S. & Loucks, D.P. eds.). IHP-VI, Series on Groundwater No. 10. Paris, UNESCO, 52.
6 Such as in the case of Spain. Adopted in June, 2001, the national hydrological plan provided for the construction of new dams, canals, pipelines and other infrastructure, and for water transfers from the Ebro river to southeast Spain and from the Rhône river to Barcelona, among others. Due to criticism by environmental and other NGOs, academics, scientists, unions and political parties, and by the European Commission, a review of the plan was initiated in 2004 and the Ebro water transfer was cancelled. A modified version of it was approved by parliament in 2005.

ecosystems, notably wetlands and watersheds, as well as at considering equity factors and the long-term sustainability of groundwater development, both in terms of availability and quality;[7]

(v) *local*: a local plan is concerned with the formulation and hierarchization of the various hydraulic works within sectorial investment programmes in such a way as to comply with the overall strategy of economic and social development plans. Within the scope of each water resources development scheme, local planning includes investigations into the viability of the various projects and, in the case a project is accepted, the construction of the corresponding hydraulic works;

(vi) *sectorial*: a sectorial plan may cover water resources development for a particular purpose (urban water supply, wastewater treatment, etc.), and infrastructure operation and maintenance. It is to be noted that even the sectorial plan is of necessity also a multisectorial plan, as it cannot ignore other resources or elements connected with the sectorial development envisaged in the plan;

(vii) *multisectorial*: a multisectorial plan affects various socio-economic activities. Therefore, it should consider all sectors concerned, as well as the prejudice that the utilization of the waters might cause. This applies generally to all plans;

(viii) *integrated*: an integrated plan should consider all water uses – municipal supply, agricultural, hydropower generation, industrial – as well as the protection from the harmful effects of water and of water quality. Furthermore, it should take into account the relationship between water and other natural resources and the environment;

(ix) *long-term*: in general, plans should envisage long-term objectives (from 25 to 50 years) for decision making, with full knowledge of the implications that a decision might cause in the future. Long-term planning aims at defining the general lines of development of the water management policy, so as to facilitate the implementation of programmes in overall terms;

(x) *medium-term*: a medium-term plan (from seven to 25 years) aims at defining the development of a water management policy in greater detail, in particular by identifying the relationship between water, both quantity-and quality-wise, and the various sectors of the economy and regions of a country, and the technical and financial means required for plan implementation;

(xi) *short-term*: a short-term plan (from four to seven years) aims at detailing and sometimes adjusting medium-term planning goals, based on how implementation progresses, and on changes in the economic situation, on changes in water availability determined by the hydrological régime, on alterations in water demand in relation to forecasts, on the effects of climate change and on the evolution of science and technology;

(xii) *international planning*: international planning is only possible when the states sharing the water resources of one basin are willing to plan these resources jointly. In all other cases, each state will make plans for that portion of the common water resources

---

7 GW-MATE (2002–2005) Groundwater Dimensions of National and River Basin Planning – Ensuring an Integrated Strategy. *Sustainable Groundwater Management: Concepts and Tools.* Briefing Note Series. Briefing Note 10. Washington, D.C., The World Bank.

located within its own national territory;[8] a new dimension has to be considered, keeping in mind the particular geographical position of the state and the legal and institutional framework existing at the international level.

Among the characteristics common to water resources plans, it is opportune to mention the following:

(i) water resources planning must take place within the framework of a continuous and permanent planning process. In other words, without prejudice to its binding force and validity in time, a water resources plan must always be flexible and dynamic, to suit to the changing circumstances determined by the progress of technology, the changing economic and social habits of a nation and the vagaries of the climate;
(ii) there is need for a single authority to regulate water uses and control polluting activities, so as to avoid the conflicts which are likely to be caused by man's action with respect to the same resource, ensure coherent measures and prevent waste of human and financial resources;
(iii) interdependence with other resources. The planning of water resources is linked to the planning of other natural resources connected with the hydrological cycle, which constitutes an indivisible whole. Partial decisions might affect the balance of an ecosystem. In the past, water resources planning was usually the task of engineers, for the simple reason that plans were resource development oriented and these professionals were deemed to have sufficient training for evaluating and comparing the tangible benefits of the solutions proposed. Recently, however, it has been realized that the discipline of engineering alone is insufficient for a correct consideration of all the problems that may arise, in particular those connected with water quality, the need to conserve water resources, climate change, environment protection and the quality of life, so that it is necessary to resort to multidisciplinary teams composed of specialists in the technologies relating to hydraulic works, and specialists in natural, social, juridical, economic and administrative sciences;
(iv) in order to implement a plan, appropriate and efficient institutional arrangements are required. In turn, institutions are set up by legislation. This is why planning is closely associated with the legal and institutional aspects of water resources management.

8.4 The relationship between water resources planning and economic and social sectors

It must not be forgotten that water resources plans are not by themselves an objective, but are part of the system of economic and social development planning at the national level. Thus, water resources are usually dealt with by means of *cross-planning*, which should be carried out in coordination with sectorial planning and regional planning.

Generally speaking, sectorial planning is intended for the preparation, in each social or economic sector, of plans concerned with the achievement of sectorial objectives established

---

8 This is the solution advocated in the EU Water Framework Directive whenever member states are not willing to embark on a common planning process. It remains mandatory, however, to plan for the national territory.

in accordance with overall national objectives. Regional planning is an attempt to define a strategy for the physical planning of the national territory, pointing out criteria for the use of the soil and natural resources, the distribution of population and activities over a given area, the ranking of urban centers, transportation networks, power distribution and other infrastructure. Cross-planning is aimed at the correct assignment of those resources, the availability of which cannot be increased significantly, including human resources and natural resources such as soil, water, air and forests.

Water resources planning is a typical case of cross-planning, which has a markedly coordinating function, due to the fact that water resources are essential to the activities of most of the sectors conditioning economic and social development, and because the execution of hydraulic works calls for investments that must be considered within the scope of sectorial planning. In addition, there are regional involvements dictated by the way in which waters are distributed throughout a territory. The cross-planning of water resources thus ensures an interconnection between the networks of sectorial planning and regional planning.

8.5 Methods for planning processes

Water resources planning should comply with the general *methods for planning processes* and take into account some specific aspects that condition it, in particular the following:

(i) the interdependence, as concerns both quantity and quality, of the different forms of occurrence of water and of the different water uses, which makes it necessary to consider the river basin as the basic unit for water resources planning, although this does not mean that other significant conditions, as for instance those of a historical, administrative, economic or political nature, can be disregarded;
(ii) when groundwater is connected to surface water within the basin, groundwater characteristics and behaviour;
(iii) the possible irregularity of the annual availability of water resources, which makes it necessary to consider, during the planning process, not only the mean values, but also other characteristic values of distribution;
(iv) the circumstance that rivers are both sources of water supply and natural collectors of polluted waters. This makes it necessary to consider the problems of both water quantity and water quality contemporarily;
(v) the distinction that has to be drawn between consumptive and non-consumptive water uses, and the uses that condition water quantity and water quality simultaneously, or only the quantity, or only the quality. In view of their importance as regards the definition of priorities among water uses and the optimization of the benefits that water may bring to the community, these aspects should be considered by water resources planners;
(vi) the need to minimize any harmful economic, social and environmental impact of hydraulic works.

8.6 Relevant administrative and institutional issues

In the past, the public administration institutions concerned with water resources have generally been oriented towards activities directed to the solution of specific problems at the sectorial level. Consequently, water resources development projects have been designed with a rather sectorial objective (for example, agricultural, industrial or water supply), without considering the possibility of multi-purpose utilizations. In parallel with this, water pollution has not been regarded as an issue deserving attention until relatively recently. Focus was placed mainly on development.

8.6.1 Administrative and institutional issues

The administrative and institutional issues associated with water resources planning are as follows:

(i) *policy and objectives*: administrative action must first be directed towards the establishment of clear-cut policy objectives. This means that every state organization must adopt a structure to achieve these objectives. The institutional apparatus must be considered as a whole of which all components play a defined role in this effort. Particular interests are subordinated to sectorial or regional interests, which are themselves subordinated to the national interest;

(ii) *authority and responsibility*: the development of a water resources policy needs institutions with well-defined functions and powers;

(iii) *coordination issue*: it is the aim of an orderly regulation of the efforts to promote joint action for the realization of a common objective;

(iv) *continuity issue*: the structure of an institution must not only serve those activities representing the object of the institution itself, but also ensure the continuity of the operations for the time necessary, and for the duration of the institution;

(v) *centralization, decentralization and deconcentration*: the existence of a single authority for the regulation of water uses implies the existence of a centralized water resources planning administration for what concerns the management of the multi-sectorial actions having common aims contained in the national development plan. *Decentralization* may be defined as the transfer of state functions and competences to other legal persons, either territorial collectivities or institutions at the river basin or other territorial level. *Deconcentration* is the transfer of functions or competences between bodies of the same institution, with a view to providing the administrative action with a wider operational flexibility;

(vi) *administrative unity*: a national water policy must be administered uniformly and in accordance with the criteria established by a modern water resources administration;

(vii) *location of the water resources planning function*: a unit for the coordination of sectorial water resources plans and/or the formulation of a master water resources plan should be created. It should be located within the regulatory institution for water management so as to facilitate national water policy implementation. Whenever feasible, institutional arrangements for resource planning should be devised at the river basin (or aquifer) level to ensure that basin-specific (or aquifer-specific) issues are adequately addressed on-site;

(viii) *upward-downward planning procedure*: in the planning process there is a continuous interrelation between the base and the top level, and vice versa.

Downward planning starts at the top level of the planning system, i.e., where the directives for the formulation of the plan are determined and the national objectives to be attained within a fixed deadline are defined. On the contrary, the upward planning starts from the base, i.e., from the bodies executing the plan at the local, regional or sectorial level. Such bodies must identify specific objectives and goals, in order to find alternative solutions to specific problems.

Compatibility between upward and downward planning may be reached in different ways. The base contributes to the creation of a policy by expressing wishes, needs, values and options (upward planning). The decision makers at the top level explain to the base the problems for which it is necessary to find a solution and indicate the consequences of the decisions to be made and the various options (downward planning).

Public participation in the planning process is now seen as essential to achieving success in plan implementation. Indeed, it becomes easier to implement plan determinations that may affect rights and legitimate interests when plan approval takes place after thorough consultations and considering the views of the various stakeholders and water users.

8.6.2 Other factors contributing to a better administration and planning of water resources

The following factors have to be considered in the administration and planning of water resources:

(i) *institutional aspects*: for a study of these aspects, it is necessary to carry out a critical and comparative analysis of the water resources administration. This makes a global evaluation of the existing institutional framework possible.

The analysis should include a description of:

- the competent bodies;
- functions and powers;
- the political-administrative situation with respect to the organigramme of the central government;
- type of institutional structure: centralized, decentralized, deconcentrated;
- interinstitutional relations;
- effectiveness in the attainment of the objectives;
- administrative conflicts, overlapping of functions, etc.;
- relationship with water users and level of participation by them in management decisions;
- organigramme;

(ii) *legal aspects*: these refer to the analysis of legal texts relating, directly or indirectly, to water resources management. The analysis will provide a complete picture of the existing legal situation, on the basis of which it will be possible to formulate further legislation, as needed;

(iii) *the need for research*: the development of research programmes in all economic (productive, services) sectors, such as agriculture, energy, public health, industry, tourism, as well as in the activities concerning the conservation, protection and improvement of water resources and the environment;

(iv) *cooperation* between different sectors, which may take place:

- through the execution of specific studies and research by each sector directly or indirectly concerned with water resources, according to its competence; and

- through a contribution of money or personnel to those assigned to carry out a study or research programme;

(v) *personnel needs*: the formation of an interdisciplinary group composed of different specialists (planners, engineers, economists, sociologists, administrators, lawyers), as well as technicians and mid-level administrative and technical personnel. The number and qualifications of those taking part in the work should be determined on the basis of integrated water resources management needs.

8.7 Water resources planning under the European Water Framework Directive[9]

Article 13 of the EU Water Framework Directive requires that member states ensure that a river basin management plan containing the information detailed in its Annex VII is produced for each river basin district lying entirely within their territories. In the case of an international river basin district, the member states concerned must make efforts to cooperate with a view to developing a single basin plan when the district falls entirely within the boundaries of the EU, or to arriving at a single basin plan for the portion of the district located in the EU when the river basin extends beyond EU boundaries, if it is not possible to produce a plan for the entire river basin district.

As may be inferred from the directive, the main steps of the planning process are:

(i) identification of the river basin districts and designation of the competent administrative authorities for coordinating activities;
(ii) assessment of the current status of surface water, groundwater and protected areas in the basin districts and preparation of an inventory of resources, including the registration of protected areas;
(iii) identification of significant pressures (water uses, land uses and other activities) and assessment of their impact;
(iv) economic analysis of water uses;
(v) setting up of environmental objectives to improve the status of water resources and prevent their further deterioration;
(vi) establishment of monitoring programmes to assess which water bodies are at risk of not meeting the objectives, identify the reasons for this failure and facilitate the design of management strategies and measures;
(vii) analysis of gaps between the current status of water bodies and the environmental objectives;
(viii) establishment of programmes of measures, and in particular of the basic measures[10] that need to be implemented in order to achieve the objectives defined by the management plans by 2015, as established in the directive;

---

9 For an overview of the European legal framework for water resources management, see Chapter 6, Section 6.4.5.
10 Under Article 11 of the Water Framework Directive, basic measures are those required to achieve the following: water protection, cost recovery, efficient and sustainable water use, protection of drinking water quality, control of abstractions, control over groundwater recharges, regulation of wastewater discharges, control of diffuse pollution, control of the hydromorphological conditions of water bodies, prohibition of direct discharges into

(ix) development of a basin plan, with the participation of the public and the active involvement of stakeholders in the production, review and updating of the plan.

Public participation is a key ingredient of river basin planning. Article 14 of the directive calls for the active involvement and consultation of all interested parties at various stages of the process, a first step being the definition of the time schedule and work programme for plan preparation, which have to be published and made available for public comments. As a second step, inputs by the public are required for the identification of significant water management issues in the basin, to be addressed through the plan. Finally, the view of the interested parties are to be sought with regard to the draft river basin management plan. The manner in which public consultation and involvement is to take place is not indicated in the directive, since each EU member state follows its own methods.

River basin planning is conceived as a dynamic process, to be kept under constant review. Therefore, the directive mandated EU member states to publish their river basin management plans in 2009 and, thereafter, to review and update them every six years, each six-year period corresponding to a planning and management cycle. The first planning cycle ended in 2015 and, by now, all member states have updated their plans.

References

Goodman, A.S. (1984) *Principles of Water Resources Planning*. Englewood Cliffs, New Jersey, Prentice Hall Inc.
Helweg, O.J. (1985) *Water Resources Planning and Management*, John Wiley & Sons.
United Nations (1964) *Manual of Standards and Criteria for Planning Water Resources Projects*. Water Resources Series No. 26. New York, United Nations.
United Nations (1956) *Multiple-Purpose River Basin Development*, Part I, *Manual of River Basin Planning*, Flood Control Series No. 8, New York, United Nations.
Vega da Cunha, L. (1983) *Water Resources Planning and Water Use*. Paper presented at the European Institute for Water, Seminar 'Water in Western Europe', 4–8 July 1983.
United Nations (1988) *Water Resources Planning to Meet Long-term Demand: Guidelines for Developing Countries*, Natural Resources Series No. 21. New York, United Nations.
Wessel, G. (1991) Water Awareness, Planning, Barriers and Communication. *Water Awareness in Societal Planning and Decision-Making*. Stockholm, Swedish Council for Building Research.

---

groundwater (except for authorized injections), abatement of surface water pollution by priority substances, prevention/reduction of leakages and accidental pollution.

Chapter 9

National water resources administration

9.1 Introduction

To a large extent, the success of a water resources management policy depends on the existence and smooth functioning of institutions within an overall government framework. These institutions carry responsibilities concerning the political, technical, economic, social, financial and legal aspects of water resources management.

Generally speaking, a 'water resources administration' is any form of institution or organization responsible for the management of water resources, either sectorially or globally. The main purpose of a water resources administration is to ensure the successful implementation of a government water resources policy and to achieve, according to the stated policy, the most 'economic' and/or 'social' and/or 'rational' use, development, protection and conservation of the water resources available in a country. Incidentally, each one of these expressions emphasizes a certain resource management aspect. The most 'economic' use is the most justifiable from an economic standpoint, but does not necessarily include 'social' or other non-economically quantifiable values (environmental, for example) and is not necessarily sustainable. The most comprehensive expression seems to be 'integrated management' of water resources, which is intended to cover surface water, groundwater, quantitative and qualitative aspects and all water uses, and considers the relationship between water and other natural resources.

9.2 Different types of water resources administration

Water resources administrations and institutions are of different types and may be considered from different standpoints according to their powers, functions, water uses and activities, territorial level of operation, legal status and degree of autonomy.

Various types of water resources administrations are indicated herebelow, with a brief description of their characteristics.

9.2.1 Institutions according to their powers

The major powers which an institution may exercise are of a (i) political, (ii) executive, (iii) technical, or (iv) judicial nature. The differences between these powers are clear. An institution may hold several of these powers:

(i) *political powers*: generally, political overall decision-making powers are reserved to the highest level of administration, i.e., a council of ministers, or a minister, state secretary or national commissioner having the rank of minister. These powers are indispensable to make high-level policy decisions, the implementation of which is the responsibility of the institutions exercising executive or technical powers.

(ii) *Executive powers*: these powers refer to executive actions which are taken as regards the sectorial aspects of water resources management, or at lower levels of administration. They are derived from the political power, which dictates the policy decisions to be implemented.

(iii) *Technical powers*: technical powers are those which are exercized by high-ranking officials of the administration, supervised by the executive power. These generally are engineers, hydrologists, hydrogeologists and other professionals within a ministry, and are often posted at various locations within the country.

(iv) *Judicial or para-judicial powers*: these are intended to include the power to monitor compliance with, and enforce, water legislation.

Often there is inadequate coordination among the ministries, departments or other autonomous authorities responsible for specific or sectorial aspects of water resources. Coordination is essential if sound management of the water resources is to be achieved.[1]

9.2.2 Institutions according to their functions

The institutions concerned with water resources management may be distinguished according to the functions they are called upon to carry out. These functions may include:

(i) inventory (of surface water, groundwater, atmospheric water, water quality), as could be the case of a meteorological department for data collection, or of a groundwater department which is often combined with a mineral department, or of a specialized department attached to a ministry of environment;

(ii) research, such as a water research centre or institute;

(iii) policy formulation, as might be the case of a national water council at the highest political level;

(iv) coordination of water-related activities, as in the case of a national water commission at the highest technical level;

(v) administration of water rights, as is the case of a water rights division or other administration responsible for the management of water use permits and concessions, whether at the national, basin, or other territorial level of administration;

(vi) water resources planning, which may be performed by a central/national water management department by itself, or in cooperation with other planning units of other ministries, or by a river basin institution;

(vii) water resources development, use and conservation, such as a resource development agency, which may also be a basin development agency, or a provider of water services;

---

1 See this chapter, Section 9.3.1.

(viii) operation, maintenance and control of waterworks (a waterworks agency, department or company);
(ix) monitoring (of surface water, groundwater, water quality), such as in the case of a meteorological department, or a specialized department of an environment or health ministry or, in some cases, that of a sectorial ministry;
(x) inspection, such as in the case of inspection units; and,
(xi) settlement of disputes, such as in the case of water tribunals or other courts, to adjudicate questions on water.

9.2.3 Institutions according to uses

A classification of institutions based on economic activities of relevance to water resources management is also possible. The activities in question may have to do with domestic and municipal uses, agricultural and industrial uses, fishing, hydroelectric production, transport, public health, land use planning, water quality control and environment protection. The same institutions may also have responsibility for dealing with the harmful effects of water, such as flood management, riverbank protection, drought and erosion control.

The institutions responsible for particular water utilizations may include a municipal water supply and/or sewerage authority or company, a national, para-national or private electricity company or authority, a land development and irrigation agency, an industrial development corporation or a river navigation company.

9.2.4 Institutions according to their territorial level of jurisdiction

The following territorial levels of jurisdiction may be mentioned:

(i) the *national level*, such as in the case of a national ministry, a national electricity agency, a national irrigation department, or a water supply agency responsible for the provision of domestic water to a whole country;
(ii) the *intermediate level*, including various levels between the national and the local ones, such as:

- the *interstate, interprovincial, interregional, or state, provincial and regional level*. This is the case of agencies or bodies having authority at the interstate or state (or interprovincial or province) level within a federation, or across two or more regions, or within a region of a unitary state. The Tennessee Valley Authority of the USA and the Kitakami River Basin Authority of Japan are examples. Other institutions of this type include the Geneva Water Supply and Sewerage Authority of Switzerland, the Mahaweli Authority and Gal Oya Development Board of Sri Lanka and other similar institutions.
- The *inter-basin or basin level*. These institutions perform water resources management or development functions with reference to a basin or a group of basins. Water resources management and development institutions such as the basin agencies of France, Spain, Nigeria, Swaziland, Tanzania and a number of other countries, and the water companies of the U.K., belong to this category. It is to be noted that at the basin level not all agencies have the same powers. For instance, the water companies of the U.K. are responsible for water supply and sewerage, while

the basin agencies of France and Spain perform all the functions necessary for effective water resources management. When dealing with these basin institutions, provision should be made for adequate coordination between them and the central government.

- The *aquifer level*. Depending on the size of an aquifer or aquifer system, the jurisdiction of these institutions may extend across administrative boundaries, such as in the case of the Great Artesian Basin (GAB) Coordinating Committee, an advisory body which covers the Australian states of Queensland, New South Wales, South Australia and the Northern Territory,[2] or be limited to more localized hydrogeological units. This is the case of the groundwater districts in the USA, of the 'technical groundwater committees' of Mexico (COTAS) and of the groundwater management committees that have been set up in Australia to deal with groundwater management planning.[3] A common feature of these institutions is that representatives of water users participate in them.
- The *sub-basin level*. While a single agency may be made responsible for the administration and management of various small basins, sometimes the management of an important deltaic area such as New York, Tokyo or Rotterdam may call for the setting-up of a specific institution based on hydrological criteria other than that of the basin. Also, in the case of very large basins such as the Mississippi in the USA, a sub-basin management institution could be a rational alternative.

(iii) The *local level*. At this level, water resources management institutions are responsible to the local authorities (municipalities, district officers or village authorities), and their powers are more or less extended. Under this category we may find municipal water supply and/or sewerage and/or urban sanitation authorities, companies or institutions, and customary water supply and management institutions (for water points, public rural wells, irrigation districts, etc.). Water users' associations and cooperatives also operate at this level.

(iv) The *international level*. Institutions may be created at the international level to handle matters relating to the utilization and protection of 'shared,' 'international' or 'transboundary'[4] water resources. To this category belong international basin commissions such as those for the Rhine and Danube in Europe, the Mekong in Asia, the Senegal, Nile, Niger and Zambesi in Africa, or the commissions for boundary waters between USA and Canada and USA and Mexico in North America, and the commissions for the Rivers Uruguay and Plata in South America.[5] Sometimes the decisions of these commissions influence water legislation at the national level.

---

2 See Chapter 6, Section 6.2.3.1.
3 Burchi, S., & Nanni, M. (2003) How Groundwater Ownership and Rights Influence Groundwater Intensive Use Management. *Intensive Use of Groundwater, Challenges and Opportunities*. Llamas, R. & Custodio, E. eds. Rotterdam, Balkema, 235 ff.
4 'International' water resources refer to water bodies which, even if located within national boundaries, are part of a system of water which at some point is intersected by a boundary, while a 'transboundary' water body is itself intersected by a boundary.
5 See Chapter 13.

9.2.5 Institutions according to their legal régime

Every institution is created and operates within a legal framework which is set up by law and which may be referred to as its 'legal régime.'

The legal régime of water institutions may be of three types, as follows:

(i) *fully governmental*; this is the case of government ministries, directorates, departments, and of their branch offices;
(ii) *semi-governmental*, or *mixed*, or *para-governmental*; in this case, the institutions are composed of government and non-government representatives and interests. Often, these institutions are set up in order to ensure financial autonomy, flexibility in operations and minimum interference by the political power in day-to-day operations. The salaries of semi-governmental or mixed company officials are generally higher than those of government officials. A large number of water companies are of this type, including water supply and sewerage companies, hydroelectric and other power companies, and navigation companies. Irrigation undertakings rarely take this form, probably because their revenues are less attractive to investors;
(iii) *private*; in some countries, sectorial, use-related aspects of water resources management are handled by private companies, associations or corporations. This may be the case of water supply, sewerage and hydroelectric companies, irrigation undertakings and the like. In many cases, individual water users join forces to form water users' associations. These associations may take different names. In this connection, it is worth mentioning customary water users' associations. Often these associations operate like private companies and enjoy the legal status of a juridical person, as opposed to that of a physical person.

These types of legal régime may be found at all levels of administration, such as in the case of water utilities, which may be national in a country of a limited size like Belize, or in a small island state, or municipal as in the majority of cases.

9.2.6 Special water development agencies

Special water resources development agencies may be created to promote or carry out the development of land, water and other related resources. This category also covers those agencies that provide water-related services, such as a water supply and sewerage entity, or an irrigation authority. These agencies may be either governmental or para-governmental, and generally enjoy a large degree of economic and financial autonomy. They are concerned with the use of water for one or more purposes in a particular region, basin or sub-basin. Water legislation should contain provisions with respect to their constitution, their relationship with the government (and/or basin) water resources administration, their degree of autonomy and their dissolution.

The practice of allowing special development agencies to grant and administer water use rights should be discouraged, because these agencies develop and use water and it follows that it is not wise to let an entity play the role of the poacher and that of gamekeeper at the same time. However, it may happen that there is no alternative other than allowing an agency to do so, due to lack of manpower or financial resources in the overall administrative apparatus, or for other causes. This has been the case of the Mahaweli Authority of

Sri Lanka, for instance. In the event, it is important to establish firm coordination with the national (or basin) water rights administration.

The national water resources administration should consider special water resources development agencies as primary water users to which a certain amount of water has been allocated in concession for one or more purposes with, eventually, the further right to grant part of that water (or hydropower produced) to secondary users. A complete autonomy of these agencies is not conducive to successful administrative control over water resources on an integrated basis.

Examples of these agencies are the Tennessee Valley Authority (USA), the Damodar Valley Authority (India), the Kitakami Development Agency (Japan), the Mahaweli Authority of Sri Lanka, and the Compagnie Générale du Canal de Provence (France).

9.2.7 Water users' associations

Water users' associations represent the best means:

(i) to combine many water users into one group at the final stage of water distribution;
(ii) to provide a solution to conflicting customary and traditional water rights and ensure equitable water distribution;
(iii) to facilitate the operation, maintenance and rehabilitation of infrastructure, and the recovery of the costs relating or incidental thereto; and
(iv) to facilitate the implementation of water legislation through the active participation of water users in the planning and management of water resources.

Water users' associations may be *de facto*, meaning traditional and customary, or *de jure*, that is, created under the provisions of water or related legislation.

De facto associations of water users are those which have been in existence since time immemorial, particularly in arid or semi-arid areas, where the available water resources must be utilized in the most efficient way.

Traditional water users' associations are governed by well known customary law, which sets detailed rules for individual water rights in terms of quantity, time, purpose of utilization, obligations as regards the construction, operation and maintenance of canals and other structures and, finally, for the adjudication of water rights disputes at the users' level. In countries where written water legislation is not known or is neglected, traditional associations are particularly relevant with respect to rural water supplies for domestic uses and irrigation.

Water or other appropriate legislation should favour the establishment, organization or reorganization of water users' associations. These associations should be recognized as autonomous bodies possessing legal personality and financial autonomy. Nevertheless, the legislation should reserve to the water administration the right to supervise and advise them in order to ensure that they comply with financial and other legal rules, and to take action in case of wrongdoing, or when it is deemed necessary to safeguard larger public interests.[6] In

---

6 However, in order to prevent possible abuse, it is advisable to provide a clear-cut definition of the concept of 'public interest.'

cases specified in the legislation, the water administration should be empowered to dismiss the management board of a water users' association and to call for new elections.

In addition to regulating the formation, dissolution and organization of water users' associations, the legislation should promote the participation and involvement of these associations in the process of water resources planning and management. To facilitate this involvement, provision should also be made for the possibility of combining two or more associations into larger ones.

Users' associations may take the form of cooperatives, consortia or unions of water users. They are established for achieving better water utilization and distribution (irrigation associations, irrigation districts or areas, farmers' associations or cooperatives, watershed management associations, groundwater management associations, farm management associations, etc.), or for cooperating jointly and more efficiently against damage caused by water (drainage boards, districts or areas, farmers' drainage cooperatives, flood control boards, associations or cooperatives, associations of hydraulic interest, etc.), or for any other purpose.

The problems connected with the use of water for grazing purposes by nomadic or semi-nomadic populations may find a solution within the framework of water users' associations. The people concerned could unite into an association to which a certain geographical area with well defined watering points is recognized in a written water use permit. However, whenever a well serves both the purpose of satisfying drinking needs of a village settlement and that of providing water for cattle during a certain period of the year, precautions should be taken in order to avoid conflicts between the villagers and the nomads.[7]

9.3 Major issues of water resources administration

The many issues arising in connection with the setting up or reorganization of the institutional framework for water resources management are a major preoccupation, and solutions are being sought both in developing and industrialized countries. Only the most relevant and recurrent issues are discussed here, with reference being made to recent literature on the subject.

It should be underlined that the undertaking or implementation of institutional framework reform depends not only on historical, ethnic, cultural, religious, geo-physical and legal factors, but also on the political will.

The issues described here refer to matters relating to the need for coordination, the degree of centralization, decentralization and deconcentration, the need to set up or strengthen an administration responsible for the administration of water use rights and, finally, the need to establish a 'regulatory' water resources institution in parallel with the existing 'development' agencies, i.e., those responsible for the actual exploitation and development of water resources for various purposes of use.

9.3.1 The need for coordination

In most countries, numerous institutions carry responsibilities concerning sectorial aspects of water resources management. These institutions may include ministries of public works,

---

[7] Conflicts of this kind sometimes arise in countries of the Sahel, such as in the case of Niger.

agriculture, industry, energy, public health, as well as ministries of environment and natural resources. Each of these ministries operates at the national and intermediate (regional, provincial) levels and represents a centre of decision. Often, sufficient coordination among institutions is lacking, the most serious consequence of this being the absence of a consistent water resources policy, together with the delays attendant upon the approval of projects.

Planning the development, protection and conservation of water resources requires an integrated approach in order to obtain as much knowledge as possible of the availability and quality of water resources (inventory), existing uses, and future needs.

Thus, it follows that water resources should be brought under integrated or coordinated management, which would include,

(i) a centralized water resources inventory;
(ii) a unit for clearing or preparing water resources plans, programmes and projects; and
(iii) a water rights administration for the issuance and administration[8] of water use permits and concessions and, possibly, also wastewater discharge permits.

9.3.2 The question of centralization, decentralization and deconcentration of the water administration

For the sake of clarity, a definition has to be given at the outset for the terms employed in this section.

While the concept of *centralization* entails that all powers are exercized from the centre, that of *decentralization* implies the transfer of powers from a central or national institution to another institution which, in this case, can act autonomously. India, Germany, Switzerland, Argentina, the USA and, in general, federal states, are examples of decentralization.

By *deconcentration* is generally meant the delegation of powers by one institution to branch offices located elsewhere, in order to facilitate water resources management, even though the principle followed is that of governmental centralization. Indonesia and the Philippines are examples of countries in which a good degree of deconcentration is appropriate.

In the case of a water resources administration, the question of the degree of centralization, decentralization or deconcentration is one of the most delicate. While the problem is ubiquitous, no clear-cut formula exists for its solution or is valid for any one country, as the choice of alternative solutions varies from one extreme to the other.

The idea of the centralization of water management activities is appealing, but cannot be applied in all cases.

The degree of centralization depends on many factors:

(i) the historical/cultural framework: the USA, for instance, was born as a fully decentralized nation where powers, originally vested completely in the individual, have been progressively handed over to the state and to the nation;

---

8 This term is meant to include, in addition to the issuance, the modification, suspension, termination and renewal of water use permits.

(ii) geographic/physical conditions: large countries such as India, Indonesia, the Philippines and Brazil are examples where full centralization is impossible, as opposed to Mauritius, Togo, The Gambia, Botswana, Andorra, Monaco and S. Marino;
(iii) the legal and institutional framework dictating the administrative organization of a country (a federal state or a unitary state);[9]
(iv) functions which necessarily have to be decentralized or deconcentrated. When discussing the degree of decentralization or deconcentration of an institution, it is important to know that certain functions may easily be – or even must be – deconcentrated (such as the screening of applications for water use permits or concessions), while others should not (such as the approval of important water resources projects);
(v) the political will.

As regards the planning function, the ideal decentralization for optimum water resources management should be at the basin level, but here, again, this is not always possible in view of:

(i) the size of a basin, either too large (Mississippi, Ganges, Amazon), or too small; a basin may include the whole country, such as in the case of the Mekong while flowing through Cambodia;
(ii) inter-jurisdictional conflicts between state governments, regional governments or federal governments;
(iii) the presence of underground basins – aquifers and aquifer systems – which may or may not correspond, or be related, to surface basins;
(iv) the presence of other relevant hydromanagement areas (New York, Tokyo, Calcutta, Rotterdam, Rome, etc.), which may require a different treatment from that of a basin.

9.3.3 The water rights administration

Within the governmental water resources management machinery, a water rights administration is a unit responsible for the granting of rights to use water and of permits for wastewater discharge and, in general, for the implementation and enforcement of the legal provisions on water rights. From an administrative viewpoint, it is possible to regulate all activities connected, directly or indirectly, with the various water uses and types of wastewater discharge through a water rights administration.

Since water is a vital natural resource, it is necessary to consider it in its entirety, but in many countries, a water rights administration does not exist, or performs sporadically, sometimes through different government departments not having an effective coordination. A water rights administration makes it possible to know the amount of water utilized or discharged, by whom and where, and about water quality.

While water resources development, protection and conservation activities may be carried out by different departments, agencies, private individuals or undertakings, the allocation of water as to its amount, purpose and timing should be centralized and deconcentrated,

---

9 The federal or unitary status of a country has to be considered with regard to management functions in the water resources sector. Federal countries include the USA, Argentina, Brazil, Canada, Mexico, the F.R. Germany, Switzerland, Nigeria, India and Australia, while most European, African and Asian countries are unitary.

according to circumstances, at the state or at the basin level, or, in the case of large basins, at the sub-basin level.

In order to facilitate water rights administration so that water users are not obliged to travel long distances, consideration should be given to the idea of delegating this responsibility to the appropriate local level. It is possible also to give to a local authority the power to grant minor permits (to be specified in the law), while in any case the central or basin administration retains for itself the power to grant permits for larger diversions. The basin and central water administration should always be informed of the permits issued at the local level and record this information in the appropriate database. The power to grant permits for inter-basin diversions should remain centralized.

Water use permits, concessions and wastewater discharge permits should be recorded in a special register of water rights, kept by the water rights administration with reference to distinct river basins within the country concerned. Subsequent modifications to permits and concessions should also be entered in the register. Together with the inventory of water resources and their quality, the register should be used as the basis for water resources planning.

Although the concept of a water rights administration is universally valid, each country has its own characteristics and needs. Therefore the question as to how to set up a water rights administration should be studied on a case-by-case basis.

9.3.4 The need for a water resources 'regulatory' institution

9.3.4.1 Definitions

Generally speaking, there are two kinds of institutions concerned with the management of water resources, of which one may be termed 'regulatory institution' and the other 'development institution.'

The distinction between these two types of institution lies in the fact that a regulatory institution is more of a policy-making body which coordinates the activities related to water utilizations undertaken by individuals, development institutions or governments, while the main objective of a development institution is the actual development and utilization of water and, in some cases, of water-related resources, such as land. This includes the construction of waterworks and hydraulic structures.

Development institutions often perform functions which are regulatory or coordinating in character, but these will be secondary to development activities and will be undertaken only when necessary to facilitate the latter. Conversely, regulatory institutions sometimes perform development activities, generally as a secondary or residual function, in limited areas or for a limited time.

9.3.4.2 Major objectives and functions of a regulatory institution

In order to achieve the most rational and sustainable use of water resources, any basic regulatory institution, whether national, international or basin, should be vested with the following functions:

(i) to formulate the overall water resources policy and plans covering all waters, whether surface, underground or atmospheric. This function would include, in particular, the preparation of water resources plans at the national and basin levels providing for the

satisfaction of the maximum possible uses and demands, depending on water availability, and for the protection of water quality, irrespective of administrative boundaries. Environment protection requirements, including the maintenance of minimum flows, should be included in the plans. Sectorial plans should be fitted into these plans;
(ii) to coordinate all projects, schemes or plans for the use of the water resources, so as to establish a rational and harmonious scheme for multiple, integrated and sustainable water use, thus economizing on costs. For instance, a dam located on the upper reaches of a river might be constructed to serve several purposes at the same time: hydropower, irrigation, regulation of the water flow downstream for the purpose of facilitating further use, navigation and fishing. The dam could also help to control floods and alleviate droughts by storing excess water as a reserve for release during dry periods;
(iii) to act as the depository and administrator of all data relating to water availability and quality, and to coordinate data collection and processing within the framework of appropriate resource monitoring programmes. This would be the equivalent of creating an inventory of the available resources, out of which future uses, and/or protection measures, may be planned. Data for this inventory should be collected and processed in accordance with standard methodologies, and should be filed by river, lake or water body, and by basin, aquifer or other hydrologic management area;
(iv) to determine the respective rights and obligations of users (individuals, public institutions, corporations or states) with regard to the use, conservation and protection of water resources. In order to plan the management of present and future water resources, it is essential to know the extent to which such resources have already been put to use, for what purposes and where. This becomes easier if a register of water rights is kept. By means of a comparison between the total amount of water available (inventory) and the existing uses (existing rights for utilizing water), it is possible to determine how much water is available for developing or permitting new water utilizations. It is obvious that a new project, if not coordinated and planned taking into account existing utilizations, may frustrate the implementation of other projects. This is the function of a water rights administration;
(v) possibly, to regulate the activities of polluters through the issuance and administration of wastewater discharge permits. These permits and their subsequent modifications should be entered in the register of water rights, so as to facilitate access by the water resources administration to the information needed for planning purposes. In many countries, however, water quality-related functions are in the hands of environment protection agencies acting autonomously, without coordination with the water resources administration, thus generating considerable difficulties whenever a water resources management plan is to be prepared. Therefore, arrangements for cooperation and the exchange of information between the environment protection agencies and the water resources administration are indispensable;
(vi) to study, with a view to their approval and subject to modification if needed, those activities, projects and plans which could significantly modify the quantity and/or the quality of the waters, or affect existing projects and rights, or the hydrologic régime of a river basin. The duty of the institutions or persons (physical or legal) intending to implement these activities, projects and plans to submit them beforehand to the regulatory institution should be provided for in the relevant legislation.

The existence of a regulatory institution does not preclude that of development institutions carrying responsibilities with regard to the actual development of certain areas, or the utilization of the basin waters and other resources. This is the case of the Gezira Board in the Sudan, of the Lake Basin (Lake Victoria) Development Authority in Kenya and of the Lake Nasser Development Authority in Egypt. It is advisable, however, that these development institutions be closely coordinated with the regulatory institution. This holds true both at national and international levels.

9.3.5 The role of water law in institution building

Water law addresses the problem of institution building or reorganization at every level and purports to define the duties, powers, functions and degree of decentralization or deconcentration of the administration down to, and including, the associations of water users and special autonomous water development agencies. Sometimes, separate legislation contains provisions to this effect, thus allowing for more flexibility in the case that institutional changes are required.

The water law or other special legislation must provide for the establishment of a regulatory water resources administration or, should this be unfeasible, at least require coordination among the various institutions and agencies dealing with sectorial aspects of water resources management. The legislation providing for the water resources administration should define the duties, powers, functions and degree of decentralization or deconcentration of the administration. Naturally, any decision would depend upon the availability of manpower and financial resources.

9.4 A possible institutional solution

There are multiple possible solutions which may be adopted to address the numerous issues arising in connection with the institutional aspects of water resources management, since, as has been indicated, these issues cross vertically and horizontally all sectors of a country's administration. For each solution there are advantages and disadvantages. In addition, detailed solutions can only be devised on the basis of the particular circumstances and possibilities prevailing in each particular country.

In order to secure efficient management and control of water resources, coordination among all those ministries, departments and agencies responsible for sectorial aspects of water resources management must be established. This coordination, however, must be not only institutionalized, but made obligatory, and must operate at all levels of the administration: national, regional, basin, aquifer, local and international.

9.4.1 Institutions at the national level

Given the multiplicity of institutions responsible for the sectorial aspects of water resources, political and technical coordination is essential. On the one hand, the involvement of the highest political level, i.e., the ministerial level, facilitates the making of decisions on important issues and interministerial cooperation without making the individual ministers feel threatened in their sectorial responsibilities over water. On the other hand, the involvement of the technical and economic levels, i.e., of the senior administrators of the sectorial aspects of water resources management, is a requisite for promoting interministerial cooperation at their level.

Thus, a possible institutional setup could be the establishment of a national water resources council and of a national committee or commission, as are described below.

The name under which these may be designated (council, commission or committee, for instance) is irrelevant and will depend on the particular legal and institutional framework of the country in which these institutions have to operate.

9.4.1.1 A national water resources council

At the political level, this council should be composed of the government ministers having a sectorial responsibility with regard to water resources. It should ensure coordination at the highest political level and constitute a kind of sub-council of ministers on water resources. The council should have the power to make decisions:

(i) on a national water resources policy;
(ii) on the adoption of a national water resources plan and/or basin water management plans;
(iii) on the allocation of funds for investment in the water sector;
(iv) on reimbursement policies, the payment of water charges or rates and other financial questions concerning water;
(v) on all questions submitted to it by the national water committee or commission; and
(vi) on any other high-level matters concerning the implementation of the government water policy, including pollution control and environment protection.

9.4.1.2 A national water committee or commission

In parallel with the national water resources council of a political nature, a national water resources committee or commission of a technical and economic nature should be established, which should be composed of high-level officials, technicians and economists with responsibilities concerning sectorial aspects of water matters within the different ministries and other institutions and representatives of major water users' associations. The committee or commission would have to ensure, at the national level, an institutionalized cooperation, as well as interministerial coordination, from the technical and economic standpoint. Its functions may be advisory or executive.

For the purpose of ensuring that the decision-making process becomes obligatory, the law establishing this institution should specify the quorum for decision making and the binding force of the decisions arrived at. The law could also specify that if agreement is not reached at this level, the matter will be submitted to the national water resources council (political body) for decision.

In many countries, interministerial water committees or commissions already exist; however, it can happen that, because of the weakness of their statutes or because of their young age, or their sectorial approach (interministerial water supply committees, for instance), their impact on overall water management coordination requirements is either inconsequential or non-existent.

9.4.1.3 A central water administration

The law should also consolidate or establish a coordinated or, preferably, a unified water resources administration at the national level. This water administration should be vested with the following functions:

- implementing, in its name and on behalf of the national water council and committee (or commission), the decisions taken by them;
- serving as a technical and economic body with the power to recognize, allocate or reallocate, and administer, rights to use water for any purpose, or to discharge wastewater, through a system of permits and concessions;
- evaluating and coordinating different projects before authorizing their execution;
- acting as the depository and administrator of hydrologic, hydrogeologic, water quality and other water-related data; correlating, standardizing and publishing all information and data relating to water resources;
- preparing the national water resources management plan and coordinating the preparation of river basin management plans, or preparing the latter plans when the institutional capacity is inexistent or insufficient at the basin level;
- promoting or preparing regional water resources plans;
- controlling, authorizing (or executing, if so decided), individual projects; and, finally,
- monitoring all water resources and water-related activities, or coordinating the implementation of the relevant monitoring programmes.

In so doing, the water resources administration is in a position to ensure existing and future uses of a clearly defined amount of water and water quality, taking into consideration the government water policy, the availability of water resources, the existing uses and individual rights, the foreseeable future requirements and the protection of the public interest and the environment in general. For this purpose, the water administration should possess all the necessary powers and discretion, within the limits established by the basic water law.

The water resources administration should constitute the executive organ of the national water resources council and committee (or commission). It could be located either in a ministry of water resources, or in a ministry of the environment or of natural resources, or in the ministry with the largest spectrum of responsibilities with regard to water management, or else, in the case of interministerial rivalry, in the prime minister's office, in the president's office or in the ministry of planning; its location, composition, chairmanship and other elements should be decided on the basis of the country's requirements.

9.4.2 Institutions at the regional, basin, sub-basin and local levels

9.4.2.1 At the regional level

Institutions at the regional level should be designated either as deconcentrated departments of the central water administration or as decentralized organizations. In any case, they will be responsible for a given zone.

9.4.2.2 At the basin or sub-basin level

A river basin agency may also be designated as a deconcentrated or decentralized entity. In some countries, river basins may be as vast as the country itself, and there may be no need to devise basin-specific institutional arrangements, as water-related interests at the basin level coincide with water-related interests at the national level. The same may apply in the case of small countries, where, in spite of the existence of several distinct river basins, the short distances and small population do not warrant the establishment of river basin entities, as basin management functions may be performed at the national level. In larger countries, however, river basin organizations may greatly facilitate water resources management and planning. Their level of autonomy from the national water administration and the functions which they are called to perform will vary from country to country, depending on circumstances. However, responsibilities relating to inter-basin water transfers, or reflecting a national interest, should always remain vested in the appropriate institution at the national level.

9.4.2.3 At the aquifer level

The jurisdiction of aquifer-level institutions may vary, depending on what is intended to be covered. In the case of the Great Artesian Basin (GAB) in Australia, for instance, the GAB Coordinating Committee covers several states and the Northern Territory. In other cases, however, a choice is made to limit the jurisdiction of groundwater management institutions to localized situations, such as in the case of the groundwater districts in the USA. Functions and powers may also vary. The GAB Coordinating Committee, for instance, is an advisory body, while some USA groundwater management districts are vested with broader powers, such as that to limit pumping.

9.4.2.4 At the local level

Local institutions may be created at the municipal, district, department or other level. Their activity should be either integrated or, at least, coordinated at the basin, sub-basin or aquifer level, or at the level of another hydrological unit. Precaution should be taken to avoid the splitting of water resources management responsibilities, which should be exercised as far as possible at the basin and aquifer level, with little or no interference from artificial administrative subdivisions.

Water users' associations exist in many countries, particularly in those where water uses have developed as a result of private initiative. In general, these associations are very effective but are slow to adjust to a regulatory framework imposed by the central government. Their creation is often useful for modernizing traditional customary water administrations with respect to water rights and, in particular, to groundwater rights.

9.4.2.5 At the international level

In the presence of an international basin or aquifer it is necessary to create an adequate institution, such as a basin or boundary or river commission, or a consultation mechanism for the aquifer for the purpose of promoting international cooperation and facilitating international financing.[10]

---

10 See Chapters 12 and 13.

9.4.3 Conclusion

The powers and functions of a water resources administration may vary from country to country, depending on government policy, technical and other local conditions, and on the availability of human and financial resources.

The first priority seems to be the need to have coordinating institutions.

The second priority should be the creation or the strengthening of a regulatory water resources administration, keeping in mind that a regulatory institution should avoid performing development functions such as the construction, operation and maintenance of hydraulic works.

In parallel with this, efforts should be made towards an operationalization of the concept of water resources management by hydrological (or hydrogeological) unit, which is often mentioned in the preamble of water laws enacted during the past decades, but may encounter difficulties in its implementation. No golden rule applies to the design of institutional arrangements for river basin (or aquifer) management, as decisions as to whether a river basin agency should be established will depend on circumstances which vary from country to country. Nevertheless, it should be recognized that river basin organizations greatly facilitate the development and implementation of plans, as well as the implementation of measures aiming at the conservation and protection of water resources and related ecosystems. Water resources management by river basin has become mandatory in the European Union.[11]

The third priority should be the establishment by law of a clear link between the regulatory water resources administration and those institutions responsible for the actual development of specific water resources projects and works.

Although any solution will depend on local circumstances, it is not advisable to augment the number of agencies and institutions dealing with water resources management, except when the operationalization of the river basin management concept so warrants. Matters relating to the regulation of water resources use, conservation and water quality control are best dealt with by a single administration responsible for water resources management. Nonetheless, it is necessary to establish a close cooperation among the appropriate authorities at different territorial levels (national, basin, regional, provincial, local). The water legislation should clearly indicate the respective responsibilities of the administrations involved in water resources management aspects, and provide these with the powers necessary to facilitate the implementation of water resources policies and plans, in accordance wih their own functions.

There is no standard framework for an institutional setup which may be adopted in all cases. The problems here stressed will help circumscribe the complexity of the administrative issues and possibly facilitate their comprehension and study before specific decisions are taken.

---

11 As mentioned in Chapter 6.

References

Hodgson, S. (2003) *Legislation on Water Users' Organizations, A Comparative Analysis*. Legislative Study No. 79. Rome, FAO.

Hodgson, S. (2009) *Creating Legal Space for Water Users' Organizations: Transparency, Governance and the Law*, Legislative Study No. 100. Rome, FAO.

Salman, M.A.S. (1997) *The Legal Framework for Water Users' Associations – A Comparative Analysis*. Technical Paper No. 360. Washington, D.C., World Bank.

Solanes, M. (1983) *Irrigation Users' Organizations in the Legislation and Administration of Certain Latin American Countries*. Legislative Study No. 24. Rome, FAO.

Teclaff, L.A. (1977) *Legal and Istitutional Responses to Growing Water Demand*, Legislative Study No. 14. Rome, FAO.

United Nations Department of Economic and Social Affairs (1974) *National Systems of Water Administration*. Doc.ST/ESA/17.

Water Deveopment and Management (1977). *Proceedings of the United Nations Water Conference*, Vol. 1–4. Pergamon Press.

Chapter 10

International water resources law in general

10.1 Introduction

Water resources, whether surface, underground or atmospheric, are not respectful of political boundaries. An appreciation of this fact will help us understand the tendentially international character of water, and the need for adequate rules to regulate water use, conservation and management at the international level.

Due to its fluidity and mobility, water cannot always be viewed in a purely national context. In fact, it may happen that the bed of a river or lake serving as the boundary between two or more states will shift as a result of erosion or avulsion, with consequences as to the exact determination of the boundary. Or it may be that the use of water belonging to an international river basin by one state jeopardizes the use that another state is making or may wish to make of the same water within its own territory.

For example, a dam built in the territory of one state for irrigation or power generation purposes may affect the natural flow of the river downstream on the other side of the border, or, conversely, may flood the territory of an upstream neighbouring state. An upstream state may render the waters of a river unsuitable for certain uses in the territory of a downstream state by discharging, or allowing the discharge of, pollutants. Many other examples can be cited.

Difficulties may arise not only on the main course of the river or around the banks of a lake, but also on the whole system of tributaries in the catchment area or international drainage basin. The substantial unity of the river/lake systems (and aquifers) forming part of a drainage basin is due to the fact that the water resources therein are nearly always component elements of one and the same hydrologic cycle, and any action taken by a state which modifies the natural water régime within its territory, either quantitatively or qualitatively, will have repercussions on the waters of the same basin located in the territory of another state.

The same negative repercussions may follow in the case of omission of an action by one state concerning the waters of an international basin, which, if it had been taken, would have alleviated or prevented an injury to the waters or other interests in the territory of another state. The best examples relate to the duty to provide early warning about the occurrence of floods and other natural calamities, or about other accidental events.

In the case of international groundwater resources, i.e., those water resources extending below the territory of two or more states, the problem is more complex. Unlike surface water, groundwater is not visible and, without a specialized survey, it is difficult to determine its exact location and characteristics. Underground water tables lie at different depths and often have an extension which does not respect the political demarcation lines and, in addition, may reach well inside the territory of different states. The abstraction and use within a state of water belonging to a water table should be allowed, but keeping in mind the international character of the underground basin.

At the beginning of its development, the law relating to international water resources was concerned mainly with the question of boundary demarcation between sovereign states; the regulation of the use of water resources for domestic purposes, irrigation, timber floating, fishing and other traditional purposes of use was limited to the national law of each state, or to bilateral agreements between bordering states. Only navigation on the main courses of international waterways was the object of consideration at the international level, in view of the commercial interests involved. Thus, it is with reference to navigation that the law of international rivers and lakes developed.

Since the beginning of the twentieth century, new forms of water utilization have evolved on international rivers, particularly for hydropower generation and irrigation; new international treaties have been concluded for the exploitation of water resources common to two or more states. Subsequently, other treaties have been concluded in order to prevent or mitigate the harmful effects of water resulting from erosion or floods, or to provide protection from natural disasters. Dating from the second half of the twentieth century, the protection of international water resources from pollution has become a major concern, and a number of treaties have been concluded for this purpose. Finally, the recent worldwide concern for the protection of the environment has fueled international action in the preparation of treaties for the protection of water resources in all of its forms and of other elements of the environment: air, space, seas, land, forests, and other natural resources directly linked with water.

It may be said that the evolution of international water resources law has closely followed the economic, technical and social development of nations.

10.2 The concept of 'international water resources' and other definitions

10.2.1 A historical review

Simply stated, international water resources law, or international river law to the exclusion of the law of the sea, is that branch of international public law governing the relationships among states, or between states and international organizations, in water resources matters.[1]

Many definitions of international water resources, or international rivers or waterways, have been adopted by state practice and proposed by learned writings in the past, following the progress in the understanding of the nature and potentiality of inland waters for many uses and the damage that such waters could cause.

It is important to point out that the expression 'international river' includes two aspects: one geographic and the other legal.

---

1 For the benefit of non-lawyers it may be useful to understand the distinction which exists between national and international law. National law regulates the relationship between man and men, between man and things and between man and the state within a given sovereign nation. International law is divided into international private law and international public law. International private law covers all the questions relating to the status of aliens in a given country, while international public law regulates the relations among sovereign states and those between states and international organizations.

According to classical international law theories, from a geographical point of view a river is 'international' when it flows through, or separates, the territories of two or more states. This definition still holds.

From a legal standpoint, always according to international law, in the past a river was considered 'international' when a riparian state ceased to have the totality of the powers or competences on it, which normally belonged to a state on the navigable portion of a river located within its own territory or under its jurisdiction.

Usually, a river which is geographically international is also legally international. However, in the past it could happen that a river geographically national became subject to an international legal régime, due to its navigability. Thus, a river geographically national could not be international from a legal standpoint if it were not navigable.

Until recently, the theories of international water law have been unanimous in recognizing that a river may be international in two ways: either because it serves as a boundary between two or more states, and in this case it is called *contiguous* international river, or because it crosses the territories of two or more states, of which some are upstream and other downstream, and in this case the international river is called *successive*.

The evolution of the expressions international 'river,' 'watercourse,' 'waterway,' 'river system' and 'drainage basin' is briefly illustrated herebelow.

In the late eighteenth century, watercourses common to several states were referred to as *common rivers or watercourses*.[2] During the nineteenth century, they were frequently described as *international rivers or lakes*, an expression enshrined in Article 108 of the Final Act of the Congress of Vienna of 1815.[3] This expression refers to navigable waterways of concern to two or more states, either successive or contiguous, or to lakes crossed by a frontier or surrounded by several riparian states ('international' or 'frontier lakes'). The Act of Berlin of 1885[4] also referred to the Rivers Congo and Niger as *international rivers*.

The peace treaties following World War I (1919) used the expression *rivers declared international*. This meant that national waterways crossing the territories of the defeated powers could be considered international waterways and consequently subjected to freedom of navigation by the winning powers.

Another expression used in the Barcelona Convention of 1921,[5] namely, *waterway of international concern*, reflected the need for extending the principle of freedom of navigation to all flowing waters, whether national or international, provided, however, that the states concerned agreed thereon. This illustrates the growing interest of the international community in the freedom of communications.

---

2 Reichsdeputationshauptschluss of 25 February, 1903, Art.39 (Martens, *Recueil des traités*, 2ème ed.); Convention between France and the Elector of Mainz of 15 August, 1804, Art.2 (Martens, *Recueil des traités*, 2ème ed., VIII, 261).
3 Text in: Herstlet, E. *Commercial Treaties, A Collection of Treaties and Conventions between Great Britain and Foreign Powers*, Vol. 1, 3. The Congress of Vienna dealt mainly with European rivers, such as the Main, the Neckar, the Moselle, the Meuse and the Scheldt. A river was considered 'international' if it was navigable.
4 Signed by the United Kingdom, Austria, Hungary, Belgium, Denmark, France, Germany, Italy, the Netherlands, Portugal, Russia, Spain, Sweden, Norway and Turkey on February 26, 1885. Text in: Herstlet, *op. cit.*, Vol.17, London 1820–25, 62.
5 Text in 7, LNTS, 37.

Later on, the expression *international rivers or lakes system* acquired relevance in international practice, making it possible to extend the international rules to tributaries, canals and secondary courses in addition to main streams. The expression was broad enough to include lakes and lake sources connected to main streams, though only concerning surface water. Underground waters were not covered.[6]

10.2.2 The drainage basin concept

Mainly as a result of studies carried out by the International Law Association (ILA), in the late 1950's it was proposed to use the expression *international drainage basin*. A precise definition was given by Article 2 of the Helsinki Rules adopted by the ILA in 1966, whereby 'an international drainage basin is a geographical area extending over two or more States determined by the watershed limits of the system of waters, including surface and underground waters, flowing into a common terminus.'[7] This concept, which is broader than the ones applied in the past, connotes the entire system of main streams, tributaries and lakes, including groundwater.

The multiple and intensive uses to which water resources can be put, together with the rapid development of hydraulic engineering techniques and the latest economic theories of integrated river basin development, have necessitated the revision of the traditional criterion of 'navigability' of a river to designate it as international. It has been realized that what is of international concern is not only the main course of a river, but all the waters belonging to the same river basin or system, and that any human interference with the waters located in one part of a basin or system may affect waters located in another part of it, directly or indirectly. If a state ignores this situation and behaves as if it had full sovereign jurisdiction over the waters located within its own territory, disputes with neighbouring states may ensue. The potential for disputes increases as the importance of water for the satisfaction of economic and social needs grows.[8]

The 'international drainage basin' concept is based on the idea that the unity of the physical fact corresponds to a community of interests. This entails the need to avoid unilateral action by individual states participating in the use of water belonging to the same international drainage basin, and the resulting fragmentary legal régime of the resource.[9] Thus, the concept offers a rational basis for planning the development of water resources, as, within the basin, all natural resources (land, water, fauna, flora) can be quantified.

The notion of 'international drainage basin' has given rise to perplexities and doubts by some states. For this reason the International Law Commission of the United Nations (ILC), charged with the study of the subject in order to prepare a codification thereof, has adopted the appellation 'international watercourses,' thereby designating rivers, lakes, glaciers and underground water. In addition, the ILC has studied the concept of 'international

---

6 Caponera, D.A. (1980) *The Law of International Water Resources*. Legislative Study No. 23. Rome, FAO, 5.
7 Text in: The International Law Association, *Report of the Fifty-Second Conference*, Helsinki, 14–20 August, 1966, 484–532.
8 Caponera, D.A. & Curti Gialdino, C. (1988) *Fiumi Internazionali*, Estratto dal Volume XIV della *Enciclopedia Giuridica*, 2.
9 *Ibid.*

water resources system,'[10] constituting, by virtue of their physical relationship, a unitary whole,[11] and that of 'shared natural resource.'[12]

These two concepts underline the substantial natural homogeneity of international water resources and the necessary interdependence of the states participating in their use. The reference to the 'system' allows to define the international character of the relations between states, not only on the basis of a physical circumstance, i.e., of the fact that the river or lake is located in the territory of two or more states, but also on the basis of the circumstance that the use by one state of international waters might have repercussions in the territory of another state or might limit the use the latter may wish to make of them. It is to the interdependence in the use of shared water resources that the attention of the lawyer is drawn.

In the final draft of the Convention on the Law of Non-Navigational Uses of International Watercourses, adopted by the United Nations General Assembly on 21 May, 1997, the ILC retained the expression 'international watercourses.' This expression designates 'a system of surface waters and groundwaters constituting by virtue of their physical relationship a unitary whole and normally flowing into a common terminus.'

In 2002, the ILC took up the concept of 'shared natural resources,' which covers oil, gas and groundwater, and included it in its long-term programme of work. Studies started with the consideration of groundwater resources, leading to the final adoption, by the commission in 2008, of a full set of draft articles on the law of transboundary aquifers. On 11 December, 2008, the General Assembly adopted a resolution that took note of the draft articles which were appended to it, and commended the draft articles to the attention of governments, pending a decision on their final form.[13]

10.2.3 The expression 'international water resources'

The foregoing review of terminology shows that in the course of history many expressions or concepts have been resorted to in order to identify the scope of international rules applicable to water resources common to two or more states. The choice between one or the other expression is closely bound to the juridical position taken by the interested states and their readiness or otherwise to accept limitations on their sovereign rights over the natural resources located on their territories. The neutral expression 'international water resources' is adopted here as 'offering a single expression that resolves and covers the traditional distinction between the problems arising under river and lake navigation and questions arising in the use, development and protection of water resources of concern to several states,'[14] and including all water resources (surface, underground, atmospheric and frozen waters).

---

10 United Nations (1976) *Management of International Water Resources: Legal and Institutional Aspects.* Report of the group of experts on legal and institutional aspects of water resources. New York, United Nations, 14.
11 International Law Commission, Forty-third session, 19 April-19 July 1991, Art.2.
12 UNEP (1975) *Report of the Executive Director on Cooperation in the Field of Environment,* Doc. UNEP/bc/44.
13 See Chapter 12.
14 Caponera, D.A. *The Law of International Water Resources; op. cit.,* 6.

10.3 The sources of international water resources law

10.3.1 Introduction

In order to ascertain the legal position of a particular issue arising in connection with the management of international water resources, it is necessary to know the sources from which this legal position derives.

The various sources of international water resources law are enumerated in Article 38 of the Statute of the International Court of Justice, which is the world court where international disputes in general, including water disputes, are handled. Article 38 indicates the basis on which the court shall decide the cases submitted to it. In the course of its proceedings and according to it, the court shall apply:

'a) international conventions, whether general or particular, establishing rules expressly recognized by the contesting States;
b) international custom, as evidence of a general practice as accepted as law;
c) the general principles of law recognized by civilized nations;
d) ... judicial decisions and the teachings of the most highly qualified publicists of the various nations, as subsidiary means for the determination of rules of law.'

It is to be noted that, as a consequence of the time when the Statutes of the International Court of Justice were drafted (just after the Second World War), Article 38:

(i) refers to principles of law recognized 'by civilized nations'; the expression 'civilized nations' can no longer be considered valid;
(ii) does not consider the law-making activities of international bodies which were created later, such as the European Economic Community (EEC), or those of other institutions possessing supranational powers, such as the Senegal River Basin Development Organization (OMVS);
(iii) does not consider the procedures constituting the quasi-legislative activities of international organizations such as the Security Council of the United Nations and the United Nations General Assembly, and the 'final acts' of *ad hoc* international intergovernmental conferences, such as the Stockholm Conference of 1972 and the UN Water Conference of Mar del Plata of 1977.

Throughout this chapter, each subject is treated following the sources of international water resources law as indicated above.

10.3.2 International conventions

The expression 'international convention' corresponds to the terms: treaty, agreement, protocol, pact, charter, compromise, exchange of notes, final act, *modus vivendi* or other international instrument. International law does not prescribe any specific form for conventions or agreements. However, international agreements are usually written and constitute the most common procedure for creating rules of conduct and binding obligations between sovereign states. According to Article 2 of the Convention on the Law of Treaties,[15] the term 'treaty' means, 'an international agreement concluded between States in written form and governed by international law, whether embodied in a single instrument or in two or more related instruments and whatever its particular designation.'

International agreements generally take one of the following forms:

(i) agreements open to signature and ratification by the contracting parties;
(ii) agreements not subject to ratification ('in simplified form'), which enter into force upon signature or upon a supervening given set of circumstances;
(iii) exchanges of notes, which enter into force on a specified date or upon the actual exchange taking place, i.e., upon receipt and confirmation by one state of the note transmitted by another state;
(iv) instruments of a less formal nature ('joint statement,' 'act,' 'procès-verbal,' 'memorandum of agreement').[16]

As regards international water resources, agreements may cover only substantive issues or they may encompass procedural aspects as well. Substantive issues include allocation of the water or regulation of use. Procedural issues might include the definition and identification of systems of control of uses, a machinery for mutual consultation on major projects, mechanisms for joint resource monitoring and for the exchange of data and information on water quality, rights of inspection, and procedures for the solution of differences arising from the interpretation and application of a convention. At times, complex processes for reaching agreements are followed in order to make room for mutually acceptable concessions.[17]

Article 38 of the Statute of the International Court of Justice distinguishes between general and particular international conventions.

10.3.2.1 General conventions

General conventions are multilateral agreements codifying rules of conduct in a given sector. They may be of universal or regional application.

Among the general conventions of universal application concerning international water resources, the following may be mentioned:

(i) Treaty of Vienna, 9 June, 1815, which internationalized the rivers of Europe;[18]
(ii) Convention and Statute on the Régime of Navigable Waterways of International Concern, Barcelona, 20 April, 1921, the purpose of which was to facilitate navigation;[19]

---

15 *Convention on the Law of Treaties*, Vienna, 23 May, 1969. Text in: *A/Conf. 39/27*, 23 May, 1969, and *corrigenda*. The convention entered into force on 27 January, 1980.
16 Caponera, D.A. *The Law of International Water Resources; op. cit.*,10.
17 Caponera, D.A. (1981) International River Law. *Proceedings of the National Symposium on River Basin Development*. Dhaka, 176.
18 Herstlet, *op. cit.*, Vol.1, 3.
19 Text in 7, LNTS, 37. The convention was ratified by only 20 states. On the other hand, according to the law of succession of states to treaties, it can be said to apply to states that attained independence from colonial rule.

(iii) Convention relating to the Development of Hydraulic Power affecting more than one State, Geneva, 9 December, 1923, which aimed at facilitating the transmission of electric power among states;[20]
(iv) Convention on Transit Trade of Land-Locked States, New York, 8 July, 1965,[21] the purpose of which is to facilitate the transit of goods of land-locked states to the sea, thus recognizing the right of such states of free access to the sea.
(v) Convention to Combat Desertification, Paris, 17 June, 1994;[22]
(vi) Convention on the Law of the Non-Navigational Uses of International Watercourses (UN Watercourses Convention), New York, 21 May, 1997, containing principles with regard to cooperation among states.[23]

The following are examples of general conventions of regional application:

(i) in *Africa*: African Convention on the Conservation of Nature and Natural Resources, Maputo, 11 July, 2003;[24] Revised Protocol on Shared Watercourses in the Southern Africa Development Community (SADC), Windhoek, 7 August, 2000;[25]
(ii) in *America*: Act of Asunción on the Use of International Rivers (Argentina, Bolivia, Brazil, Paraguay, Uruguay), 3 June, 1971;[26]
(iii) in *Asia*: ASEAN Agreement on the Conservation of Nature and Natural Resources, Kuala Lumpur, 9 July, 1985;[27]
(iv) in *Europe*: European Agreement on the Restriction of the Use of Certain Detergents in Washing and Cleaning Products, Strasbourg, 16 September, 1968;[28] Convention on the Protection and Use of Transboundary Watercourses and International Lakes (UNECE Water Convention), Helsinki, 18 March, 1992.[29]

The UNECE Water Convention applies to the region of the UN Economic Commission for Europe (UNECE), which also includes all the countries of the former Soviet Union. Because of an amendment adopted by the parties in 2003, which entered into force in 2013, all UN member states may now become parties to it.[30] The convention has thus become a global instrument.

---

20 Text in: 36, LNTS, 77. Like the Barcelona Convention, this convention was ratified by only 11 states.
21 Text in: 597, UNTS, 3.
22 Text in: 33, ILM, 1328 (1994).
23 Text in: 36, ILM, 710 (1997).
24 FAOLEX (FAO legal database online).
25 Reprinted in: Burchi, S. & Mechlem, K. eds. (2005) *Groundwater in International Law: Compilation of Treaties and other Legal Instruments*. Legislative Study No. 86. Rome, FAO, 102.
26 Text in: *Rios y lagos internacionales (Utilizacion para fines agricolas e industriales)*, 4 ed. rev. OEA/Ser.I/IV, CIJ-75, rev.2, Organization of American States, Washington, D.C., 1971, 183–186.
27 Reprinted in: Burchi, S. & Mechlem, K. eds. *Op. cit.*, 35.
28 Text in: 788, UNTS, 182–190.
29 Text in: 31, ILM, 1312 (1992).
30 Chad and Senegal acceded to the convention in 2018.

10.3.2.2 Particular conventions

Particular conventions are those tending to regulate a specific aspect of international water resources management. They may be multilateral or bilateral.

From the standpoint of form and substance, multilateral agreements may be divided into the following groups:

'(i) agreements relating to the general development of an international water resource (river, basin, aquifer);
(ii) agreements relating to a specific utilization or development of an international water resource or basin;
(iii) agreements resulting from cooperation between States within the framework of institutions established for the purpose of utilizing international water resources;
(iv) agreements concerning technical and financial assistance between donors (States, international organizations or institutions) on the one side, and riparian States on the other, for the development of international water resources.'[31]

Most existing agreements are bilateral. A United Nations study classifies them as follows:

'(i) framework agreements, i.e., agreements usually concluded in respect of contiguous watercourses, setting up a joint commission designed to facilitate the exchange of information and consultation;
(ii) agreements for the integrated management of an international basin or water resource;
(iii) agreements for the study of potential uses and development of an international basin or water resource;
(iv) agreements for a specific use (navigation, timber floating, fishing, irrigation, hydropower production, etc.) of an international basin or water resource;
(v) agreements for the control of harmful effects of water (flooding, erosion, salinization) of an international basin or water resource;
(vi) agreements for the control of water quality (pollution, contamination) and environmental protection of international waters;
(vii) technical and/or financial assistance agreements between donor States or international organizations and basin States;
(viii) agreements calling for the harmonization of national laws governing water resources with a view to avoiding discrimination against users of different nationalities. Normally, in such cases, municipal legislation, i.e., the national law, is introduced and referred to as 'parallel' legislation, and the preparatory work is often done by a joint institution appointed by the States concerned.'[32]

---

31 United Nations (1976) *Management of International Water Resources: Legal and Institutional Aspects*. Report of the group of experts on legal and institutional aspects of water resources. New York, United Nations; see also Caponera, D.A., *The Law of International Water Resources; op. cit.*, 11.
32 *Ibid.*

10.3.3 International customary water law

International custom is mentioned in Article 38 of the Statute of the International Court of Justice as the second source of international law. It is generally recognized that international custom is constituted by:

(i) the constant and uniform conduct by states (*inveterata consuetudo – usus*), together with
(ii) their conviction as to the obligatory nature of such conduct as being in conformity with a juridical norm (*opinio juris sive necessitatis*).

The two elements just mentioned have undergone criticism by some authors who have stated that *opinio juris* may conceivably not be one of necessity, and that the time taken for custom to establish itself may not necessarily be a matter of centuries; in fact, a number of international customary rules have come into being in a very short period of time.[33]

International custom may be general, i.e., binding upon all states, or particular, binding only upon a given group of states.

The identification of a rule of customary law of international validity governing international water resources can be elusive. International state practice (conventions concluded between states, declarations of principle of international organizations, international judicial decisions and the most recent learned opinion) must be analyzed in order to assess the existence of a conformity in the states' conduct. Such conformity may be taken as a proof of the existence of a general rule of conduct.[34]

In this connection, there is a clear affirmation of the general customary rule whereby the rights of states are limited in relation to any shared resource. The Permanent Court of International Justice, in its decision concerning the territorial competence of the River Oder Commission of 10 September, 1929, recognized this rule by noting that 'when consideration is given to the manner in which States have regarded the concrete situations arising out of the fact that a single waterway traverses or separates the territory of more than one State, and the possibility of fulfilling the requirements of justice and the considerations of utility which this fact places in relief, it is at once seen that a solution of the problem has been sought not in the idea of a right of passage in favour of upstream States, but in that of the community of interests of the riparian States. This community of interests in a navigable river becomes the basis of a common legal right, the essential features of which are the perfect equality of all riparian States in the uses of the whole course of the river and the exclusion of any preferential privilege of any of the riparian States in relation to the others.'[35]

International custom has provided some of the most important rules for the use and protection of shared water resources. The following may be mentioned:

(i) duty to cooperate and to negotiate with a genuine intention of reaching an agreement;

---

33 As the International Court of Justice was able to affirm in its judgement of 20 February, 1969, concerning the North Sea Continental Shelf Case between the Federal Republic of Germany, Denmark and the Netherlands. *International Court of Justice Reports*, 1969, 3 ff.
34 Caponera, D.A. *The Law of International Water Resource; op. cit.*,12.
35 Text in: *Permanent Court of International Justice*, Series A, No.23, Series C, No.17 (II).

(ii) prohibition of management practices likely to cause substantial and lasting injury to other states;
(iii) duty of prior consultation;
(iv) equitable and reasonable utilization of shared water resources.[36]

10.3.4 The codification of international water resources law

By 'codification' is usually meant that process by which, by act of authority, certain customary or unwritten rules are translated into a written body of rules (codified rules), so that as such they can be brought to the knowledge of the community by which they will have to be observed.

Article 13, 1(a) of the United Nations Charter entrusts the General Assembly with the task of 'encouraging the progressive development of international law and its codification.' For this purpose, in 1947 the General Assembly appointed the International Law Commission as a permanent subsidiary organ.[37]

In 1970, the United Nations General Assembly recommended that the International Law Commission undertake the study of the law of non-navigational uses of international watercourses with a view to the progressive development and codification of this law.[38]

The work of the commission started in 1971,[39] culminating with the adoption of the Convention on the Law of the Non-Navigational Uses of International Watercourses (UN Watercourses Convention) by the General Assembly in 1997.[40] Between 2002 and 2008 the commission focused its work on the development of draft articles on the law of transboundary aquifers, which were eventually adopted and submitted to the General Assembly in 2008. In its resolution 63/124 of 11 December, 2008, the General Assembly took note of the draft articles, the text of which was annexed to the resolution. It commended them to the attention of governments, inviting them to make appropriate bilateral or multilateral arrangements for the management of their transboundary aquifers. A decision as to the final form of the draft articles is still to be made.

10.3.5 The law-making activity of the European Union

Although not considered by Article 38 of the Statute of the International Court of Justice as sources of international law, the enactments of the European Communities,[41] and later of the European Union (EU), deserve special mention, as they have effect not only within the sphere proper to the institution, but also on the national law of each member state. These enactments prescribe rules of conduct to be observed by the member states themselves, by their organs and by public and private entities and individuals operating within their territories.

---

36 Caponera, D.A. International River Law, *op. cit.*, 176.
37 General Assembly, *Resolution No. 174 (II)* of 21 November, 1947.
38 General Assembly, *Resolution No. 2669 (XXV)* of 8 December, 1970.
39 *Yearbook of the International Law Commission*, 1971, Vol. II (Part I), 370, Doc. A/8410/Rev. 1, paras. 119–122.
40 The convention entered into force on 17 August, 2014.
41 The European Coal and Steel Community, instituted by the Treaty of Paris of 1951; the European Economic Community and Euratom, instituted by the Treaty of Rome of 1957.

According to the Treaty of Rome of 1957, which established the European Economic Community, 'for the achievement of their aims . . . the Council and the Commission shall adopt regulations and directives, make decisions and formulate recommendations or opinions.' The regulations of the EU are immediately binding upon the member states, while the directives require national legislation for their entry into force.

Generally speaking, while the list of existing regulations is relatively short, a number of directives have been adopted in the field of water resources, among which the most important are the following:

(i) Council Directive 91/271/EEC of 21 May, 1991, concerning urban waste water treatment (Urban Wastewater Treatment Directive);[42]
(ii) Council Directive 91/676/EEC of 12 December, 1991, concerning the protection of waters against pollution caused by nitrates from agricultural sources (Nitrates Directive);[43]
(iii) Council Directive 98/83/EC of 3 November, 1998, on the quality of water intended for human consumption;[44]
(iv) Directive 2000/60/EC of the European Parliament and of the Council of 23 October, 2000 establishing a framework for Community action in the field of water policy (Water Framework Directive);[45]
(v) Directive 2006/7/EC of the European Parliament and of the Council, of 15 February, 2006, concerning the management of bathing water quality and repealing Directive 76/160/EEC;[46]
(vi) Directive 2006/118/EC of the European Parliament and of the Council, of 12 December, 2006, on the protection of groundwater against pollution and deterioration;[47]
(vii) Directive 2007/60/EC of the European Parliament and of the Council, of 23 October, 2007, on the assessment and management of flood risks;[48]
(viii) Directive 2008/05/EC of the European Parliament and of the Council, of 16 December, 2008, on environmental quality standards in the field of water policy, amending and subsequently repealing Council Directives 82/176/EEC, 83/513/EEC, 84/156/EEC, 84/491/EEC, 86/280/EEC and amending Directive 2000/60/EC of the European Parliament and of the Council;[49]
(ix) Directive 2009/90/EC of the European Parliament and of the Council, of 31 July, 2009, laying down, pursuant to Directive 2000/60/EC of the European Parliament and of the Council, technical specifications for chemical analysis and monitoring of water status;[50]

---

42 *Official Journal*, L 135, 30/5/1991, 40.
43 *Ibid.*, L 375, 31/12/1991, amended in 2003.
44 *Ibid.*, L 330, 5/12/1998, 32–54. On 1 February, 2018, the Commission adopted a proposal for a revision of this directive. The proposal seeks to improve the quality of drinking water by considering new and emerging substances that may adversely impact such water. A further aim of the proposal is to improve access to drinking water, especially for vulnerable and marginalized groups.
45 *Ibid.*, L 327, 22/12/2000, 1.
46 *Ibid.*, L 64, 4/3/2006, 37–51.
47 *Ibid.*, L 372, 27/12/2006, 19.
48 *Ibid.*, L 288, 6/11/2007, 27.
49 *Ibid.*, L 348, 24/12/2008, 84.
50 *Ibid.*, L 201, 1/8/2009, 36.

(x) Commission Directive (EU) 2015/1787 of 6 October, 2015, amending Annexes II and III to Council Directive 98/83/EC on the quality of water intended for human consumption;[51]

In particular, the Water Framework Directive consolidates the provisions of a number of previous directives, which have been repealed in accordance with specified schedules. The Nitrates Directive and the Urban Wastewater Treatment Directive complement it. EU member states have taken steps to approximate their water legislation to the provisions of the directives adopted so far, since they are under a duty to do so.[52] Candidate states and a number of European states intending to join the EU have also engaged in this process.

The instruments listed above are by no means exhaustive, as certain important issues such as drought management are not covered sufficiently. Emphasis is placed on water quality issues and pollution control.

10.3.6 General principles of international water resources law

'General principles of law' represent the third source of international law, to be resorted to in the absence of international conventions or customary rules.

An analysis of general principles helps in the task of ascertaining whether or not international rules exist. Such a reconstruction is made through judicial decisions and in learned writings, which, in their attempt to affirm limitations to the sovereignty of states sharing common water resources, are based on certain principles outlined below, namely:

(i) the principle that there shall be no abuse of rights (*sic utere tuo ut alienum non laedas*). Whenever a state makes use of its own territory in an arbitrary fashion, thereby causing unjustified loss or damage in another state, such action should be deemed to be contrary to international law. Presently, almost all national laws recognize this principle. Differences may arise, however, as to the degree and scope of the rights acknowledged, and the degree to which abuses are prohibited. The laws of most countries prohibit intentional or culpable harm to others in the exercise of their rights;

(ii) the principle that co-basin states shall act in a way consonant with good neighbourly relations. Under the good-neighbourliness principle, no state may engage on its own territory in activities likely to have negative repercussions on the territory of another state. Territorial propinquity obviously facilitates greater collaboration;

(iii) the principle that the national water laws of the basin states should be applied harmoniously in mutual disputes.[53] Almost all national water laws provide for a balancing of rights between competing users. Such principle informs, though without any marked determining force, the criteria of equitable apportionment and use of waters among the states concerned.

---

51 *Ibid.*, L 260, 7/10/2015, 6.
52 See Chapter 6, Section 6.4.5.
53 Caponera, D.A. International River Law, *op. cit.*, 177; *The Law of International Water Resources*; *op. cit.*, 16.

10.3.7 Resolutions of intergovernmental organizations

Having been adopted prior to the emergence of the phenomenon, Article 38 of the Statute of the International Court of Justice takes no account of the resolutions of intergovernmental organizations containing 'declarations of principles' of conduct in the relations between states. Since then, many resolutions have been adopted by the United Nations General Assembly in the field of natural resources, of which water resources are a component.

Similarly, declarations, resolutions and recommendations of major significance have been adopted at the conclusion of intergovernmental conferences convened by the United Nations. Among these, the following may be cited:

(i) Declarations of the United Nations Conference on the Human Environment, Stockholm, 16 June, 1972;[54]
(ii) Declarations and Resolutions of the United Nations Water Conference, Mar del Plata, March, 1977;[55]
(iii) Declarations of the United Nations Conference on Desertification, Nairobi, 9 September, 1977.[56]
(iv) Declaration of the International Conference on Water and the Environment, Dublin, 31 January, 1992;[57]
(v) Agenda 21, output of the United Nations Conference on Environment and Development (UNCED), Rio de Janeiro, 14 June, 1992;[58]
(vi) United Nations Millennium Declaration, New York, 8 September, 2000.[59]

In addition, a great number of declarations, resolutions and recommendations have been adopted by the United Nations Economic and Social Council (ECOSOC), the Economic Commission for Europe (UNECE), and, outside the United Nations system, by the Organization for Economic Cooperation and Development (OECD), by the Organization of the American States (OAS), by the Council of Europe and by the Asian-African Legal Consultative Committee.

Given the universal vocation of the United Nations, the legal value of these declarations, resolutions, and recommendations, especially where 'declarations of principles' are concerned, has been the subject of lengthy doctrinary discussions. Neither actual state practice nor the preparatory works leading to them warrant an affirmative reply as to their legal status as full-fledged legislative or quasi-legislative sources of international law. However, these declarations and the like have considerably influenced the processes of formation of general

---

54 *Report of the UN Conference on the Human Environment*, United Nations Publication, 4–7, 17, 20, 22–23.
55 *E/CONF.70/29*, 51.
56 *Report of the United Nations Conference on Desertification*, A/CONF.74/36, 55.
57 Text in: International Conference on Water and the Environment, *The Dublin Statement and Report of the Conference*, 26–31 January, 1992.
58 Text in: *Report of the United Nations Conference on Environment and Development*, A/CONF.151/26 (Vol. II), 169 and 176.
59 UN General Assembly Resolution 55/2 of 8 September, 2000. Available from www.un.org/millennium/declaration/ares552e.htm

rules of international law. Likewise, 'they have had the function of crystallizing opinions and state practice whence international customary rules take their origin and develop.'[60]

10.3.8 Judicial decisions

Water law principles regulating the relations between co-basin states have also been developed through judicial decisions by courts and arbitral tribunals.

Judicial decisions may be classified into:

(i) judgements and advisory opinions of the international courts;
(ii) awards rendered by arbitral tribunals;
(iii) decisions of national tribunals.

10.3.8.1 Decisions of international courts

The expression 'judicial decisions' contained in Article 38 of the Statute of the International Court of Justice covers, *inter alia*, the decisions of this court and those of its predecessor, the Permanent Court of International Justice.

However, Article 59 of the Statute provides that 'the decision of the Court has no binding force except between the parties and in respect of that particular case.' Thus, the decisions of international tribunals have no force of precedent. In spite of this, the court's decisions not only expressly recognize the evolution of international law when determining the law to apply in a specific case, but, through the interpretation of the international law in force and its application, they clarify that law, thereby paving the way for its progressive development.

Moreover, the jurisprudence of the court plays an important role in the codification of international law, since the International Law Commission submits its draft articles to the General Assembly of the United Nations together with a commentary which includes a full summary of the precedents and other relevant material, including the judicial decisions of the court.[61]

Among the cases submitted to international courts in the field of water resources, the following deserve to be mentioned:

(i) the case relating to the territorial jurisdiction of the International Commission of the River Oder, judgement of 10 September, 1929.[62] It was decided that the powers of the commission extended to the sections of the tributaries of the Oder, Warthe and Netze, situated in Polish territory;
(ii) the Oscar Chinn Case, judgement of 12 December, 1934.[63] This case, between the United Kingdom and Belgium, dealt with an alleged violation by Belgium of the principle of freedom of navigation;

---

60 Caponera, D.A. *The Law of International Water Resources, op. cit.*, 17.
61 The International Court of Justice (2014) *Handbook*. 6th Edition. The Hague, 100.
62 *Permanent Court of International Justice*, Series A, No. 23, Series C, No.17. Parties to the dispute were Germany, Denmark, France, Great Britain, Sweden, Czechoslovakia and Poland.
63 *Ibid.*, Series A/B, No.63, Series C, No.75. Parties: Great Britain, Belgium.

(iii) the diversion of water from the Meuse, judgement of 28 June, 1937;[64]
(iv) the case relating to the Gabcikovo-Nagymaros Project, between Hungary and Slovakia; judgement of 25 September, 1997.[65] This case concerned the suspension, by Hungary, on environmental grounds, of a joint dam construction project on the Danube. The project had been agreed upon together with Czechoslovakia on the basis of a treaty concluded in 1977. Following the suspension of the joint project, Czechoslovakia undertook the construction of a dam at Gabcikovo, inside its own territory, entailing unilateral diversions of Danube water;[66]
(v) the Pulp Mills Case, between Argentina and Uruguay, judgement of 20 April, 2010.[67] This case, which was submitted by the Government of Argentina to the International Court of Justice on 4 May, 2006, related to the alleged violation, by Uruguay, of legal obligations stemming from the 1975 Statute of the Uruguay River. In substance, Argentina charged Uruguay with having unilaterally authorized the construction of two pulp mills on the Uruguay River without complying with the obligatory prior notification and consultation procedure. It claimed that these mills would damage the environment of the river and the areas affected by it.[68]
(vi) Two cases, the first concerning certain activities carried out by Nicaragua in the border area (Costa Rica vs. Nicaragua), and the second the construction of a road in Costa Rica along the San Juan River (Nicaragua vs. Costa Rica); joint judgement of 16 December, 2015. The first case started in 2010, with the claim by Costa Rica that Nicaragua had occupied the territory of Costa Rica in connection with the construction of a canal from the San Juan River in Costa Rican territory and had conducted dredging operations in the river. One year later, Nicaragua instituted proceeding against Costa Rica, contending that road construction works by Costa Rica along the border were leading to grave environmental consequences.[69]

While, in the case concerning the diversion of water from the Meuse, in that relating to the Gabcikovo-Nagymaros Project and in the Pulp Mills Case, the decisions were based on particular treaties concluded between the parties, in the case relating to the territorial jurisdiction of the International Commission of the River Oder, the court invoked the *principle of community of interests* of riparian states, which could be considered as one of the customary rules of international law.[70] In the Oscar Chinn Case, the decision was based on the Convention of Saint Germain (1919), which confirmed the principles of freedom of navigation and freedom of trade.

---

64 *Ibid.*, Series A/B, No.70, Series C, No.81. Parties: Belgium, the Netherlands.
65 The text of the ruling may be found on the website of the International Court of Justice, at www.icj-cij.org/icjwww/idocket/ihs/ihsjudgement/ihs_ijudgment_970925.html. Parties: Slovak Republic, Hungary.
66 'Variant C.'
67 International Court of Justice, Case concerning Pulp Mills on the River Uruguay (Argentina v. Uruguay) Judgment of 20 April, 2010, text available at: www.icj-cij.org/files/case-related/135/135-20100420-JUD-01-00-EN.pdf.
68 Developments with regard to this matter are illustrated on the website of the International Court of Justice, at www.icj-cij.org/icjwww/idocket/iau/iauframe.htm.
69 The two cases are summarized in International Court of Justice, *Yearbook 2015–2016*, The Hague.
70 Caponera, D.A. *The Law of International Water Resources*; *op. cit.*, 18.

In the Gabcikovo-Nagymaros Case, the court invoked the principle of equitable utilization to conclude that Czechoslovakia had committed an internationally wrongful act by operating the project,[71] thus depriving Hungary of its reasonable and equitable share in the use of the natural resources of the Danube. By referring to the case relating to the River Oder Commission, it reasserted the principle of community of interest in the waters of an international watercourse. In parallel with this, the court placed emphasis on an obligation for co-basin states to cooperate, and in this specific case to negotiate in good faith in order to achieve the objectives of the 1977 treaty, taking into account current environmental standards.

Finally, in its decision on the case concerning the construction of a road in Costa Rica along the San Juan River, the court found that Costa Rica had not complied with its obligation under general international law to carry out an environmental impact assessment (EIA).

Another case was submitted by the Government of Chile to the International Court of Justice on 6 June, 2016,[72] relating to the status and use of the waters of the Silala/Siloli, which is a water resource situated at the border between Bolivia and Chile. Chile maintains that the Silala is an international river according to the UN Watercourses Convention and long-standing practice of the two states. Bolivia contends that the Silala is not transboundary, since it would not flow into Chile if a canal had not been constructed to such effect in the early 1900's. The filing of a counter-memorial by Bolivia is pending.

As regards advisory opinions, those of the International Court of Justice are open only to international organizations; they have consultative nature and are not binding on the requesting bodies. An example of advisory opinion requested by the Council of the League of Nations to the Permanent Court of International Justice in the field of water resources is the one concerning the jurisdiction of the European Commission of the Danube between Galatz and Braila,[73] delivered on 8 December, 1927.

10.3.8.2 Arbitral awards

The contribution to international water resources law made by arbitral awards is significant. Among the cases submitted to arbitration, the following may be cited:

- Helmand River Delta Case, arbitral awards of 19 August, 1872, and 10 April, 1905;[74]
- San Juan River Case, arbitral award of 22 March, 1888;[75]
- Kushk River Case, arbitral award of 22 August-3 September, 1893;[76]

---

71 It is interesting to note that, at the same time, the court considered acceptable the construction by Czechoslovakia of Variant C – the Gabcikovo Dam.
72 The application is available on the website of the International Court of Justice, at www.icj-cij.org/files/case-related/162/162-20160606-APP-01-00-EN.pdf.
73 *Permanent Court of International Justice*, Series B, No.14, Series C, Nos.13-IV(V), (II), (III), (IV). Parties: France, Great Britain, Italy and Romania.
74 Mayors St.John, Lovett and Evan Smith and Mayor-General Sir Frederick John Goldsmid, *Eastern Persia, An Account of the Journeys of the Persian Boundary Commission*, 1870–71–72 (London, 1876), Vol.I, 413. Parties: Afghanistan, Persia.
75 Moore, *History and Digest of International Arbitration to which the United States has been a Party*, Washington, 1898, Vol.V, 4706. Parties: Costa Rica, Nicaragua.
76 De Martens, G.F., *Nouveau recueil général de traités*, 1888, 2ème série, Vol. XIII, 566. Parties: Great Britain, Russia.

- Faber Case, award of 1903;[77]
- Tacna-Arica Case, award of 4 March, 1925;[78]
- Trail Smelter Case, awards of 16 April, 1938 and 11 March, 1941;[79]
- Zarumilla River Case, arbitral award of 14 July, 1945;[80]
- Lake Lanoux Case, award of 16 November, 1957;[81]
- Gut Dam Case, awards of 15 January, 1968, 12 February, 1968 and 27 September, 1968.[82]
- Milestone 62-Mount Fitz Roy Case, award of 21 October 1994;[83]
- Indus Waters Kishenganga Arbitration, partial award of 18 February, 2013, and final award of 20 December, 2013.[84]

The awards rendered in these cases reveal a tendency towards the construction of the rights of riparian states in terms of limited territorial sovereignty in respect of shared water resources. The principle of the *innocent use of rivers* which emerged during the Faber Case can be regarded as yet another principle of international water resources law that has gained currency.[85]

10.3.8.3 Decisions of national tribunals

National courts have also contributed to the evolution of international law. These courts, particularly those of federal states in their decisions concerning water disputes, have invoked a variety of highly relevant concepts.[86]

In one of its first cases involving a water dispute (Kansas v. Colorado), the Supreme Court of the United States upheld the principle of equality of rights between the two states.[87] In other cases, the Supreme Court has applied the theory of equitable apportionment.[88] This theory was also upheld in India in 1941, in the case of Sind against Punjab concerning the Indus waters[89] and in the Narmada dispute between the states of Madhya Pradesh, Maharashtra and Gujarat.[90] In the River Krishna dispute, the tribunal, constituted by the central government to settle the dispute between the states of Maharashtra, Karnataka and Andhra Pradesh, decided that groundwater is a relevant factor to be taken into consideration for

---

77 United Nations, *Reports of International Arbitral Awards*, Vol.X, 466. Parties: Germany, Venezuela.
78 *Ibid.*, Vol.II, 921–958. Parties: Chile, Peru.
79 *Ibid.*, Vol.III, 1905 ff. Parties: Canada, United States.
80 *Informe del Ministro de las Relaciones Exteriores a la Nacion*, Quito, 1946, 623. Parties: Ecuador, Peru.
81 *International Law Reports*, 1957, 101. Parties: Spain, France.
82 8 ILM, 118–143 (1969). Parties: USA, Canada.
83 Tribunal Arbitral Internacional, Sentencia del 21 de octubre de 1994, *Controversia sobre el recorrido de la traza entre Hito 62 y el Monte Fitz Roy*. Parties: Argentina, Chile.
84 Both texts are available on the website of the Permanent Court of Arbitration, at https://pca-cpa.org/en/search/?q=Kishenganga. Parties: India, Pakistan.
85 Caponera, D.A. *The Law of International Water Resources; op. cit.*, 20.
86 Caponera, D.A. International River Law; *op. cit.*, 179.
87 See *185 US 125* (1902), *206 US 46* (1907).
88 Nebraska v. Wyoming, *325 US 589* (1945), New Jersey v. New York, *283 US 336* (1931), Connecticut v. Massachusetts, *282 US 660* (1931).
89 *Report of the Indus (Rao) Commission*, 1942, 10–11.
90 *Gazette of India Extraordinary*, 6 October, 1969, pt. II, S.3(ii).

equitable distribution.[91] It is also worth mentioning the dispute between the Argentinian provinces of La Pampa and Mendoza, over the sharing of the waters of the Atuel River, in which the Argentina Supreme Court applied the principle of equitable utilization.[92]

As regards Europe, in the judgement of the German Staatsgerichtshof on the *Donauversinkung* case (Württemberg and Prussia v. Baden)[93] the German Supreme Court based its decision on the principle of equitable utilization. In Switzerland, the equality of rights between cantons for the use of public watercourses was affirmed,[94] and in Italy the Court of Cassation recognized the principle of community of ownership (Electricité de France v. Compagnia Imprese Elettriche di Liguria).[95]

The question arises as to whether municipal law, i.e., the law applied by national tribunals, can be transformed into customary international law concerning water resources. It seems obvious that the only method by which municipal law can be transformed into international law is through its recognition as 'general principles of law.' Therefore, in order to establish this, its consistent application in international practice must be proven.

In some countries such as Germany, a federal state, the constitution recognizes international law as part of the law of the country, but even in such cases the decisions of national courts do not constitute a source of international law. If in municipal disputes the court expresses the principle of the equitable apportionment of water, it might imply that the solution was based on law and equity (*ex aequo et bono*). In almost all national water law systems will be found the principle that one state, by using water, must take into account the needs of neighbouring states. This principle, in fact, is identical to the principle of customary law.

10.3.9 Contributions of publicists and international non-governmental organizations

The law of international water resources has been particularly enriched through scholarly contributions. Convincing proof of this may be seen in the work of the Institute of International Law, the International Law Association, the Inter-American Bar Association and the International Association for Water Law (AIDA).

10.3.9.1 The work of the Institute of International Law[96]

Since the end of the last century, the Institute of International Law has given attention to international river law. Among its resolutions, there are the following:

- International Regulation on River Navigation, Resolution of Heidelberg, 9 September, 1887;

---

91 *Ibid.*, 10 April, 1969, pt.II, S.3.
92 República Argentina, *Fallos de la Corte Suprema de Justicia de la Nación*, 1987, 310, Vol. 3, 2577.
93 *Annual Digest of Public International Law Cases 1927–28*, Case No. 86, A. McNair and H. Lauterpacht eds., London, 1931, 128.
94 *Recueil officiel des arrêts du tribunal fédéral*, IV, 34–37.
95 *Annual Digest*, 1938–40, 120.
96 See *Annuaire de l'Institut de Droit International*, 1887, 535; 1911, Vol. 24, 365; 1934, 713–719; 1961, Vol. 49, II, 381–384; 1979, Vol. 58,I, 197.

- International Regulation regarding the Use of International Watercourses for Purposes other than Navigation, Declaration of Madrid, 20 April, 1911;
- Regulation governing Navigation on International Rivers, Resolution of Paris, 19 October, 1934;
- Resolution on the Use of International Non-Maritime Waters, Salzburg, 11 September, 1961;
- Resolution on the Pollution of Rivers and Lakes and International Law, Athens, 12 September, 1979.

10.3.9.2 The work of the International Law Association[97]

The largest professional organization devoted to international law, the International Law Association (ILA), has made a notable contribution to the development of international water resources law through:

- Statement of Principles, Resolution of Dubrovnik, 1956;
- Resolution on the Use of the Waters of International Rivers, New York, 1958;
- Recommendations on the Procedures concerning Non-Navigational Uses, Hamburg, 1960;
- The Helsinki Rules on the Uses of the Waters of International Rivers, Helsinki, 1966;
- Articles on Flood Control, New York, 1972;
- Articles on Marine Pollution of Continental Origin, New York, 1972;
- Articles on Maintenance and Improvement of Naturally Navigable Waterways separating or traversing several States, New Delhi, 1974;
- Resolution on the Protection of Water Resources and Water Installations in Times of Armed Conflict, Madrid, 1976;
- Resolution on International Water Resources Administration, Madrid, 1976;
- Articles on Regulation of the Flow of Water of International Watercourses, Belgrade, 1980;
- Articles on the Relationship between Water, Other Natural Resources and the Environment, Belgrade, 1980;
- Articles on Water Pollution in an International Drainage Basin, Montreal, 1982;
- Rules on International Groundwaters, Seoul, 1986;
- Complementary Rules Applicable to International Water Resources, Seoul, 1986;
- Articles on Cross-Media Pollution resulting from the Use of the Waters of International Drainage Basins, Helsinki, 1996;
- Articles relating to Private Law Remedies, Helsinki, 1996;
- Berlin Rules on Water Resources, Berlin, 2004.

---

97 See Manner, E.G., & Metsalampi, V.M. eds. (1988) *The Work of the International Law Association on the Law of International Water Resources*. Helsinki, Finnish Branch of the ILA. See also International Law Association, *Report of the Sixty-Seventh Conference held in Helsinki, 12–17 August, 1996*. London, 401; and, *Report of the Seventy-First Conference held in Berlin, 16–21 August, 2004*. London, 334.

Special attention has to be paid to the Helsinki Rules, as they embody principles that have been formally accepted by many countries cooperating in the integrated development of international basins in Asia, Africa and Latin America. In addition, they have been used by the International Law Commission as the basis for the codification of the law of international water resources, leading to the adoption of the UN Watercourses Convention by the United Nations General Assembly in 1997.

It is in the Helsinki Rules that the concept of equitable utilization has received its definitive formulation. Article 4 stipulates that 'each basin State is entitled, within its territory, to a reasonable and equitable share in the beneficial uses of the waters of an international drainage basin.' In quantifying this reasonable and equitable share, due weight must be given to all relevant factors in each specific case. These factors are indicated in Article 5; the choice, however, is not limited to them. No use, or group of uses, must be granted priority according to subjective criteria.

In accordance with the principle of equitable utilization, states 'must prevent any new form of water pollution or any increase in the degree of existing water pollution in an international drainage basin which would cause substantial injury in the territory of a co-basin State.' Article 9 of the Helsinki Rules defines the term 'water pollution' as 'any detrimental change resulting from human conduct in the natural composition, content, or quality of the waters of an international drainage basin.' Basin states should take all necessary steps to eliminate existing pollution, so as not to cause substantial damage in the territory of a co-basin state.

Finally, the Helsinki Rules recommend that information be exchanged between states, and establish procedures for the prevention and settlement of disputes. The rule recommending the exchange of information has become a duty under the UN Watercourses Convention, and has been incorporated into several other international legal instruments relating to international water resources.

The ILA Berlin Rules of 2004 are an attempt at revising the Helsinki Rules, but while the latter were based on prevailing international water law (*lege lata*), the Berlin Rules have placed emphasis on recent developments and trends (*lege ferenda*), and in particular on the environmental context into which water resource management is increasingly being absorbed.

10.3.9.3 The work of the Inter-American Bar Association

The Inter-American Bar Association has also contributed to the development of the law of international water resources through its resolutions and declarations. The Declaration of Buenos Aires of 1957[98] contains general principles applicable to a system of international waters, while the Resolutions of San José of 1967 provide certain suggestions to the Permanent Committee on the use of international rivers and lakes.[99] In the resolution of Caracas

---

98 Inter-American Bar Association, *Proceedings of the Tenth Conference*, Buenos Aires 14–21 November, 1957, Vol.I, 246–248.
99 Inter-American Bar Association, *Resolutions, Recommendations and Declarations approved by the XV Conference*, San Jose, Costa Rica, 10–15 April, 1967, 1–2, 190.

of 1969, the Association recommended the unification of laws in American countries on the industrial and agricultural utilizations of rivers and lakes.[100]

10.3.9.4 The work of the Asian-African Legal Consultative Committee[101]

The Asian-African Legal Consultative Committee has undertaken the study of the legal aspects of water resources. This body, which includes specialists in water law from Asian and African countries, has produced a set of very relevant 'propositions'. These propositions, formulated in New Delhi on 18 January, 1973,[102] contain rules which were inspired by the Helsinki Rules of the International Law Association.

After defining the term 'drainage basin', as a result of which a basin state is entitled, within its territory, to a reasonable and equitable share in the beneficial uses of the waters of an international drainage basin, the propositions affirm that the reasonable and equitable share is to be determined on the basis of all the relevant factors in each particular case.

10.3.9.5 The work of the Pan American Union

During its Seventh International Conference, the Pan American Union, now called the Organization of the American States (OAS), prepared a declaration concerning the industrial and agricultural uses of international rivers.[103] While proclaiming concern for navigation, this declaration states that announcement shall be made by any state intending to construct works which are likely to cause substantial injury to other riparian states, and that within three months an objection can be raised.

The Resolution of the Inter-American Economic and Social Council of 1966 (Resolution of Buenos Aires)[104] recommends that member countries, with the technical and financial assistance of international agencies, continue joint studies for the economic utilization of the hydrographic basins of which they are a part.

10.3.9.6 The work of the Council of Europe

The Consultative Assembly of the Council of Europe has adopted a recommendation on fresh water pollution control in Europe.[105] This recommendation provides extended guidelines for the defence against pollution. In 1967, a European Water Charter was prepared at Strasbourg, which is also concerned with the protection of water quality. The charter calls

---

100 Inter-American Bar Association, *Resolutions, Recommendations and Declarations approved by the XVI Conference*, Caracas, Venezuela, 1–8 November, 1969.
101 Now the Asian-African Legal Consultative Organization.
102 Text in: Asian-African Legal Consultative Committee, *Report on the 14th Session*, New Delhi, 10–18 January, 1973, 7–14.
103 Pan American Union, Seventh International Conference of American States, Plenary Session, Minutes and Antecedents, Montevideo, 24 December, 1933.
104 Pan American Union, Final Report of the Fourth Annual Meeting of the Inter-American Economic and Social Council, Vol. I (OEA/Ser. H/XII-11), Washington, D.C., 1966, 48.
105 Consultative Assembly of the Council of Europe, *Report on Fresh Water Pollution Control in Europe*, Doc. 1965, part III.

for viable water policies and recommends that international problems arising from the use of water be settled by agreement between the states concerned, to conserve the quantity and the quality of water. Then, in 1971 at Strasbourg, the Council of Europe made recommendations concerning the pollution of the Rhine water table.

10.3.9.7 The work of the International Association for Water Law

The International Association for Water Law (AIDA/IAWL) was created in 1967 for the purpose of promoting the evolution, study, understanding and application of water law, national and international. Its members have contributed, and continue to contribute, to the progressive development of legal rules on water management within both domestic and transboundary contexts. The Association has held three international conferences with published proceedings,[106] as well as regional conferences.[107] In addition, it has taken a leading role in the delivery of the water law and governance track in the programmes of the three latest World Water Congresses of the International Water Resources Association (IWRA),[108] by soliciting papers and presentations from within and outside its membership covering a variety of water law and governance topics and by contributing to the relevant debates.

Despite the fact that the International Court of Justice has had no occasion to rely on writings of international jurists, these have directly influenced the law of international water resources, thus helping in the identification of international customary rules on the subject. Sometimes, the rules proposed in such writings have been accepted favourably by states. For instance, the Helsinki Rules formulated by the International Law Association have been incorporated into several international agreements and, even in the absence of general recognition, are likely to receive further appreciation in future treaty-making activities. The UN Watercourses Convention, which was adopted by the General Assembly in 1997, is largely based on the work of the International Law Association.

References

Bogdanovic, S. (2001) *International Law of Water Resources – Contribution of the International Law Association (1954–2000)*. The Hague, Kluwer Law International.

Caponera, D.A. (1976) *Outline for the Preparation of an Inventory of the Legal and Institutional Aspects of International Water Resources Basins*, Background Paper No. 11. Rome, FAO.

Caponera, D.A. (1980) *The Law of International Water Resources*. Legislative Study No. 23. Rome, FAO.

Caponera, D.A. (1981) International River Law. *Proceedings of the National Symposium on River Basin Development*, Dhaka, Bangladesh, 4–10 December, 1981.

---

106 See the proceedings of the IAWL Conferences: *Annales Juris Aquarum I*, Buenos Aires, 1968; *Annales Juris Aquarum II*, Caracas, 1976; *Annales Juris Aquarum III*, Alicante/Valencia, 1989. In the recommendations of the second conference held at Caracas, Venezuela, in 1976, the Association suggested the elaboration of norms pertaining to the use of international water resources.

107 Teslic, Bosnia and Herzegovina, 2001.

108 Which were held in Porto de Galinhas (Brazil), Edimburgh (UK) and Cancun (Mexico) on 25–29 September, 2011, 25–29 May, 2015, and 29 May – 3 June, 2017, respectively.

Caponera, D.A. (1985) Patterns of Cooperation in International Water Law, Principles and Institutions. *Natural Resources Journal*. Albuquerque, University of New Mexico School of Law.

Caponera, D.A. & Curti Gialdino, C. (1988) *Fiumi Internazionali*. Estratto dal Volume XIV, Enciclopedia Giuridica. Rome.

Chauhan, B.R. (1981) *Settlement of International Water Disputes in International Drainage Basins*. Berlin, Erich Schmidt Verlag.

FAO (1998) *Sources of International Water Law*. Legislative Study No. 65. Rome, FAO.

FAO (1978 and 1984) *Systematic Index of International Water Resources Treaties, Declarations, Acts and Cases by Basin*. Legislative Studies No. 15 and No. 34. Rome, FAO.

Manner, E.G., & Metsalampi, V.M. eds. (1988) *The Work of the International Law Association on the Law of International Water Resources*. Helsinki, Finnish Branch of the ILA.

Many Authors (1967) *The Law of International Drainage Basins*, Institute for International Law, NY University School of Law. Dobbs Ferry, New York, Oceana Publications, Inc.

Pan American Union (1974) *Rios y Lagos Internacionales (Utilización para Fines Agricolas e Industriales)*, 4a ed. rev. Washington D.C., OAS.

Teclaff, L. (1967) *The River Basin in History and Law*. The Hague, Nijhoff.

The International Court of Justice (2014) *Handbook*. 6th Edition. The Hague.

United Nations (1976) *Management of International Water Resources, Legal and Institutional Aspects*. New York, United Nations.

Chapter 11

International water resources law
Major issues

11.1 Boundary demarcation

11.1.1 Introduction

The question relating to the demarcation of boundaries on rivers which separate or cross the territories of two or more sovereign states is one of the first to have arisen in the development of international water law and has given birth to a number of disputes between states, sometimes leading to armed conflict.[1]

These disputes may arise in the absence of formal international agreements and for a number of reasons, among which, the modification of an original border, the dubious validity of treaties establishing a boundary, the colonial character of some of the agreements and the caducity of some of the provisions contained in the treaties.

International court decisions and arbitral awards have been rendered also on questions of boundary demarcation.[2] The learned jurists who have dealt with these questions are not always in agreement with the solutions proposed for the issues at stake.

State practice and the various solutions which have evolved in connection with boundary demarcation are briefly analyzed here below.

11.1.2 The boundary on a successive river

The fixing of a boundary on a successive river or watercourse, i.e., a river crossing the territory of two or more states, does not pose great problems as regards the actual determination of the boundary itself. The frontier runs along an imaginary straight line which cuts the river or watercourse and joins at the two banks the extreme limits of the land frontier between the two states. The river in this case is cut in as many sections as there are states crossed by the river.

Legal difficulties may come about in connection with the right of passage on the river for navigation purposes, or on the right to cross the boundary for other purposes, such as

---

1 The conflicts which have arisen out of the question of boundary demarcation include those between Iran and Iraq as regards the border on the estuary of the Shatt-el-Arab river and between USSR and China concerning the Ussuri river. Many other similar situations exist today.
2 Arbitral awards between Guatemala and Honduras of 1933, which determined the boundary on the right bank of the Rivers Tinto and Montagua. UN *Recueil des Sentences Arbitrales*, II, 1368.

timber floating. Other questions may arise concerning the level of the water flow at the river boundary and the quantity or quality of the water crossing the river boundary. However, these questions do not belong to the issue of boundary demarcation on rivers, but to those arising from the uses of the river for navigation or for other purposes where the interests of the upper and lower riparian states, or those of the riparian and non-riparian states in the case of navigation, are not similar.

11.1.3 The boundary on a contiguous river

The fixing of a boundary on a contiguous river, i.e., on a river separating the territories of two or more sovereign states, has always had controversial aspects for a number of reasons. These may be of a physical-geographic nature, as the size, the nature and the behaviour of the river water vary considerably, and of a legal nature, in that the issue can only be settled on the basis of an agreement between the states concerned, as no clear-cut rule of international law exists in this respect.

Problems may arise even when an international boundary agreement exists; in this case, the dispute may relate to questions such as the inflexibility of a treaty as compared to a new physical situation, the interpretation, the application or the particular relevance of a treaty or indeed to its own total or partial validity, or to a number of other legal questions.

As to the ways in which the boundaries on a contiguous river have been fixed throughout history, these are indicated here below.

11.1.3.1 The boundary at the banks (river res nullius)

A medieval theory set the boundaries at the banks of the river, while the river itself was considered *res nullius*, i.e., belonging to nobody. This was stated in the Latin maxim, *Rhenus est una ripa Galliae et altera Germaniae limes*, one bank of the Rhine belonged to Gaul and the other to Germany. However, this practice only prevailed during the Middle Ages, as the Rhine then constituted the border between the Roman and Germanic worlds.

11.1.3.2 The boundary at the banks (river res communis)

There have been cases in which the boundary has been fixed at the banks of a river, while the river itself was considered *res communis*, i.e., the common and undivided property of the two riparian states. This was the case, *inter alia*, of a treaty between Prussia and the Netherlands of 7 October, 1816, dealing with the Rhine, the Moselle and other rivers. This régime is also applied to that portion of the Moselle River separating Luxembourg and Germany.

11.1.3.3 The boundary at one of the banks

In this case the boundary of one state extends up to the bank of the opposite riparian state, and its territorial jurisdiction includes the whole river.

Many treaties have fixed this type of river boundary, of which one of the first was the Treaty of Osnabruck of 24 October, 1648, concerning the Oder River. The same situation applies to various international rivers and lakes on the basis of treaties signed, generally, between a strong colonial power and a less strong independent state. A case in point is that of the boundary fixed on the Shatt-el-Arab, which gave Iraq sovereignty over the whole river up to the bank of Iran, thus allowing Iraq to control and collect tolls from the traffic entering the Shatt-el-Arab. A series of negotiations, followed by the signature of a treaty in 1975, brought about a change in this situation, and the boundary was fixed on the thalweg (deepest channel). The adoption of the thalweg criterion, however, has remained largely on paper, as disputes arose and a war started in 1980 between the two states. Apparently, boundary issues have not been resolved, even after the war.

An arbitral award was also rendered in 1933 in a dispute between Guatemala and Honduras, which fixed the boundary on some sections of the Rivers Rio Tinto and Montagua at the right bank.

11.1.3.4 The boundary at the median line

Under this practice, the state boundaries are fixed at the middle of the river, i.e., on the 'median line' or on the imaginary line equidistant from either riverbank, also corresponding to the geometric centre of the river.

One of the first treaties fixing the boundary at the median line was the Treaty of Paris of 10 November, 1763, concerning the Mississippi River, signed between England, France and Spain. There are a number of agreements concluded between states establishing this type of boundary. However, in the case of navigable rivers, the median line boundary has not been a satisfactory solution, in that, in following the median line, navigation is not always possible. In addition, this line tends to move when the water level augments, or in case of floods or whenever the opposite banks of a river are of different heights. Due to natural occurences, therefore, in many cases the median line cannot be firmly determined, and continuous revision may be required.

11.1.3.5 The boundary at the thalweg

'Thalweg' is a German word meaning deepest channel. The fixing of the boundary at the thalweg has been introduced in many international agreements, particularly in the case of navigable rivers, in order to facilitate their use for navigation purposes.

The first agreement fixing the boundary at the thalweg may have been the Treaty of Luneville of 9 November, 1801, concerning the Rhine. However, an older record seems to exist in a boundary controversy between the Abbot of St.Gallen and the Bali of Feldkirchen of 1560, when it was stated that the boundary between Austria and Switzerland was where 'a feather with a calm weather follows the Rhine flow . . . .' The Treaty of Westphalia of 1648 also fixed the boundary between France and the German Empire at the thalweg. The criterion of the thalweg has almost always been followed for the determination of river boundaries in Africa, Asia and the Americas.

The exact meaning of the expression 'thalweg' varies from convention to convention. Sometimes it indicates the deepest part of the river, sometimes that following the main navigable channel, other times the median line of the main navigable channel. The Treaty of 16 November, 1825, between France and Switzerland concerning the Rhone defines the thalweg as 'the line according to which the last watercourse would flow without interruption in all of its extension, assuming a gradual lowering of the level which would end by its drawing up.'

The question as to whether the thalweg criterion in the case of navigable contiguous rivers and the median line criterion in the case of non-navigable contiguous rivers are to be considered as general principles or rules of international law is still the subject of discussion.

Some authors consider them as positive rules of international law, while others uphold a different view.

11.1.4 Natural modifications of the boundary on a contiguous river

It is possible that a natural modification of a contiguous river causes a disruption in the existing boundary.

Ancient Roman law distinguished between sudden modifications due to violent causes and gradual modifications due to a process of sedimentation; in the first case, the boundary remained fixed on the old line of demarcation, while in the second case it followed the slow shifting of the watercourse. The same principles have been upheld in various judiciary cases[3] and in an arbitral award rendered between the USA and Mexico for the Chamizal territory, of 15 June, 1911.

The legal effects of the shifting of the bed of a contiguous river as a consequence of natural factors can only be determined on the basis of the will of the concerned states, as expressed in an agreement. If changes occur after a boundary has been conventionally fixed, the states have to resort to a second convention if they want to amend the provisions concerning the boundary.

Any modification or interpretation of an existing boundary treaty should be the subject of an agreement between the states concerned. Thus, Article 30 of the Treaty of Peace of Versailles following World War I states: 'It shall be the duty of the boundary commissions established by the present Treaty to indicate if the boundary line shall follow, in the case of shifting, the course of the channel now so defined, or if it shall be determined in a permanent manner according to the position of the watercourse or the canal at the moment of the entry into force of the present Treaty.'[4]

11.1.5 The boundary on a bridge over a contiguous river

The legal questions raised by the ownership, construction, operation and maintenance of bridges crossing contiguous rivers are many. Here we are concerned only with the determination of the boundary on the bridge in relation to the existing boundary on the river, median line or thalweg.

Generally speaking, it is the median line which determines the border on the bridge.[5] However, difficulties arise when the boundary on the river is determined by the thalweg. In this connection, the Treaty of Versailles of 28 June, 1919, at Article 66 declared that the sovereignty over the bridges crossing the Rhine belonged to France up to the German bank, although the border on the river was fixed on the thalweg. This clause was modified in 1930 by an agreement which determined the border at the middle of the bridge. Other agreements have been concluded concerning bridges which cross contiguous rivers.

---

3 Resolution of 1888 concerning the River Drewens separating Russia and Prussia, and Exchange of Notes of 8–15 July, 1891, concerning the River Juba between British Somaliland and Italian Somalia.
4 Treaty of Peace of Versailles, 28 June, 1919. Text in: 29, Herstlet, C.T., 603.
5 See Verdross (1964) *Volkerrecht*, 5th ed. Vienna, 272.

No rule of international water law exists concerning the boundary to be fixed on international bridges; the riparian or basin states should determine among themselves the legal régime of the bridge by means of an international agreement.

11.2 Navigation

11.2.1 Origins

Ancient civilizations did not consider the possibility of navigating on rivers, which were considered as sacred; express prohibitions to navigate on rivers were contained in the laws of Persia and Egypt.

According to Roman law, rivers were *res communis*, i.e., common to everyone. The state had only the right to control and to collect taxes for the maintenance of watercourses; navigation was free.

During the Middle Ages navigation was made subject to all sorts of harassments and fiscal measures which hindered development, even to the point of certain waterways being closed by treaty, as in the case of the Scheldt by the Treaty of Münster of 30 January, 1648.

After the feudal period, in the sixteenth century, Grotius, who is known as the father of international law, introduced the principle of freedom of innocent passage,[6] i.e., the principle according to which rivers should be open for transit for legitimate purposes; however, a person intending to exercise the right of innocent passage had to apply for permission. A permission could only be denied for well-stated reasons; otherwise a violent reaction could be justified.

This conception was restricted by Ziegler, a commentator of Grotius, who asserted that unless a servitude allowed for it expressly, an agreement was necessary to establish a right of passage.[7] According to Pufendorf, a famous Swedish jurist and follower of Grotius, a right of transit must be granted when requested for honest and necessary reasons, in conformity with the law of nature.[8] Another author, Wattel, asserted that the right of passage was an imperfect right, since its request left to the grantor the freedom of deciding whether or not the right had to be accorded.[9]

11.2.2 The internationalization of navigation

It was with the peace Treaty of Westphalia (Münster) of 1648 after the war between France and Germany that navigation became a matter of international concern. Article 9 of the treaty provided that rivers be given their ancient freedom and security and be so maintained forever, as they had been before the war, although before the treaty there had been neither

---

6 Grotius, Hugo, *De Jure Belli ac Pacis, libri tres, in quibus jus naturae et gentium, item juris publici praecipua explicantur*. Book 2, Cap. 2, 3.
7 Ziegler, C., *Hugonis Grotii de Jure Belli ac Pacis Libros quibus naturae et gentium jus explicavit*. Notae et animadversiones subitariae. Book 2, Chapter 2E.
8 Pufendorf, S., *De Jure Naturae et Gentium*. Libri octo, Book 3, para 5–6 (233–6).
9 Wattel, *Le Droit des Gens*, Principes de la loi naturelle appliquée à la conduite et aux affaires des nations et des souverains. Book 2, Cap. 9.

freedom nor security on international rivers. The declared principle of freedom was contradicted by the closure of the River Scheldt by the United Provinces by virtue of the very same treaty.

A large number of bilateral treaties according freedom of navigation to the contracting parties were concluded between 1648 and 1800.

The inhibitory system prevailing throughout the Middle Ages was strongly shaken by the French Revolution. A decree of the interim Executive Council of the French Republic of 16 November, 1792, declared: '1. That the difficulties and obstacles to which navigation and commerce, both on the Scheldt and on the Meuse, have been subject are contrary to the fundamental principles of natural law that the French people have undertaken to respect. 2. That the course of rivers is common, inalienable property of all riparian land . . .' Based on these considerations, the council instructed the chief general of the French army during the Belgian expedition to take all measures and to employ all means available to ensure freedom of navigation and transport on all courses of the Scheldt and the Meuse Rivers.[10]

During the following years, this principle was incorporated into various conventions, among which the treaty of 16 May, 1795, between France and the Batavian Republic concerning navigation on the Rhine, Scheldt and Moselle, and the Campoformio Peace Treaty of 17 October, 1797, between France and the Austro-Hungarian Empire. Here, also, freedom of navigation was restricted to riparian states only.

In 1798, at the Congress of Rastadt concerning the peace between France and the German Empire, the French plenipotentiaries proposed absolute freedom of navigation on the Rhine for all flags, including those of non-riparian states, subject to the condition that consent be granted thereto by France and the Empire. The proposal was not accepted, and the congress ended with the murder of the French plenipotentiaries. This shows the animosity surrounding this issue, which continued to be debated during the successive two centuries.

A convention relating to the right to navigate on the Rhine was concluded between France and the German Empire at Paris on 15 August, 1804. This convention, which at least implicitly excluded non-riparians from the enjoyment of such right, was inspired by the concept of community of interests to the benefit of riparians and created a Franco-German administration for the collection of taxes.

Progress in the regulation of international rivers was made with the conclusion of the Treaty of Paris of 30 May, 1814, which established that navigation and commerce on the Rhine be open (free) to all flags, with the possibility for riparian states to levy certain tolls. In addition, the treaty provided that the extension of this régime to other international rivers be studied during the future Congress of Vienna.

11.2.3 The Congress of Vienna (1815)

The Congress of Vienna re-established the political régime which had existed before the French Revolution and the Napoleonic period. During the congress, the question of international rivers was again discussed by means of a navigation commission, formed first to deal with the Rhine and its tributaries and then with other rivers. Three main theories were proposed: the first, a French position, would limit the freedom of navigation and commerce

---

10 Decree of the Provisional Council of the Convention, 16 November, 1792.

to riparians; the second, proposed by the British, propounded such freedom to all flags, subject to their respect of police regulations to be established; the third, propounded by Russia, would extend to non-riparians the freedom of commerce only, that of navigation being excluded and reserved to the riparians. Hence, the commerce of non-riparians would have had to be effected by means of vessels belonging to riparians, and by riparians only.

The outcome of the congress was the approval of 32 articles concerning navigation on the Rhine and of seven articles relating to navigation of the Rivers Main, Neckar, Moselle, Meuse and Scheldt. Nine general articles (108–116) were incorporated in the Final Act of Vienna of 9 June, 1815.[11] These general articles are very relevant, as they set up and still constitute the basis of international law with respect to navigation on international rivers.

According to Article 108, the riparian states undertake to regulate all matters relating to navigation on international rivers by common agreement. Article 109 establishes the principle of freedom of navigation on international rivers, but, as far as non-riparians are concerned, they are only accorded freedom of commerce with the exclusion of navigation.

Articles 110 to 115 deal with the régime applicable to the assessment and collection of tolls and taxes for navigation and related services. These have to be uniform, invariable, moderate (covering real operation and maintenance of services costs) and easy to collect by well-identified officers. In short, these procedures, including customs, should not interfere with navigation. For the undertaking of maintenance works, the riparian states must consult each other (Article 113).

The Congress of Vienna constituted a step backwards as compared to the liberal principles proclaimed by the French Revolution and during the Napoleonic period. While absolute freedom of navigation and commerce for riparians and non-riparians was not established, the following principles did become crystallized:

(i) the principle of freedom of navigation and of commerce for the riparian states;
(ii) the principle of freedom of commerce, but not of navigation, for the non-riparian states;
(iii) the duty to consult on, and settle by common agreement among the riparian states, all questions affecting navigation.

It may be said that these have become general principles of international river navigation law. Support for this opinion was given also by the Permanent Court of International Justice in the River Oder Case in stating that 'this community of interest in a navigable river becomes the basis of a common legal right, the essential features of which are the perfect equality of all riparian states in the use of the whole course of the river . . .'[12]

The principles established in Vienna were applied to a number of international rivers, but always in a particular agreement concerning, separately, each river and were always limitative of the interests of third states as to freedom of navigation and commerce. Between 1815 and 1831, particular conventions concerning the Rivers Moselle and Vistula (1815), the Oder (1818), the Neman (1818), the Elbe (1821), the Weser (1823) and the Rhine (1831) were concluded. These conventions provided for freedom of commerce for all flags, but

---

11 Text in: 1, Herstlet, C.T., 3.
12 P.C.I.J. Ser. No. 23, 26–27, 1929.

reserved cabotage, i.e., the transport of passengers and goods on the rivers within one riparian state's jurisdiction, to that riparian state only.

This is why, under the provisions of the Convention of Mayence of 1831[13] concerning the Rhine, anyone could load goods on vessels directed to any one of the ports of the navigable network of the Rhine, under the condition, however, that such vessels belonged to riparians. The Central Commission for Navigation on the Rhine was established by the convention of 1831. This commission, which was composed of one representative for each riparian state, had a consultative role; it was responsible for studying and preparing solutions as concerned both navigation regulations and works, a final decision thereon being left to the governments. The commission, however, had judiciary powers: violations of navigation regulations were brought before the national courts first; appeals against the judgements of these courts could then be brought before either the national courts of second instance or before the central commission.

11.2.4 The Treaty of Paris (1856): the régime of the Danube

A step towards the affirmation of the principle of freedom of navigation was made after the Crimean War with the Treaty of Paris of 1856,[14] concerning the Danube. Despite its importance because of its length and the many states it crosses or separates, this river had not been considered during the Congress of Vienna, as Turkey, i.e., the Ottoman Empire, riparian to the downstream section of the Danube, including its mouth, was not regarded as a part of the European community of interests. Already during the War of Crimea, conferences had taken place in Vienna between 1853 and 1856 to study the problem of navigation on the river, with a view to ensuring its freedom under the control of a permanent authority in which both riparian and non-riparian states would have been represented. In the course of these conferences, Austria expressed the intention to establish different régimes for the upper Danube, in respect of which agreements between Austria and other riparian states limited the freedom of navigation to riparians only, and the lower Danube, which would have to be open to all flags. After the War of Crimea, Turkey was admitted to benefit from European public law. The Treaty of Paris established freedom of navigation on the entire course of the river for all flags. In addition, the treaty provided for equality of treatment of all nations as concerned the payment of tolls at the mouth of the Danube. These would be fixed and meant for covering the cost of the execution of works. No duties or charges based solely on the fact of navigation or on the type of goods transported would be levied. The Treaty of Paris extended the principle of equality of treatment of all nations to other aspects, including cabotage.

Two commissions were established: the Commission of the Riparians, and the European Commission. While the former, composed of representatives of the riparian states and conceived as a permanent body, was put in charge of the preparation of the navigation act for the Danube, the latter, composed of representatives of the European powers and of a temporary nature, was made responsible for the supervision of the works to be executed on the delta in order to improve navigation.

---

13 Text in: 10, Herstlet, C.T., 471.
14 *Ibid.*, 421.

However, the Navigation Act of 1857, prepared by the Commission of the Riparians, contained clauses which restricted the freedom of navigation by reserving cabotage to riparians only. For this reason, it was not applied and was the reason for the abolition of the commission. The European Commission, the duration of which was to have been limited to the duration of the works on the delta, was therefore charged, in 1865, with the responsibility of preparing navigation regulations for the entire river. This was carried through satisfactorily, and the European Commission has been in function up to present times.

11.2.5 The navigation régime after 1856

The régime established for the Danube in respect of freedom of navigation was more liberal than that established for the Rhine. This induced the states riparian to the Rhine to introduce changes. A new convention was concluded at Mannheim on 17 October, 1868, to substitute the régime established in the convention of 1831 with a new régime. The new convention provided for absolute freedom of navigation, besides that of commerce, for all nations, prohibited the establishment of tolls and excluded the payment of duties on goods in transit. In addition, it extended this régime to the tributaries of the Rhine.

This absolute freedom was, however, only apparent. In fact, it was subject to the obtention of pilot and vessel licences. As regards pilot licences, these were only accorded to persons who could prove to have practiced navigation on the Rhine for a given period of time and to have obtained from the government of the state in which they had established domicile a licence authorizing them to exercise the independent profession of pilot (Article 15). As to licences for vessels, these had to be issued pursuant to an inspection by the competent authorities of one of the riparian states and renewed after any important repair or change.

It is easy to see how, although the Convention of Mannheim proclaimed absolute freedom of navigation, the two above-mentioned requirements rendered the effective exercise of navigation difficult for non-riparians.

11.2.6 The Act of Berlin (1885)

In 1884 the European powers met in Berlin in order to regulate their respective spheres of influence in the occupied territories of Africa and the development of commerce there. On 26 February, 1885, the Act of Berlin,[15] which established rules for the partition of these territories, was concluded. The act provided for the 'internationalization' of two main African rivers, the Congo and the Niger, and proclaimed freedom of navigation and commerce on these rivers, their tributaries and their mouths for all states. As regards the Congo only, the act stipulated freedom of access for armed vessels, provided that the general provisions concerning neutrality in times of war were respected. The régime of freedom was extended to land in the sections where the course of the river was interrupted by rapids. In addition, the act prohibited the establishment of tolls for purposes other than in the interest of navigation. Two particular acts were formulated, one for the Congo and the other for the Niger.

Following this, a large number of treaties proclaimed the principle of freedom of navigation in Asia and Latin America.

---

15 Act of Berlin of 26 February, 1885. Text in: 17, Herstlet, C.T., 62.

11.2.7 The régime after World War I

The First World War caused enormous changes in the territorial and political set-up of the world. With particular reference to Europe, the interest of the Allied and Associated Powers, i.e., the winning powers, to establish a régime of freedom of navigation on the rivers flowing within the territories of the defeated states caused the introduction of a series of liberal principles concerning international rivers by means of the peace treaties following the war.

The Treaty of Peace with Germany, concluded at Versailles on 28 June, 1919, established a system of integral internationalization of the waterways to which it expressly applied, i.e., the Rivers Elbe, Oder, Neman and Danube. In addition, it contained clauses relating to the Rhine and the Moselle. The main characteristics of the system were as follows:

(i) it also applied to national waterways, given that these belonged to Germany, which had lost the war, and particularly to the Oder and to the waterway connecting the Rhine to the Danube, should this be constructed (Article 33);
(ii) it applied to the benefit of all flags, riparian and non-riparian (Article 332);
(iii) it implied the resort to an international jurisdiction instituted by the League of Nations (the Permanent Court of International Justice) as concerns the obligation of the riparian states to execute maintenance and improvement works and in the case of execution of works hindering navigation (Articles 336–337);
(iv) it implied the establishment of river commissions composed of representatives of both the riparian and non-riparian states;
(v) it confirmed the rules already established as regards tolls and duties by limiting these to what was necessary to cover the costs incurred to maintain or improve navigability conditions.

As regards the Rhine, the Treaty of Versailles introduced modifications geared towards a more liberal régime of navigation on the river; the rights enjoyed by riparians were extended to non-riparians. In addition, the composition of the central commission was modified to include additional members as well as representatives of non-riparian states (Great Britain, Italy and Belgium), and it was made responsible for preparing a new statute of the Rhine to replace the Convention of Mannheim of 1868.

The principles introduced by the Treaty of Versailles were also applied to navigation on the Elbe by the Convention of Dresden of 22 February, 1922.

A step backwards as regards the régime introduced in Africa by the Act of Berlin of 1885 was made by the Treaty of St. Germain-en-Laye of 10 September, 1919. The absolute freedom of navigation for all nations on African rivers was restricted, to the advantage of the members of the League of Nations only, or for health purposes.

11.2.8 The régime established at Barcelona (1921)

The liberal régime introduced by the Treaty of Versailles was reconsidered in the course of the Barcelona Conference convened by the League of Nations in March, 1921, to elaborate a general and uniform navigation régime for international rivers.

This conference prepared an act composed of a convention by which the contracting parties undertook to respect given rules, a statute containing these rules and an additional protocol.[16] The act was signed on 20 April, 1921, and entered into force on 31 October, 1922.

While the régime introduced by the Treaty of Versailles only applied to specified rivers, that of Barcelona regarded all international rivers. The statute, entitled 'Statute on the Régime of Navigable Waterways of International Concern,' designated by this expression (i) navigable waterways of international concern by nature, i.e., geographically, and (ii) navigable waterways of international concern by declaration, i.e., declared as such by way of convention or unilateral acts of the states under whose jurisdiction they are placed (Article l). Waterways belonging to these two categories must be naturally navigable; their navigability must not have been the result of man's action.

This new expression is much broader than the expression 'international rivers.' Besides rivers, it covers all waters, including lakes, to the exclusion of the sea. In addition, it allows for the extension of the international régime to waterways geographically national. The adoption of the expression 'of international concern,' on the other hand, reveals the intention to leave a riparian state free to regulate the waterways situated within its territory, provided that the principles introduced by the statute are respected.

The régime introduced by the Act of Barcelona was characterized by the principle of freedom of navigation, but this freedom was accorded only to the contracting parties. A step backwards was therefore made with respect to some of the provisions of the Treaty of Versailles, which extended the régime to all nations. In addition, the statute reserved cabotage to riparian vessels.

As regards dues, Article 7 confirmed the principles introduced by the Act of Vienna, by establishing that no dues of any kind be levied, other than those 'in the nature of payment for services rendered and intended solely to cover in an equitable manner the expenses of maintaining and improving the navigability of the waterway and its approaches, or to meet expenditure incurred in the interest of navigation.'

11.2.8.1 The administration of international waterways

Article 14 of the Barcelona Convention provided for a régime of international administration by means of international river commissions, which could include the following attributions:

(i) the power to draw up navigation regulations;
(ii) the power to indicate to the riparian states the works advisable for the maintenance of navigability conditions;
(iii) the right to receive from the riparian states information concerning improvement projects;
(iv) the power to approve dues and charges, when these are not provided for by the act of navigation concerning the waterway.

The duty of riparian states to maintain navigable waterways and to remove any obstacle to navigation was kept by the Barcelona statute.

---

16 Convention and Statutes on the Régime of Navigable Waterways of International Concern, Barcelona, 20 April, 1921. Text in: 7, LNTS, 36.

If not otherwise stipulated by agreement, a reasonable contribution towards upkeep costs could be requested to a co-riparian state. However, the state could free itself from such obligation if a co-riparian state agreed to execute the works.

11.2.8.2 Evaluation

The Convention of Barcelona included principles which were too abstract and rigid, in that, in reality, each international river presents its own particular characteristics.

However, it introduced a new principle, that of the prevailing economic function of a river. Such function was still navigation, but provision was made in Article 10.6.7, that a riparian state could close all or part of a waterway to navigation if it could 'justify its action on the ground of an economic interest clearly greater than that of navigation.' This exceptional provision underlines the fact that already at that time utilizations other than navigation were acquiring relevance. In addition, it crystallized the principle of freedom of navigation.

11.2.9 Developments after Barcelona

After the Barcelona Convention another navigation convention was concluded in 1921 as regards the Danube.[17] It should be noted that when the First World War broke out this river was divided into four sections:

(i) the upstream section, from Ulm to Orsova, was administered by a consortium of four riparians (Bayern, Württemberg, Austria-Hungary and Serbia). These states had maintained in force among themselves the Navigation Act of 7 November, 1857, which reserved cabotage to riparians only;
(ii) the second section, from the Iron Gates (after Orsova) to Turnu Severin, was dangerous but navigable. Works had been undertaken by Hungary, which, had, however, by its own regulations, levied taxes which were not commensurate to the tonnage of vessels, but to the kind of goods;
(iii) the third section, from the Iron Gates to Galatz (middle Danube), was not subject to uniform regulation. Each state had its own regulations;
(iv) the fourth section (lower Danube) was under the authority of the European Commission, which exercised strong powers with respect to navigation and works.

For a period of time after the victory of the Allied Powers, the river was administered by inter-allied military commissions.

The Navigation Convention of 23 July, 1921,[18] divided the Danube into two sections, the Maritime Danube (up to Braila) and the Fluvial Danube.

The Maritime Danube section was placed under the administration of the European Commission, re-established with a modified composition (France, the United Kingdom, Italy and Romania); the Fluvial Danube section was subjected to the control of an international

---

17 Text in: 26, LNTS, 173.
18 Danube Statutes. Text in 26, LNTS, 17.

commission composed of the riparian states and three non-riparian states (the U.K., France and Italy). This commission was given the following powers:

(i) to control and supervise the freedom of navigation;
(ii) to draw up an annual programme of work;
(iii) to prepare navigation and police regulations.

The German Reich, in a note of 14 November, 1936, denounced the clauses of the Treaty of Versailles. Another aspect of this regressive trend is represented by the Agreement of Sinaia of 18 August, 1938, which reduced the powers of the European Commission of the Danube with respect to navigation on the Maritime Danube; these powers were transferred to an autonomous Romanian service.

11.2.10 The régime after World War II

The Paris Peace Treaties of 10 February, 1947,[19] established freedom of navigation on the Danube for the vessels and merchandise of all nations, to the exclusion of cabotage. The USSR and the Danubian states, however, were oriented towards a limitation of the freedom of navigation to riparian states only. The Belgrade Conference initiated a new Navigation Convention for the Danube, signed on 18 August, 1948, which reserved to the riparian states the right to control and regulate navigation. The creation of an international commission composed only of representatives of the riparian states was provided for.

An additional protocol expressly abrogated the pre-existing navigation régime and, in particular, the convention of 23 July, 1921. However, the USA, France and the United Kingdom (this also on behalf of Belgium, Italy and Greece) did not accept the new régime, which limited their rights, and they continued to honour the convention of 1921 and the European Commission.

As regards the Moselle, freedom of navigation and equal treatment for all nations were provided for by the Convention of 27 October, 1956, between France, the Federal Republic of Germany and Luxembourg. The same applied to the canals between Metz and Koblenz up to its confluence with the Rhine. This convention created a commission composed of representatives of the three riparian states, national river tribunals and a court of appeal. Canalization works are to be financed by the Societé Internationale de la Moselle, a German limited company.

With respect to the Rhine, the Convention of 20 November, 1963, between Belgium, France, Great Britain, the Netherlands, the Federal Republic of Germany and Switzerland introduced some procedural amendments to the Convention of Mannheim of 1868. It is to be noted that Switzerland had never participated in the Rhine system.

Other agreements have been concluded for the Scheldt (1957 and extension of the régime of the Convention of 1963 for the Rhine), the Saar and the Ems (1960 and 1962).

---

19 Text in: 41, LNTS, 21.

11.3 Non-navigational uses of water

11.3.1 Introduction

Since ancient times, civilization and development have been based on the availability and use of water resources for domestic, agricultural and commercial purposes. Many ancient civilizations developed along rivers: the Tigris and the Euphrates, the Nile, the Huang Ho, the Indus, the Ganges and the Tiber. Even then, conflicts arose on the allocation and use of waters flowing through two or more political units. Societies felt the need for a body of rules governing water use. At the same time, two principles were recognized, i.e., that of the sovereignty of the states on the territory where water resources of interest to other states were located, and that of international cooperation and solidarity as a basis on which to organize the joint use of water resources.

The historical development of international water law has followed closely that of political, economic, technical and social needs. According to such needs, on which depends the prevailing of one water use over another, the development process is more or less marked.

If, in the past, conflicts were primarily related to navigation and small-scale water uses, today they cover a much wider spectrum. Water diversions by one riparian state may have negative effects on the uses that another riparian state intends to make of the same water.

The non-navigational uses of water include irrigation, floating, domestic uses, fishing, aquaculture, hydroelectric power production and industrial uses.

The river basin concept is particularly relevant for the latest theories concerning these uses.

11.3.2 The theory of absolute territorial sovereignty

Nowadays, few accept the thesis according to which a state is the absolute master of its own territory, empowered to use the resources it encounters therein without consideration for the effects it may cause beyond its frontiers. This thesis was sustained by the Attorney General of the United States, M. Harmon, in 1895, in a controversy between the United States and Mexico concerning the diversion and use of water from the Rio Grande. Rejecting the Mexican claim to the effect that prior agreement between the two countries was necessary on the grounds that the United States could not make use of the river water in such a way as to markedly reduce the flow, Harmon stated, 'The fundamental principle of international law is the absolute sovereignty of every Nation as against all others within its own territory ... all exceptions, therefore, to the power of a Nation within its own territory must be traced up to the consent of the Nation itself. They can flow from no other legitimate source.' On this premise, 'the rules, principles and precedents of international law impose no liability or obligation upon the United States.'[20] In the 1950's the USA renounced the principle as not conducive to cooperative arrangements, and the principle was never applied.

This theory is known as 'absolute territorial sovereignty' and is favored by upstream states. It is an extreme theory which ignores the rights of the downstream states. It fails to

---

20 Moore (1906) *Digest of International Law* 654. See also Caponera, D.A. (1980) *The Law of International Water Resources*. Legislative Study No. 23. Rome, FAO, 7.

appreciate the dual character of a state, namely, that territorial sovereignty is a source of obligations as well as of rights.

The Advocate General of East Punjab, India, referred to the principle in the wake of the Indus dispute. However, the 1960 Indus Water Treaty does not reflect the notion. In the case of the Ganges dispute, for quite some time India subscribed to the notion that the Ganges, if not an Indian river, is 'essentially an Indian river.'

11.3.3 The theory of absolute territorial integrity

The theory of 'absolute territorial integrity' is upheld by lower riparian states and corresponds to the 'theory of natural flow,' whereby a state is entitled to expect that the same volume of water, uninterrupted in quantity, flows into its territory.[21] The concept of the natural flow theory likewise awards 'rights without duties' to the lower riparian states. It includes a right of veto against any upstream water utilization that is likely to disturb the natural flow. This theory was never accepted in international law and practice, since it was conceived as unworkable. In the 1950's it was accepted by Egypt, Pakistan and Bangladesh, which are downstream states.

11.3.4 The theory of limited territorial sovereignty and integrity

The two above-mentioned theories, territorial sovereignty and territorial integrity, are unworkable and may lead to lengthy discussion among the co-riparians. Instead, 'limited territorial sovereignty and integrity' is a theory based on the assertion that every co-riparian is free to use the waters of shared rivers within its territory on condition that the rights and interests of the other co-riparian states are taken into consideration. In this case, sovereignty over shared waters is relative and qualified. The co-riparians have reciprocal rights and duties in the use of the waters of common rivers. Physical unity creates a unique legal unity leading to the formation of a 'community of interests,' and the waters of the shared rivers so become *res communis*.

This theory has been well accepted in international law, as it was recognized by the Permanent Court of International Justice in 1929, in its judgement on the territorial competence of the River Oder Commission,[22] when it affirmed that states have a common legal right to the resources of a shared river, not only a right of passage, the essential characteristic being the community of interests of all the parties in the use of the river and the exclusion of preferential privilege of any one riparian state in relation to the others.

A further refinement of the court's theory was called for to deal with water uses other than navigation, and in particular with complex multi-purpose projects which required large water diversions and substantial alterations of the water flow. In the case concerning the diversion of water from the Meuse, the court held that while the two parties (Belgium and the Netherlands) could modify, enlarge, transform, fill the canals and increase the volume

---

21 For a discussion of the common law principle, see Kiechel, J.R., and Green, M. (1976) Riparian Rights Revisited: Legal Basis for Federal Instream Flow Rights. 16, *Natural Resources Journal*, 969.
22 Text in: *Permanent Court of International Justice*, Series A, No.23, Series C, No.17 (II). Parties to the dispute were Germany, Denmark, France, Great Britain, Sweden, Czechoslovakia and Poland.

of water in them, this could only be done on condition that 'the volume of water was not affected.'[23] The same principle was upheld in the arbitral award concerning the Helmand River Delta Case.[24] In the award between France and Spain concerning the utilization of the waters of Lake Lanoux in 1957,[25] it was held that the protection of the interests of all riparian states is an important principle.

11.3.5 The shared natural resources concept

This theory, which emerged during the second half of the twentieth century, stems from the United Nations General Assembly (UN General Assembly) resolutions on the permanent sovereignty over natural resources by the states. This sovereignty is limited by the similar rights of countries sharing the same basin. The theory reinstates the community of interest approach and attributes a positive duty to engage in active cooperation in the rational development and utilization of the shared water resources.

11.3.6 Equitable and reasonable utilization and participation

The latest theory concerning the use of international waters for purposes other than navigation has been referred to as 'equitable and reasonable utilization and participation.' It is embodied in the Convention on the Law of the Non-Navigational Uses of International Watercourses (UN Watercourses Convention), which was adopted by the UN General Assembly in 1997 and is the result of the work of the International Law Commission. *Inter alia*, the focus of this work has been on how better to define, in the draft articles that were then embodied in the convention, the substance of international cooperation in the use and development of the waters of international watercourses. The convention deals with surface water, groundwater being covered only in so far as it is connected to surface water.

The equitable and reasonable utilization theory derives from the Helsinki Rules of the International Law Association (ILA) of 1966,[26] which, in turn, derive it from the theory of limited territorial sovereignty and integrity. It asserts that the right of a co-basin state is to be regarded in the light of the similar right of another co-basin state. It does not purport to identify the respective share of a basin state in the waters, which has to be ascertained on a case-by-case basis, based on the relevant factors. This flexibility represents its strength, as it allows states to accommodate a variety of interests. Equitable and reasonable utilization may also provide a response to the question as to whether there is a human right to water in international law, which has been raised on a number of occasions in recent years.[27]

---

23 The diversion of water from the Meuse, Judgement of 28 June, 1937. Text in: *Permanent Court of International Justice*, Series A/B, No. 70, Series C, No. 81.
24 Arbitral Awards of 19 August, 1872, and 10 April, 1905. Text in: Mayors St. John, Lovett, and Evan Smith and Mayor-General Sir Frederick John Goldsmid, *Eastern Persia, An Account of the Journeys of the Persian Boundary Commission*, 1870–71–72, London, 1876, Vol.I, 413. On 7 September, 1950, the two governments signed the 'Terms of Reference of the Helmand River Delta Commission.'
25 Text in: *International Law Reports*, 1957, 101.
26 See Chapter 10, Section 10.3.9.2.
27 See Section 11.8. For a discussion of this issue, also see Salman, A.M.S., & McInerney-Lankford, S. (2004) *Human Right to Water: Legal and Policy Dimensions*. Law, Justice and Development Series. Washington, D.C.,

Equitable apportionment, on the contrary, is the version of the equitable and reasonable utilization theory favoured by engineers, who like to 'apportion' water. At a certain stage of the codification effort, it was mentioned by the International Law Commission, together with equitable utilization and participation. However, the apportionment approach is more difficult to achieve in view of the various interests involved and the need to quantify the water of a basin in order to reach agreement. Among others, this approach was opted for by Egypt and Sudan through the Agreement for the Full Utilization of Nile Waters, signed on 8 November, 1959.[28]

In 2010, six of the Nile basin countries signed the Agreement on the Nile River Basin Cooperative Framework,[29] of which a cornerstone principle is equitable and reasonable utilization. Three of them, Ethiopia, Rwanda and Tanzania, have ratified it. In spite of an impasse over going forward with the ratification of the agreement because Egypt and Sudan are reluctant to relinquish what they define as their 'historical water rights,' these two countries, together with Ethiopia, endorsed the Joint Declaration of Principles on the Grand Ethiopian Renaissance Dam in 2015.[30]

The expression 'equitable participation' corresponds to that of 'equitable utilization.' This is an obvious development of a legal principle appreciating the physical facts of water, the sovereignty claims, the shared natural resources concept and the community approach. It is a flexible rule which provides scope to establish justice and fairness on the basis of circumstantial factors. The principle was initially developed by courts in federal countries.

The reciprocity of respective rights and duties of states sharing a common basin acquires the force of a rule of conduct generally applicable to the relations between these states. The rule includes the obligation of a state to refrain from causing significant harm to other states concerned with the same water resources and the obligation to cooperate.

11.3.7 Obligation not to cause significant harm

Article 7 of the UN Watercourses Convention requires watercourse states to 'take all appropriate measures to prevent the causing of significant harm.' This obligation also derives from the theory of limited territorial sovereignty and integrity, and entails the duty of a state to take all necessary preventive measures in order that the question of harm shall not arise where international water resources are concerned. Thus, it is an obligation of conduct based on the standard of due diligence; it is not an obligation of result. The appropriateness of the measures that a state is bound to take to prevent the harm will depend on its capability in terms of human, financial and other resources. Emphasis is placed on the 'significance' of the harm, as it is only in such case that the violation of a rule of international may be envisaged. 'Significant' means that the harm must neither be trivial, nor substantial, and must be capable of being determined on the basis of objective evidence.

The World Bank; and McCaffrey, S. (1992) A Human Right to Water: Domestic and International Implications. *Georgetown International Environmental Law Review*, Vol. V, Issue 1, 1–238.
28 Text in: ST/LEG/SER.B/12, 143.
29 Available from www.nilebasin.org/index.php/nbi/cooperative-framework-agreement.
30 See this chapter, Section 11.3.9.

The obligation to prevent harm was highlighted in the Trail Smelter Case arbitral awards,[31] in the Lake Lanoux case,[32] and was reiterated in Principle 21 of the Declaration of the United Nations Conference of Stockholm in 1972. Under the UNECE Water Convention, 'The Parties shall take all appropriate measures to prevent, control and reduce any transboundary impact' (Art. 2, Para. 1). Although not related to water resources, the United Nations Convention on Biological Diversity (1992) recognizes the sovereign right of states to exploit their own resources, provided that they ensure that 'activities within their jurisdiction or control do not cause damage to the environment of other states or of areas beyond the limits of national jurisdiction' (Art. 3).

The ILA Helsinki Rules do not make specific reference to the obligation not to cause significant harm. Rather, they list 'the degree to which the needs of a basin state may be satisfied, without causing substantial injury to a co-basin state' among the factors that states must consider to determine their reasonable and equitable share in the use of the waters of an international drainage basin. Under the UN Watercourses Convention, the equitable and reasonable utilization and the no significant harm rules go hand in hand. If a state deems to have suffered significant harm as a result of activities taking place on the territory of another watercourse state, it shall enter into consultation with such other state, with a view to finding a negotiated solution of the matter that is equitable and reasonable for both states, having regard to the provisions of Articles 5 and 6.

11.3.8 Floating

Floating corresponds to the use of waters for timber floating and rafts. This particular type of utilization is practiced along the rivers of the northern European and American continents. An agreement is necessary for floating in the case of shared waters, because this use comes into conflict with that of navigation.

The Helsinki Rules of the ILA recognize that this use always requires the agreement of the sharing states and recommends that the right of floating be granted in the case of rivers which are not used for navigation, thus determining the prevalence of navigation over floating.[33]

11.3.9 Production of energy and industrial uses

The exploitation of water resources for the production of energy is another use which is relevant in international water resources law. Switzerland offers various examples of international cooperation. The first legal agreement concerning hydroelectric power production with Germany was stipulated in 1879. In 1919, the German-Swiss Permanent Commission for the production of energy was formed. In parallel with this, agreements allow a German corporation with headquarters in Germany to manage an installation situated in Switzerland at Albbruck-Dogern, with a Swiss share of 22 percent, while another installation located at Birsfelden is managed by a corporation with exclusively Swiss capital and administration.

---

31 See Chapter 10, Section 10.5.2.
32 *Ibid.*
33 International Law Association, Helsinki Rules, 1966, Arts. 21–25.

There are also examples of Franco-Swiss and Italo-Swiss cooperation, among which can be cited the installation of electricity plants in the Val di Lei and for the exploitation of the Spöl in Italian territory.[34] Obviously the types of agreement and the arrangements entered into vary according to the natural characteristics of the river basin.

The industrialization of different countries worldwide has augmented the development of water resources for hydropower generation purposes considerably. Often, in developing countries international financing institutions have participated in such undertaking, particularly when it refers to the exploitation of the water resources common to several states. It suffices to recall the cases of the Congo (Inga I and II Dams), Niger, Zambesi (Cahora Bassa) and Omo-Turkana (Gibe I, II and III Dams) basins in Africa, the Amazon and the Rio de la Plata (Salto Grande) in South America and the basins of the Ganges, the Indus and the Mekong Rivers in Asia, not to mention similar developments on the Rhine and Danube in Europe.

In the majority of these cases, the production of electricity is compatible with other traditional uses as regards quantitative aspects, provided that certain operational rules are respected. For instance, if the waters of an international watercourse are diverted and stored upstream during the summer (dry season) for the purpose of producing hydropower, it may be difficult, if not impossible, to use those waters to irrigate crops downstream. This has been the case of a hydropower reservoir located on the Syr Darya River (a major tributary of the Aral Sea) in Kyrgyz territory, into which water was stored during the growing season, thereby giving birth to a conflict with Uzbekistan, a downstream state, in the late 1990's.[35] In the more recent case of the Grand Ethiopian Renaissance Dam (GERD), on the Blue Nile in Ethiopia,[36] Egypt and Sudan showed their concern for diminishing flows in the river, particularly during the filling period, and voiced their strong opposition to the dam. After studies on the impact of the dam, a number of tripartite meetings at the technical and ministerial levels, and growing support for the dam by Sudan, on 23 March, 2015, the three countries reached an agreement on the GERD. They signed a joint declaration of principles,[37] which includes equitable and reasonable utilization, the obligation not to cause significant harm, the obligation to cooperate, the duty to exchange information and the peaceful settlement of disputes, as well as principles specific to the dam, its operation and safety. The joint declaration also calls for the setting up of tripartite coordination mechanisms. The sharing of the resource and of the benefits deriving from its exploitation on the basis of criteria of equity is therefore possible.

For industrial uses, one cannot say the same thing as regards water quality. It is not always possible to keep this quality at the level of purity needed for fish culture or for

---

34 Agreements of 18 June, 1949, 25 November, 1952, and 27 May, 1957. Text in: 155 BFSP, 731; ST/LEG/SER.B/12, 852, 859.
35 The dispute was resolved, at least on paper, through an agreement concluded in 1998 between Kyrgyzstan, Uzbekistan and Kazakhstan.
36 Construction started in 2011 and was entirely funded by Ethiopia.
37 Agreement on Declaration of Principles between the Arab Republic of Egypt, the Federal Democratic Republic of Ethiopia and the Republic of Sudan on the Grand Ethiopian Renaissance Dam Project (GERDP), signed at Khartoum on 23 March, 2015. Available from www.internationalwaterlaw.org/documents/regionaldocs/Final_Nile_Agreement_23_March_2015.pdf.

agricultural uses after the waters have been utilized for industrial purposes. The question has been felt with increasing frequency by the tourist industry.

This consideration underlines the fact that it is not possible to predetermine the priorities in the use of water resources common to two or more state in the abstract.

It is to be noted that the rule of reciprocal information in this regard is contained in the Convention of 9 December, 1923, relating to the Development of Hydraulic Power Affecting more than one State.[38]

Another utilization of water resources for industrial purposes is that of the cooling installations connected with nuclear power production.

11.3.10 Procedural rules

The general rules governing the conduct of states summarized so far constitute substantive law. Together with the general obligation to cooperate set out in Article 8 of the UN Watercourses Convention, they form the basis of contemporary international customary water law. For their implementation to be possible, procedural rules have to be taken into account. A procedural rule of general applicability, not limited to hypothetical situations where damage might arise, requires that *states shall inform and consult each other*. The rule refers both to the data and information that are to be exchanged among co-basin states on a regular basis and to the advance notice that is to be provided to the co-basin states by the state planning an activity, programme or project which may have adverse transboundary repercussions.

The duty to exchange regular data and information serves the purpose of enabling the states concerned to make informed decisions as to equitable and reasonable water utilization and to introduce resource conservation and protection measures whenever necessary in order to prevent significant harm. The substance of the data and information may vary, depending on the situation on the ground, but the minimum requirement, as provided for in the UN Watercourses Convention, which codifies rules set forth in numerous international agreements as evidence of consistent state practice, is that they must cover the condition of the watercourse concerned and, in particular, be 'of a hydrological, meteorological, hydrogeological and ecological nature and related to the water quality,' as well as include 'related forecasts.'[39] The data and information to be exchanged must be 'readily available,' but a state may be requested by another to provide not readily available data, subject, however, to payment by the requesting state of the costs of collection and processing. Finally, under the convention watercourse states are under an obligation to 'employ their best efforts to collect and, where appropriate, to process data and information in a manner which facilitates its utilization by the other watercourse States to which it is communicated.'[40] This obligation translates, *inter alia*, into positive action to be taken by watercourse states in order to monitor water resources (and water uses) in accordance with agreed standards or harmonized procedures and methods, leading to at least comparable results.

The obligation to inform and consult on projects and activities with possible adverse transboundary effects was the subject of consideration by the ILA in 1956. According to

---

38 Signed by 17 states at Geneva on 9 December, 1923. Text in: 36, LNTS, 77.
39 Article 9.
40 *Ibid*.

Article 6 of the Resolution of Dubrovnik of that year, 'a State which proposes new works (construction, diversion, etc.) or change of previously existing use of water which may affect the utilization of water by another State, must first consult with the other State. In case agreement is not reached through such consultation, the States concerned should seek the advice of a technical commission; and if this does not lead to agreement, resort should be had to arbitration.'[41] Ten years later this position became less rigid. The ILA Helsinki Rules of 1966, in fact, recommended that each basin state 'furnish relevant and reasonably available information to the other basin States' concerning the waters of a drainage basin within its territory, the relevant water uses and the intended projects. The fixing of a deadline for submitting views or raising objections was also provided for.[42] Similarly, in 1979 the Institute of International Law affirmed the existence of a duty to cooperate, through exchange of information and data, consultation and scientific cooperation, in international water resources pollution matters.[43]

In the ILA's Complementary Rules of 1986, the provision by a basin state of advance notice with regard to projects 'which may substantially affect the interests of any co-basin State' has been construed as a duty. The rules provide guidance as to the procedures to be followed in order to fulfill the duty, and to the effects of the notice. This approach has been adopted by the International Law Commission, with the result that the UN Watercourses Convention devotes an entire chapter to procedural rules on notification.[44] It is obvious, however, that a project may 'substantially affect' the interests of one basin state, but may not affect the interests of another. Moreover, each international river basin has its own characteristics. Thus, implicitly the rules – and the convention – leave to the basin states concerned the task of defining, through an agreement, the projects that are subject to notification.[45] The convention fixes the deadline for replying to a notification and raising objections, if any, at six months.

---

41 Text in: International Law Association, *Report of the Forty-Seventh Conference*, held in Dubrovnik, 1956, London, 1957, 241–243.
42 Article 29. Text in: The International Law Association, *Report of the Fifty-Second Conference; op. cit.*
43 Institute of International Law, *Resolution on the Pollution of Rivers and Lakes and International Law*, Athens, 12 September, 1979. Text in: *Annuaire de l'Institut de droit international*, Vol.58, T.I., Athens Session, September, 1979, Basel, Munich, 1980, 197 ff.
44 Part III, 'Planned Measures,' contains nine articles illustrating the minimum requirements of the notification procedure and the obligations of the watercourse states concerned when the procedure is going on. Article 12 indicates the minimum content of the notification, while Article 13 sets the deadline that has to be met by the notified state(s) replying to the notification. Article 15 is devoted to the modalities for replying to a notification; Articles 14 and 16 deal with the obligations of the notifying states in the period for replying, and with the effects of an absence of reply to notification, respectively. Article 17 calls for consultation and negotiations in the event that the notified states communicates its disagreement as to the measures planned by the notifying state. Finally, Article 18 deals with procedures in the absence of notification, while Article 19 allows the implementation of a planned measure when this measure is urgent in order 'to protect public health, public safety or other equally important interests.'
45 See, for instance, the Agreement on Cooperation for the Sustainable Development of the Mekong River Basin, signed by Cambodia, Laos, Thailand and Viet Nam on 4 April, 1995. Its articles on notification and consultation take into account seasonal requirements. The Convention on the Sustainable Management of Lake Tanganyka, signed by Burundi, the Democratic Republic of Congo, Tanzania and Zambia on 12 June, 2003, indicates in an annex the projects that are subject to notification (and environmental impact assessment).

The rule of reciprocal information and consultation, together with that governing the conduct of states with regard to the management of international water resources, has been embodied in a number of international instruments, such as the Geneva Convention of 1923 relating to the Development of Hydraulic Power Affecting more than one State, the Declaration of Montevideo of 1933,[46] the Charter of the Economic Rights and Duties of States, which, at Article 3, reads, 'In the exploitation of natural resources shared by two or more countries, each State must cooperate on the basis of a system of information and prior consultations in order to achieve optimum use of such resources without causing damages to the legitimate interests of others,'[47] and many international conventions relating to specific river basins, such as the Danube, the Rhine, the Mekong and others.

It should be noted that the obligation to inform and consult does not imply the corresponding powers of veto. It does not mean that one state is obliged to obtain the consent of all interested states, and by that token to conclude an agreement with them before it may proceed. Such an obligation would conflict with the sovereignty principle and with the principle of equality of rights and community of interests, both of these being looked upon as an expression of the priorities of today's international community.[48]

11.3.11 Conclusions

The law concerning the non-navigational uses of international water resources is still going through a process of elaboration. Nevertheless, the International Law Commission has completed its work of codification of the international rules on the subject, producing an international legal instrument, the 1997 UN Watercourses Convention, which is largely based on the work of the ILA and reflects prevailing state practice. The convention entered into force on 17 August, 2014, following the deposit of the 35th instrument of ratification, as required by its Article 35.

The Helsinki Rules of the ILA were formally accepted by many countries oriented towards the joint development of international river basins and were incorporated, in one way or another, in a number of international agreements, such as those relating to the Senegal, Chad, Kagera, Gambia and Mekong basins. In Europe, also, the principle of equitable sharing is clearly to be found in many international agreements and even in some preceding the adoption of the Helsinki Rules. In addition, the Helsinki Rules have been followed by the Supreme Court of India in deciding the Krishna River dispute.

Likewise, many of the principles reflected in the UN Watercourses Convention have been embodied in international treaties or have been followed in the solution of international disputes. For example, the principle of equitable and reasonable utilization has been incorporated into the Framework Agreement for the Sava River Basin of 2002.[49] In addition, the

---

46 *Declaration concerning the Industrial and Agricultural Use of International Rivers*, Seventh Inter-American Conference, Montevideo, 24 December, 1933. Text in: Pan American Union (1933) *Seventh International Conference of American States*, Plenary Sessions, Minutes and Antecedents, Montevideo, 114.
47 United Nations General Assembly, *Resolution 3281 (XXIX) on Charter of Economic Rights and Duties of States*, New York, 12 December, 1974. Text in: United Nations, *Resolutions adopted by the General Assembly during its Twenty-Ninth Session*, Vol. I, 17 September – 18 December, 1974, 50, 52, 55.
48 Berber, F.J.A. (1972) *Flood Control*. Report of the International Law Association. New York.
49 Available from www.savacommission.org/dms/docs/dokumenti/documents_publications/basic_documents/ fasrb.pdf

provisions of the convention are reproduced almost integrally in the Revised Protocol on Shared Watercourses in the Southern African Development Community (SADC) of 2000. Finally, reference to the convention has been made in the Gabcikovo-Nagymaros dispute.

The practical implementation of the principles contained in the Helsinki Rules and the UN Watercourses Convention requires close cooperation among states sharing the same international river basin, which can be established only through appropriate international administrative machinery. Nowadays, close cooperation is particularly required because of the multiple uses to which water is put. Moreover, it is increasingly being recognized that in order to satisfy ecological requirements a certain amount of flow is to be left in the basin.

Cooperation, however, must be established on such principles as good faith, good neighbourly relations, equality and reciprocity, keeping in mind not only national interests, but also those of the basin community as a whole.

11.4 Harmful effects of water

11.4.1 Definition

By harmful effects of water it is generally meant the damage produced by water, either from natural causes or induced voluntarily or involuntarily by man's action, or lack of it. Harmful effects include, *inter alia*, those caused by floods, soil erosion, damages deriving from poor drainage, dumping of waste, misuse of untreated wastewater, salinization, siltation, waterlogging, flow obstruction, salt water intrusion and land subsidence caused by the overpumping of water or of other liquid or gaseous substances from the subsoil. Of these, however, it would appear that international law has mostly addressed flood control.

11.4.2 Evolution

The problem of controlling floods was already present in ancient times, when the so-called hydraulic civilizations utilized water during the high water period for irrigation purposes. In fact, it is in Egypt, Mesopotamia, India and China that the need arose to control the floods caused by the high water periods of the Nile, the Tigris, the Euphrates, the Indus and the Huang Ho.

Floods are rather complex hydrologic occurrences, which may be the result of exceptional rainfalls, the rapid melting of snow, hurricanes or other natural phenomena. Examples are the high waters of the Mississippi in 1927, the floods caused in China by the Huang Ho in 1931, the inundation of Florence by the Arno in 1966, the floods occurring in the State of Bengala, India, in 1968 and 1970, and the floods which have affected many parts of Europe in the late 1990's and continue to affect them in the present century due to changing climate conditions.

The methods most commonly used for the defence against floods are:

(i) construction of dykes, flood walls, levees or embankments to protect land from flood water and to keep it within the usual main channels;
(ii) increasing the discharge capacity of the main channel by strengthening, widening or deepening it, or by a combination of the three;
(iii) diverting part or the whole of the flood waters in excess of the carrying capacity of the main channel;

(iv) constructing reservoirs to hold flood waters temporarily and releasing them later as the channel is capable of carrying them;
(v) taking steps to decrease the rate of discharge by improved land use practices, i.e., afforestation, substitution of erosion inducing crops with soil protecting crops;
(vi) using flood forecasting and early warning systems to minimize loss of life and property;
(vi) identifying and mapping flood-prone areas, in order to subject them to a special régime.[50]

It appears at first sight that flood control is primarily a problem of science and technology, and that its implementation is a matter for municipal legislation and administration. According to one author,[51] 'It is doubtful whether customary international law contains any special rules at all about flood control. The same is true of general principles of law. If they exist at all they are so vague that they are incapable of immediate application as, e.g., the general principle of good neighbourliness is a general guideline for all international water problems.'

Until recently, publicists have treated flood control in short, insignificant passages; by the same token, various bilateral treaties of the past, many of which are still in force, have dealt with flood control, but only incidentally. In addition, these agreements were mainly restricted to frontier waters and the obligations of the parties are almost always limited to the maintenance of flood discharge conditions.

With time, states have felt the need for minimum requirements and standards to be met in relation to the control and mitigation of the effects of flood occurrences within international river basins. Thus, a number of multilateral[52] and bilateral[53] agreements entered into in the past decades provide, *inter alia*, for:

(i) a duty of basin states to exchange regular data and information on water levels, rainfall, snow and ice conditions, among others;
(ii) a duty of basin states to inform each other, without delay, of any flood emergency originating in their territory and, to this effect, to set up appropriate communication, warning and alarm systems;
(iii) a duty of basin states to take all appropriate steps to prevent, mitigate or reduce the harmful transboundary effects of floods.

By the same token, the UN Watercourses Convention states the general duty of watercourse states, whether individually or jointly, to take 'all appropriate measures to prevent or mitigate conditions related to an international watercourse that may be harmful to other watercourse States, whether resulting from natural causes or human conduct, such as floods or ice conditions, waterborne diseases, siltation, erosion, salt water intrusion, drought or desertification.' This duty is all-encompassing, as it also covers events deriving from human action, such as

---

50 Normally, new construction is prohibited within designated flood control or management areas, and certain activities are prohibited or limited in the interest of public safety and health.
51 Berber, F.J.A. (1972) *Flood Control*. Report of the International Law Association. New York.
52 Among which the 1992 UNECE Convention on the Protection and Use of Transboundary Watercourses and International Lakes, the 1994 Danube Agreement, the 1999 Rhine Convention, the 2002 Framework Agreement for the Sava River Basin.
53 Such as those entered into by the Tisza basin states in the late 1990's. The Tisza is a tributary of the Danube.

industrial accidents. It translates into positive obligations to provide timely notification of emergency situations and to take appropriate measures to prevent, mitigate and eliminate the harmul effects of these situations. The convention also calls for the development of joint contingency plans.

In 2007, the European Union (EU) issued a directive on the assessment and management of flood risks, which supplements and completes the provisions of the Water Framework Directive.[54] This directive requires EU member states to assess and map flood risks within each river basin district and to prepare flood risk management plans on the basis of the data collected. In the case of an international river basin district, the member states concerned shall ensure the exchange of the relevant information and the production of a single plan, or coordinated plans, for the river basin district. When a river basin district extends beyond EU boundaries, member states are required to endeavor to coordinate their planning efforts with those of non-member states. The directive recognizes that cooperation in flood risk management planning might not always be possible, and, if this is the case, it allows each member state to plan for the portion of the basin located in its territory. If the basin crosses EU boundaries, this may result in a plan being produced only for that portion of the basin located within the EU, because flood risk management planning, like basin planning, is an obligation for EU member states but not for non-members. Under Article 7 of the directive, 'flood risk management plans established in one member state shall not include measures which, by their extent and impact, significantly increase flood risks upstream or downstream of other countries in the same river basin or sub-basin, unless these measures have been coordinated and an agreed solution has been found among the member states concerned.'

In addition to floods, a natural phenomenon which may produce harmful effects connected with the use of shared water resources relates to the erosion of embankments. This particular aspect has also been the subject of conventional norms. It suffices to recall Article 4 of the treaty between the USA and Mexico of 23 November, 1970,[55] concerning the maintenance of the embankments of the Colorado and Rio Grande.

11.4.3 The emerging rule

As may be inferred from the provisions of the UN Watercouses Convention, there is a general duty of basin (or watercourse) states to take action, as appropriate, to prevent or mitigate the effects of emergency situations that may be harmful to other basin states. For this purpose, co-basin states shall provide regular and timely exchange of data and information and consult for the planning and implementation of joint measures whenever they would be more effective than individual measures.

In addition, states shall take all measures necessary to ensure that activities under their jurisdiction or control are so conducted as not to cause water-related hazards or other adverse conditions which may result in significant harm to other states. The concept of 'significant' harm is important, as a minor damage (*de minimis*) is not relevant.

---

54 See Chapter 6, Section 6.4.5.1.
55 Treaty to Resolve Pending Boundary Differences and Maintain the Rio Grande and Colorado River as the International Boundary between the United States of America and Mexico. Text available from www.ibwc.gov/Files/1970_Treaty.pdf

The diversity of conditions in different river basins precludes the formation of uniform, detailed regulations for all countries or all river basins. However, from the study of conventional practice a rule can be formulated. This rule:

(i) recognizes the existence in the international community of an obligation to cooperate for the control of floods, on the basis of the principle of good neighbourliness;
(ii) makes it an obligation for the states to communicate, as soon as possible, information on heavy rainfalls, sudden melting of snow or other events which are liable to create floods and dangerous rises of water levels in their respective territories; and
(iii) upholds that the states should set up an effective system of transmission and priority for communicating flood warnings.

The UN Watercourses Convention has taken a similar course of action. At Article 28, it makes it an obligation for a watercourse state to notify other potentially affected states and competent international organizations of any emergency situation originating in its territory, or of which it has knowledge. The article provides that a state within whose territory a water-related danger or emergency situation originates shall immediately take all practical measures to prevent, neutralize or mitigate the harmful effects of such situation, in cooperation with the potentially affected states and, when applicable, with the competent international organizations. In order to fulfill their obligations effectively, watercourse states, in cooperation with the other potentially affected states and, where appropriate, competent international organizations, shall develop contingency plans for responding to emergency situations.

It is interesting to note that a basin state is not liable to pay compensation for damage caused to another basin state by floods originating in its own territory unless it has acted contrarily to what could be reasonably expected and unless the damage caused is significant.

The above underlines the correct interpretation of international responsibility and foresees the fault of the state for not having observed the reasonable diligence that circumstances call for to avoid significant harm.

11.5 Quality control of water

11.5.1 Definition

Water may be polluted for one purpose and not for another. Thus the term 'pollution' varies according to the receiving body of water. Therefore it is important in defining pollution to refer to the future use of that water.

Generally speaking, the origins of pollution are three:

(i) that deriving from organic sources, i.e., organic material formed by the discharge into the water of the effluents from domestic sources;
(ii) pollution originating from inorganic sources, i.e., that pollution which has its origin in inorganic materials, mainly discharges from industrial and chemical sources;
(iii) pollution which has its origin in radioactive substances; this is the case of pollution originating from nuclear sources.

As regards a definition of pollution in international law, the authorities have various solutions. The best definition is given by the ILA, which in Article 9 of the Helsinki Rules refers to the term water pollution as follows: 'Any detrimental change resulting from human conduct in the natural composition, content or quality of the waters of an international drainage basin.'[56] A similar definition is provided by the UN Watercourses Convention.

Another definition of pollution is that of the United Nations Economic Commission for Europe (UNECE), which defines a watercourse as being polluted when the condition of the water 'is directly or indirectly modified as a consequence of the activities of man to such an extent that these waters are less apt to be used in their natural state . . .'[57]

A relevant consideration on water pollution to keep in mind is the effects that such pollution causes, if and when it causes them, which determines state responsibility.

The consequences of water pollution are the progressive destruction of the flora and fauna (for example, the disappearance of salmon in the Rhine), the degradation of potable water, a phenomenon very worrysome for the Netherlands, and, generally speaking, a modification of the equilibrium of the water.

The problems relating to water pollution are being examined by numerous agencies and programmes of the United Nations, including the Food and Agriculture Organization (FAO), the United Nations Education, Science and Culture Organization (UNESCO), the World Health Organization (WHO), the World Meteorological Organization (WMO), the United Nations Environment Programme (UN Environment) and the United Nations Economic Commission for Europe (UNECE). Furthermore, they are being dealt with by the Organization for Economic Cooperation and Development (OECD), the European Economic Community (EEC) and, more recently, the European Union, the Council of Europe, the International Maritime Organization (IMO) and, finally, by the North Atlantic Treaty Organization (NATO), which has created a special office for the study of this problem. The COMECON also elaborated regulations on the subject in the past.[58]

11.5.2 Evolution

Generally speaking, the concept of pollution refers explicitly to the criteria of reasonableness or to that of normal tolerability, or finally, to that of 'consistency' of the damage.

Various agreements address the question of water pollution within the framework of the general utilization of water resources. However, in the past decades several agreements have dealt specifically with this question reaching, at times, a certain level of detail.

In Europe and in the UNECE region the problem of water pollution has received much attention and has been addressed through several agreements, including:

(i) the UNECE Convention on the Protection and Use of Transboundary Watercourses and International Lakes, signed at Helsinki on 17 March, 1992 (UNECE Water Convention);[59]
(ii) the Protocol on Water and Health to the UNECE Water Convention, signed at London on 17 June, 1999,[60] which places emphasis on public health;

---

56 International Law Association, Helsinki Rules, Art.9.
57 European Conference on Water Pollution (1958) UN Doc. E/ECE/311, 4.
58 The Regulation of 23 November, 1961, elaborated by COMECON, states the fundamental principles recommended to the riparian states.
59 Text in: 31, ILM, 1312 (1992).
60 UN Document MP.WAT/2000/1 EUR/ICP/EHCO 020205/8Fin., 18 October, 1999.

(iii) the Convention on the protection of the Rhine of 12 April, 1999, between the Federal Republic of Germany, France, Luxembourg, the Netherlands and Switzerland and the European Community.[61] The convention repeals the Berne Convention of 1963 and the 1976 Convention for the protection of the Rhine against chemical pollution;

(iv) the Convention on cooperation for the protection and sustainable use of the River Danube (Convention for the Protection of the Danube), between Austria, Bulgaria, Croatia, the Czech Republic, Germany, Hungary, Moldova, Romania, Slovakia, Slovenia, Ukraine and the EU, signed at Sofia on 29 June, 1994;[62]

(v) the Agreement on the International Commission for the Protection of the Oder from Pollution, between Germany, the Czech Republic, Poland and the European Community, signed at Bratislava on 11 April, 1996;[63]

(vi) the Convention of Bonn of 3 December, 1976, between the Rhine basin states, on the protection of the Rhine against pollution from chlorides;[64]

(vii) the Convention Concerning Fishing in the Waters of the Danube, between the USSR, Bulgaria, Hungary, Romania and Yugoslavia, signed at Bucharest on 29 January, 1958 (Article 7);[65]

(viii) the Convention on the Protection of Italo-Swiss Waters Against Pollution, between Italy and Switzerland, signed at Rome on 20 April, 1972.[66]

The UNECE Water Convention mandates the parties to adopt measures to prevent, control and reduce water pollution, based on the following principles:

(i) these measures are to be taken at source, if possible;
(ii) the precautionary principle, by virtue of which action to avoid the potential transboundary impact of the release of hazardous substances is not to be postponed on the grounds that scientific research has not fully proved a causal link between those substances, on the one hand, and the potential transboundary impact, on the other;
(iii) the 'polluter pays' principle;
(iv) inter-generational equity, meaning that water resources are to be managed so that the needs of the present generation are met without compromising the ability of future generations to meet their own needs.

The convention provides an indication of the measures to be taken by the parties, which include the control of wastewater discharges through licencing mechanisms, the setting of emission limits based on best available technologies, the setting of water quality objectives and criteria, the imposition of strict water quality requirements when conditions so warrant, the prevention of groundwater pollution and the monitoring of water resources and of wastewater discharges.

---

61 Text available from www.eda.admin.ch/dam/eda/fr/documents/aussenpolitik/voelkerrecht/rhin/conv 1999_fr.pdf.
62 Text in: *Official Journal of the European Communities*, L 342, 12/12/1997, 19.
63 *Ibid.*, L 100, 21.
64 Text in: 1404, UNTS, 91.
65 Text in: 339, UNTS, 23.
66 It concerns Lakes Lugano and Maggiore. Text available from www.fao.org/faolex/results/details/en/c/LEX-FAOC031925/.

In addition, the parties sharing specific water bodies have the duty to conclude agreements which provide for the establishment of joint institutional mechanisms, and to cooperate in the implementation of programmes and measures which are listed in the convention. In brief, these are the major features of the convention relevant to water pollution.

The convention has served as a model for a number of particular conventions, both in Europe and in the UNECE region. In fact, elements of it may be found in the Rhine and Danube conventions, and in others, including those for the Dniester, Meuse, the Sava, the Scheldt and the Oder River basins, and for Lake Prespa. What is more, by virtue of an amendment of 2003, which entered into force in 2013, the convention is now open to the participation of all UN member states.

The EU Water Framework Directive details and implements provisions of the UNECE Water Convention. As was mentioned earlier,[67] EU member states are bound to translate into domestic law the provisions of this directive and those of the other directives addressing water quality and pollution control issues. The same applies to the states applying for membership in the EU, which are required to transpose EU legislation into their legal systems. The directive provides guidance with regard to the classification of water bodies, the setting of environmental quality objectives and the standardization of monitoring, sampling and analysis methods. Furthermore, it provides for a definition of quality standards with reference to priority substances.

It is worth recalling that the directive requires the designation, by the states concerned, of an international river basin district whenever a river basin spans an international boundary, and the production of a single management plan for the basin. When an international river basin extends beyond EU boundaries, the states which are members of the EU are required to endeavour to cooperate with non-members in the achievement of the objectives of the directive, i.e., the production of a basin management plan and the identification of environmental objectives. The directive recognizes that cooperation in river basin planning might not always be possible, and, if this is the case, it allows each member state to plan for the portion of the basin located in its territory. If the basin crosses EU boundaries, this may result in a basin plan being produced only for that portion of the basin located within the EU, because basin planning is an obligation for EU member states but not for non-members. Examples of successful cooperation between EU member states and non-members are offered by the Danube basin, where Moldova and Ukraine are not members of the EU, and by Lake Peipsi and the other water resources shared between Finland and the Russian Federation, for which a commission was established in 1997.

In North America, mention may be made of the 1978 Agreement between the USA and Canada on Great Lakes Water Quality, amended on 16 October, 1983, on 18 November, 1987 and on 7 September, 2012.[68] This agreement expands the scope of work of the International Joint Commission established under the 1909 Boundary Waters Treaty to include water pollution issues affecting the Great Lakes.[69] Also to be mentioned is the progressive

---

67 See Chapter 6, Section 6.4.5.
68 Text available from https://binational.net//wp-content/uploads/2014/05/1094_Canada-USA-GLWQA-_e.pdf.
69 The Commission assists the two governments by collecting, analyzing and disseminating data, carrying out public information activities and other means. It is supported by the Water Quality Board and the Science Advisory Board. See Bourne, C.B. (1997) *International Water Law, Selected Writings*. International and National Water Law and Policy Series. The Hague, London, New York, Kluwer Law International, 354 ff.

broadening in this direction of the mandate of the International Boundary and Water Commission between the USA and Mexico, which was created on the basis of the treaty of 3 February, 1944, concerning the utilization of the waters of the Colorado and Tijuana Rivers and of the Rio Grande. As for South America, it is worth mentioning the obligation to prevent pollution contained in Art. 41 (a) of the Statute of the Uruguay River (1975).[70]

Until recently, there were very few examples of treaties dealing with water quality issues in Africa. However, a number of agreements concluded in the not so distant past demonstrate that the obligation to prevent, reduce and control pollution is becoming well consolidated all over the world. Many transboundary water agreements have been entered into in Southern Africa under the umbrella of the SADC Revised Protocol, which reflects the provisions of the UN Watercourses Convention. Among others, we may quote the agreements concerning the Okavango (1994), the Orange-Senqu (2000), the Incomati-Maputo (2002), the Limpopo (2003) and the Zambesi (2004). Agreements concluded elsewhere in Africa lead to the same conclusions. The agreements on the Nile (2010), Lake Victoria (2003) and the Congo (2007) are further cases in point, while the water charters of the Senegal (2002), the Niger (2002) and Lake Chad (2012), adopted to supplement the existing agreements, enter into detail as to the measures to be implemented by the states concerned. These agreements recognize the need for common approaches and call for the harmonization of actions by the states concerned.

Asian water treaties are mainly concerned with water resources development and use, and with the prevention of the harmful effects of water. Only a few of them deal with water quality issues. These are, *inter alia*, the 1995 agreement between Mongolia and the Russian Federation on the protection and management of transboundary waters[71] and the 2011 agreement between China and Kazakhstan on the protection of water quality in transboundary rivers[72]

The UN Watercourses Convention addresses water quality aspects by requiring watercourse states, whether individually or jointly, to 'prevent, reduce and control the pollution of an international watercourse that may cause significant harm to other watercourse States or to their environment, including harm to human health or safety, to the use of the waters for any beneficial purpose or to the living resources of the watercourse.' To this end, the states have the duty to harmonize their respective policies on the subject, and, upon the request of any of them, to 'consult with a view to arriving at mutually agreeable measures and methods to prevent, reduce and control pollution of an international watercourse, such as:

(a) setting joint water quality objectives and criteria;
(b) establishing techniques and practices to address pollution from point and non-point sources;[73]
(c) establishing lists of substances the introduction of which into the waters of an international watercourse is to be prohibited, limited, investigated or monitored.'

---

70 Text in: 1295, UNTS, 331.
71 Text available from www.ecolex.org/details/treaty/agreement-between-the-government-of-the-russian-federation-and-the-government-of-mongolia-on-protection-and-management-of-transboundary-water-tre-151518/
72 Text available from www.fao.org/faolex/results/details/en/c/LEX-FAOC110874/ (in Russian).
73 For a definition of non-point pollution, see Chapter 7, Section 7.13.5.

The provisions of the convention reflect current state practice, but in Europe this practice, which has been absorbed into a number of international agreements including those listed above, goes well beyond the obligations just mentioned. European international agreements often indicate the concrete steps that the parties must take in order to meet the legal obligations stemming from them. *Inter alia*, they require the parties to monitor and assess the qualitative status of water bodies on the basis of common or compatible methods, and provide for the establishment of international basin commissions. It is to be noted that while the UNECE Water Convention envisages a duty for the states sharing international water resources to establish international commissions, under the UN Watercourses Convention watercourse states are not obliged to do so.

As far as judicial decisions are concerned, reference may be made to the Trail Smelter Case between the USA and Canada, which concerned smoke emission connected with the activity of a smelting company in Canadian territory into United States territory. The United States requested from Canada the payment of damages for atmospheric pollution. The internal tribunal stated that, 'under the principle of international law as well as the law of the United States, no state has the right to use or permit to use its territory when the case is of serious consequences and the injury is established by clear and convincing evidence.'[74] The Corfu Channel Case is interesting because of the principles that it imposes for the responsibility for extra-territorial damages: 'It is every state's obligation not to allow knowingly its territory to be used for acts contrary to the rights of other states.'[75] In the Lake Lanoux Case the arbitral tribunal admitted that there is an 'interdiction prohibiting a state upstream to alter the water of a river in such condition as to cause substantial damage to the downstream states.'[76] Finally, in the Pulp Mills Case the obligation to prevent pollution set forth in the Statute of the Uruguay River was invoked by the International Court of Justice when it asserted the obligation of Uruguay to notify Argentina of the Pulp Mills undertaking and to carry out an environmental impact assessment.

As regards the principles of international law, it may be said that an absolute interdiction of pollution is not admissible, since it would constitute an obstruction to useful economic activities and therefore would become an abuse of right. Some authorities derive the prohibiton of pollution from the principle of good neighbourliness.

As to doctrine, in the Declaration of Madrid, 1911, of the Institute of International Law, we find a clause according to which, 'Every alteration harmful to water, every discharge of residual matter coming from industries, is prohibited.'[77] This was at the beginning of the last century when pollution was not such a grave issue. On the same line is the Declaration of Montevideo, adopted in 1933 at the Seventh International Conference of American States. Article 2 reads, 'No state may, without the consent of the other riparian states, introduce into watercourses of an international character for the industrial or agricultural exploitation of their water supply any alteration which may prove injurious to the banks of the other states.'[78]

---

74 See the Trail-Smelter Dispute in the Canadian Yearbook of International Law, 1973, 213 ff.
75 ICJ, Corfu Channel Case, 1949, 18 ff.
76 Luzzano, *Responsabilita' e colpa in diritto internazionale*.
77 Article 2.2 of the text.
78 Declaration of Montevideo Concerning the Industrial and Agricultural Uses of International Rivers, Art. 2.

Greater attention to the problem of pollution in shared water resources appears in the texts of resolutions of the ILA. During the session of Dubrovnik of 1956, the association approved the statement, 'Preventable pollution of water which does substantial injury to another state renders the former state responsible for the damage done.'[79] In the Helsinki Rules, 1966, the association made a substantial contribution. These rules, supplemented in 1982, impose on basin states the obligation to maintain water quality and the duty to compensate the prejudice caused by the increase in the existing pollution. Basin states must prevent any new form of water pollution and any increase in the degree of existing water pollution which would cause substantial injury in the territory of other basin states. Furthermore, they must take all reasonable measures to abate existing pollution in an international drainage basin to such an extent that no substantial damage is caused to the territory of a co-basin state. In the case of a violation of the rules, the state responsible shall be required to cease the wrongful conduct and to pay compensation.[80]

11.5.3 The emerging rule

International river or lake basins must be protected in their entirety. Any attempt to define the legal régime of international water resources without respecting the unity of the river basin is doomed to failure. Therefore, it is important to create basin institutions to deal with water quality and pollution issues. A number of bilateral and multilateral commissions have been created for this purpose.[81]

Whether these commissions should be made responsible for water pollution control only, or for integrated water resources management, including pollution, is a matter for discussion. Recent developments, however, show a trend to vest international basin institutions with functions encompassing all aspects whenever feasible.

There is a customary rule which recognizes the illegality of pollution. Such rule establishes the principle according to which states cannot alter the composition of a watercourse to the detriment of other states. There is an obligation of the states to prevent, control and reduce the pollution of shared water resources while these are flowing through their territory, and, if there has been pollution, there is the obligation of *restitutio in integrum*, i.e., to reconstitute the preceding situation. Naturally, this principle becomes operative only if the pollution produces 'relevant,' 'substantial' or 'significant' damage or harm to other countries and if it is a consequence of illicit behaviour. Thus, the obligation of the states sharing transboundary water resources to prevent, reduce and control water pollution is a corollary of the general obligation not to cause significant harm, which is codified in the UN Watercourses Convention. To regulate possible conflicts between states, consultation and negotiation are necessary, possibly through a basin institution.

Various principles have been called upon to justify this rule, such as the principle of the abuse of rights, the theory of good neighbourly relations, the principle of equality of rights, the principle of equitable utilization and the principle that states shall not cause substantial

---

79 ILA, Principle 7 of the text. See ILA Report of the Dubrovnik Conference, 1956.
80 For a discussion of the ILA rules relating to pollution, see Bourne, C.B, *op. cit.*, 107 ff.
81 See Chapter 13.

injury in the territory of others.[82] There is no single justification, and the principle of the prohibition of pollution can rely on each one of these considerations. Recently, emphasis has been placed on the principle of precaution (*Vorsorgeprinzip* in German), which is stated in many international environmental instruments and is now embodied in a number of international agreements relating to European river basins.[83] Outside Europe, it is also accepted in North America[84] and increasingly in Africa, but there is still no evidence of its consolidation as a principle of international water law in other regions beyond the industrialized world. The precautionary principle places emphasis on a duty of states to take cost-effective action in order to prevent environmental degradation, even in the absence of full scientific evidence.

If it cannot be denied that a state causing pollution in an international context is responsible for damages caused by its behaviour, there is no unanimity in the doctrine as to the foundation on which this responsibility is based. Authors have spoken of the violation of a conventional obligation, of the violation of a general obligation derived from general international law or from the obligation to neighbourly relations.

The 'polluter pays' principle constitutes an economic directive rather than a legal principle. However, the fact that a number of countries have introduced it into their domestic legislation and that it is increasingly being referred to by international conventions are strong arguments in favour of its recognition as a general principle of law.

11.6 Armed conflict

11.6.1 Definition

In times of armed conflict or similar circumstance, it may occur that water resources be utilized as destructive weapons. In particular, activities may include:

(i) the destruction of water installations with harmful consequences for the life and property of the population of the downstream states;
(ii) the cutting of the water supply;
(iii) the poisoning of waters, making them unfit for human consumption;
(iv) the diversion of water for military purposes to damage or destroy the minimum conditions for survival.

Unfortunately, existing conventional law and international customary law do not offer any rule to prohibit these actions.

---

82 Principle 21 of the Stockholm Declaration, 16 June, 1972.
83 The principle was reflected in the 1982 World Charter for Nature, and in the 1992 Rio Declaration on Environment and Development.
84 See the Great Lakes – St. Lawrence River Basin Sustainable Water Resources Agreement between the states of Illinois, Indiana, Michigan, Minnesota, New York, Ohio, Pennsilvanya and Wisconsin (USA) on the one hand, and the Province of Ontario and the Government of Quebec (Canada) on the other; 13 December, 2005.

11.6.2 Precedents

Various precedents purport to codify the law of war with reference to the protection of the civilian population in times of armed conflict. However, they do not contain specific reference to the protection of water and of water installations. These precedents include:

(i) the Fourth Hague Convention of 1907, which contains express rules on the law of war;
(ii) the Geneva Protocol of 1925, dealing with the same subject; and
(iii) the Fourth Geneva Convention of 1949, relating to the protection of civilian persons.

The customary rules and general principles that may be derived from these conventions affirm the following:

(i) that the right of belligerents to adopt means of injuring the enemy is not unlimited (Fourth Hague Convention, Article 22);
(ii) that it is prohibited to employ poisonous arms (Fourth Hague Convention, Article 23 lit.a);
(iii) that it is prohibited to destroy or seize enemy territory unless the destruction or seizure be imperatively demanded by the necessity of war (Fourth Hague Convention, Article 23 lit.g);
(iv) that it is prohibited in war the use of asphyxiating, poisonous or other gasses, and all analogous liquids, materials or devices, as well as the use of bacteriological methods of warfare (Geneva Protocol of 1925);
(v) that it is prohibited to destroy by the occupying powers in the territory under military occupation of any real or personal property belonging individually or collectively to private persons or to the state, unless destruction is absolutely necessary for military operations (Fourth Geneva Convention of 1949, Article 53).

Finally, Resolution 2444 (XXIII dated 13 January, 1969) of the UN General Assembly contains some rules:

(i) that the rights of the parties to a conflict to adopt means of injuring the enemy are not unlimited;
(ii) that it is prohibited to launch attacks against the civilian population;
(iii) that a distinction must be made at all times between persons taking part in the hostilities and members of the civilian population.

The Institute of International Law, at its session of Edinburgh, 1969, adopted a resolution culminating in Statement No. 7, that international law prohibit 'the use of all weapons which by their nature affect indiscriminately both military and non-military objectives, or both armed forces and civilian populations.'

In the light of this situation, the following five rules can be considered:

(i) the use of arms, materials and measures likely to cause unnecessary suffering is prohibited;
(ii) military acts of warfare may not be directed against non-combatants and civil subjects;
(iii) the application of prohibited acts is justified as legitimate reprisal only if the proportion is appropriate and if it is not directed against especially protected persons or objects;

(iv) military necessity justifies the application of prohibited acts only in cases in which this exception has been expressly reserved (Fourth Hague Convention of 1907);
(v) the use of poisonous means of combat is forbidden.

11.6.3 The emerging rule

From the above it may be said that no specific rule, either conventional or customary, exists concerning the use of water resources as a destructive means during armed conflict. The Helsinki Rules of the ILA attempt to underline that water, indispensable for the health and survival of civilian populations, should not be poisoned or rendered otherwise unfit for human consumption.

The Helsinki Rules go further: the diversion of watercourses should be prohibited if this act should be cause of 'disproportionate suffering of the civilian population or substantial damage to the environment.'

The rules also state that the water supply installations which are indispensable for the minimum condition of survival for the civilian population should not be cut off or destroyed (Article 2), and that 'the destruction of water installations containing dangerous forces such as dams and dykes should be prohibited' (Article 5).

Naturally, these measures should apply not only during military action but also during occupation of territories whenever the occupying authorities manage the resources of the population.

Finally, it is interesting to note that the ILA proposal on the effect of the outbreak of war and validity of treaties concerning the use of water resources is not termination of these treaties but only suspension (Article 7), and that a peace treaty should not deprive the population of its water resources, which are necessary to its economy and physical condition for survival.

If, as a result of war, there is a fixing of a new frontier, the hydraulic system on the territory of one state thus becoming dependent on the works within the territory of another state, arrangement should be made for the safeguarding and uninterrupted delivery of water supplies indispensable for the vital needs of the people.

Under the UN Watercourses Convention, 'International watercourses and related installations, facilities and other works shall enjoy the protection accorded by the principles and rules of international law applicable in international and non-international armed conflict and shall not be used in violation of those principles and rules.'

This is the state of the law with respect to the use of water resources during armed conflict. It can be seen that, in spite of armed conflict in various parts of the world, during the First and Second World Wars and in more recent conflicts such as the Gulf War, states have generally not resorted to measures involving water considered heinous in international relations.

11.7 Environmental aspects

11.7.1 Definitions

The relative interdependence of water with other natural resources within a shared basin has given rise to special consideration by environmental lawyers. Starting with the United Nations Conference on the Human Environment (Stockholm, 1972), international law on the environment has developed within the framework of the UN and aims at ensuring the most rational exploitation of natural resources and their conservation for the benefit of present and future generations.

Before dealing with the topic 'environment' and to make its comprehension easier, it is necessary to give some basic definitions of what is generally meant by 'natural resources' and the 'human environment.'

Natural resources are physical natural goods. They include:

(i) space and its content, i.e., air (atmosphere), mineral content, wind, climate, (Hertz) waves;
(ii) energy (solar, nuclear, wind, hydraulic, thermic);
(iii) land and topography (including slopes);
(iv) soil;
(v) panoramic/scenic resources;
(vi) minerals;
(vii) water (surface, underground, atmospheric);
(viii) the sea and the seabed, including the resources below the seabed;
(ix) plants;
(x) animals;
(xi) geothermal resources.

Natural resources may be *renewable*, i.e., replenished naturally, or *non-renewable*, i.e., depleted by use.

Natural resources are only one element of the human environment. The human environment also includes the created environment, i.e., things or institutions created by man, and the induced environment, i.e., the products of agriculture, pisciculture, forestry, etc., which are induced, not created by man. Hence, the concept of human environment is much broader than that of natural resources.

11.7.2 Evolution

International environmental law is a set of international legal rules drawn up for the purpose of protecting the environment. Its sources are the same as those of international law, i.e., according to Article 38 of the Statute of the International Court of Justice:

(i) international conventions, whether general or particular;
(ii) international custom;
(iii) general principles of law;
(iv) judicial decisions and the teachings of the most qualified publicists, as subsidiary means.

Already during the past century a number of conventions dealt with the protection of single natural resources such as fish, certain birds or certain other animals. It was then that conventions were concluded to establish protected areas (reservations, natural parks) for the conservation of fauna and flora (London Convention, 1933, Washington Convention, 1940, etc.). After World War II, in Europe various bilateral conventions concerning the use of frontier rivers were concluded. Around 1960, conventions on the protection of certain rivers or bodies of water from pollution (the Rhine, the Moselle, Lake Constance) were also concluded.

By the end of the 1960's, the industrial boom had created an increasing concern for the state of the environment. Initiatives for the defence against water and air pollution were taken under the auspices of the Council of Europe. Also, conventions for the protection of natural resources were concluded, including the African Convention on the Conservation of Nature and Natural Resources of Algiers, 1968,[85] and the global Convention on Wetlands of International Importance Especially as Waterfowl Habitat concluded at Ramsar, 1971,[86] which recognizes the ecological function of wetlands and requires the parties to formulate and implement their planning so as to promote their conservation and, as far as possible, their wise use.

The Stockholm Conference on the Human Environment of 1972, resulting in a Declaration and an Action Plan, represents an important milestone in the development of international environmental law, since it had the merit of stimulating further thinking and action in the field of the environment. The principles set out in the declaration, although not legally binding as such, have been absorbed into various international conventions. Principle 2 asserts that 'The natural resources of the earth, including the air, water, land, flora and fauna and especially representative samples of natural ecosystems, must be safeguarded for the benefit of present and future generations through careful planning or management, as appropriate.' Principle 21, approved unanimously by the UN General Assembly, concerning the duty of states not to cause or let cause damage to the environment of other states or of areas beyond the limits of national jurisdiction, is considered as an established principle of international law.

Further to the Stockholm Conference, the UN General Assembly decided to create the United Nations Environment Programme (UNEP),[87] which is now known as UN Environment, to act as a focal point for action and coordination in the field of the environment within the United Nations system. In 1978, the Governing Council of UNEP adopted Draft Principles of Conduct in the Field of the Environment to guide states in the conservation and harmonious utilization of shared natural resources. Among other considerations, the draft principles call upon states to cooperate with a view to controlling, preventing, reducing or eliminating adverse environmental effects that may result from the utilization of these resources, consistent with the principle of equitable utilization. They further recommend that states, before engaging in activities that may significantly affect the environment of other states, carry out an environmental impact assessment. These principles, although not legally binding, have been absorbed into various international conventions.

A number of UN conferences of relevance to environment protection took place after the Stockholm Conference, namely, the UN Conference on Human Settlements (Vancouver, 1976), the UN Conference on Desertification (Nairobi, 1977), the UN Water Conference (Mar del Plata, 1977) and the UN Conference on Environment and Development (Rio de Janeiro, 1992). The latter conference produced a declaration reiterating the duty of states to ensure that activities within their jurisdiction and control do not cause damage to the

---

85 Replaced in 2003 with a convention by the same title.
86 Text in: 996, UNTS, 245.
87 Resolution 2997 (XXVII) of 15 December, 1972.

environment of other states (Art. 2). A major achievement of the Rio conference of 1992 was the opening for signature of two important global conventions, namely, the Convention on Biological Diversity (Rio de Janeiro, 1992)[88] and the Framework Convention on Climate Change (New York, 1992).[89]

The Convention on Biological Diversity requires the parties to integrate considerations relating to the conservation and sustainable use of their biological resources into decision making at the national level, and to adopt measures to prevent or minimize impacts on biological diversity (Art. 10). It further calls for the protection of ecosystems, as well as for their rehabilitation and restoration once they are degraded (Art. 8). An ecosystem is defined as 'a dynamic complex of plant, animal and micro-organism communities and their non-living environment interacting as a functional unit' (Art. 2). Under the Framework Convention on Climate Change, the parties undertake to 'cooperate in preparing for adaptation to the impacts of climate change; develop and elaborate appropriate and integrated plans for coastal zone management, water resources, agriculture, and for the protection and rehabilitation of areas, particularly in Africa, affected by drought and desertification, as well as floods' (Art. 4.1.e). A further offspring of recommendations tabled during the Rio conference is the Convention to Combat Desertification (Paris, 1994),[90] which calls for cooperation, among the states affected by desertification, in environment protection and the conservation of land and water resources, as they relate to desertification and drought (Art. 4.d).

These conventions recognize that water resource management may not to be viewed in isolation from the management of other natural resources. Rather, it must be considered from a broader perspective that takes into account the interaction between water, other natural resources and other elements of the environment, not only from the standpoint of the (economic) benefits accruing to the society from the physical development of water resources. This approach recognizes that social and economic development, and the health and well-being of humankind, are closely related to the sustainable management of natural resources and the environment. Thus, environmental considerations must be integrated into decision making at all levels, and environmental requirements must be considered alongside water requirements for economic development. The concept of ecosystem, appearing in the Stockholm Declaration and defined through the Convention on Biological Diversity, has been employed in several water agreements to highlight the need to consider water and other related natural resources as a whole when making management decisions. Among other things, the obligation of states to take appropriate measures to conserve and, when necessary, restore ecosystems is provided for in Article 2 of the UNECE Water Convention. Similarly, the UN Watercourses Convention establishes the duty of states to protect and preserve ecosystems (Art. 20) and the related duty to prevent the introduction of alien and new species that may have adverse effects on the ecosystem of an international watercourse, thereby causing significant harm to other watercourse states (Art. 22).

---

88 Text in: 1760, UNTS, 79.
89 Text in: 1771, UNTS, 107.
90 Text in: 1954, UNTS, 3.

Besides the global approaches just mentioned, the following actions have been taken since Stockholm:

(i) at the regional level, the Regional Seas Programme has been implemented, leading to the development and adoption of a number of multilateral conventions;[91]
(ii) by sectors, as regards the seas,[92] the atmosphere[93] and animal species;[94]
(iii) to solve specific problems arising from emergency situations.[95]

In 1986, the Experts Group on Environmental Law of the World Commission on Environment and Development met for the second time to make recommendations for further action in the field of environment. This group developed, *inter alia*, elements for a draft global convention on environment protection, based on principles supported by various existing international agreements, both bilateral and multilateral. In 1987, the so-called 'Bruntland Report' elaborated upon the concept of sustainable development, or intergenerational equity,[96] which has been incorporated into several international water agreements, including the 1992 UNECE Water Convention and its Protocol on Water and Health, the 1994 Danube Convention, the 2002 Framework Agreement on the River Sava, the 1998 Convention between Portugal and Spain on Cooperation in the Protection and Sustainable Development of the Waters of the Hispano-Portuguese Hydrographic Basins[97] and the 2010 Nile Basin Cooperative Framework Agreement. According to some authors, the notion of sustainable development has attained the status of principle of international environmental law. According to others, rather than being a principle, sustainability must represent an objective. It is difficult to define the substance of the concept, since what is sustainable for one state may be unsustainable for another.

Other relevant documents include UN General Assembly Resolution 2995 (XXVII), on Cooperation between States in the Field of the Environment, which places emphasis on the duty of states to avoid the causing of significant transboundary environmental harm, Resolutions 2996 (XXVII) and 3129 (XXVIII), on the obligations deriving therefrom, and Resolution 3281 (XXIX), adopting the Charter on the Economic Rights and Duties of States.

---

91 Such as the Convention for the Protection of the Marine Environment and the Coastal Region of the Mediterranean, Barcelona, 1976 (amended in 1995), and the Convention for the Protection, Management and Development of the Marine and Coastal Environment of the Western African Region, Abidjan, 1981.
92 UN Convention on the Law of the Sea, 1982.
93 UNECE Convention on Long-Range Transboundary Air Pollution, 1979, and other conventions.
94 Convention on International Trade in Endangered Species of Wild Fauna and Flora, Washington, 1973; Convention on the Conservation of Migratory Species, Bonn, 1979.
95 Two conventions relating to notification of nuclear accidents and to assistance in the case of nuclear accidents, Vienna, 1989; the Convention on the Protection of the Ozone Layer, Vienna, 1985, and the Montreal Protocol, 1990; the UNECE Convention on the Transboundary Effects of Industrial Accidents, 1992, and the Liability Protocol, 2003 (not yet in force).
96 It defined it as 'development that meets the needs of the present without compromising the ability of future generations to meet their own needs.' See World Commission on Environment and Development (WCED) (1987) *Our Common Future*. Oxford, 43.
97 Text available from www.cawater-info.net/bk/water_law/pdf/spain_portugal_en.pdf.

These resolutions have introduced some principles on state responsibility.[98] Under Article 30 of the charter, the protection, preservation and enhancement of the environment for present and future generation is the responsibility of all states, which translates into the obligation of each state to ensure that activities carried out under their jurisdiction or control do not cause damage to the environment of other states, and into the obligation of states to establish appropriate policies and norms and to cooperate.

The obligation to prevent the causing of significant transboundary harm and the duty of states to cooperate in the field of the environment generate further obligations of a procedural character. These include the obligation for states to exchange information, and the duty to provide advance notification of planned activities with possible adverse effects. A further obligation, which is closely connected to the duty to notify, relates to the assessment of environmental impacts. The obligation to conduct an environmental impact assessment (EIA), which is listed in the UNECE Water Convention (Art. 3) as one of the measures to be implemented by states to prevent, reduce and control transboundary impact, is the specific subject of the UNECE Convention on Environmental Impact Assessment in a Transboundary Context (1991),[99] which enters into details as to EIA procedural requirements, the content of an EIA and, in an indicative manner, the activities subject to EIA. Under Article 12 of the UN Watercourses Convention, the notifying state shall provide 'available technical data and information, including the results of any environmental impact assessment, in order to enable the notified states to evaluate the possible effects of the planned measures.'

The obligation to conduct a transboundary EIA has been incorporated into numerous water agreements, including the Agreement on the Protection and Sustainable Development of the Prespa Park Area (2010),[100] the Treaty on Cooperation in the Field of Protection and Sustainable Develoment of the Dniester River Basin (2012),[101] the Water Charter of the Senegal River (2002),[102] the Convention on the Sustainable Management of Lake Tanganyka (2003),[103] the Niger Basin Water Charter (2008)[104] and the Nile Basin Cooperative Framework Agreement (2010). Some of these agreements reach a certain level of detail by outlining EIA procedures and indicating the activities that must undergo an EIA. The obligation has also been recognized by the International Court of Justice in 2010, when it asserted, in its judgement on the Pulp Mills Case, that '. . . it may now be considered a requirement under general international law to undertake an environmental impact assessment where there is a risk that the proposed industrial activity may have a significant adverse impact in a transboundary context, in particular, on a shared resource', and in 2015, in its judgement on the case concerning the construction of a road in Costa Rica along the San Juan River.

---

98 The texts of these resolutions, which were adopted by the General Assembly at its 27th, 28th and 29th sessions, held in 1972, 1973 and 1974, respectively, may be retrieved from www.un.org/documents/ga/resga.htm.
99 Espoo, 25 February, 1991. Text in: 30, ILM, 800 (1991).
100 *Official Journal of the European Union*, L 258, 4/10/2011, 2.
101 Text available from www.unece.org/fileadmin/DAM/env/water/activities/Dniester/Dniester-treaty-final-EN-29Nov2012_web.pdf.
102 Text available from www.portail-omvs.org/sites/default/files/fichierspdf/charte_des_eaux_du_fleuve_senegal.pdf.
103 Text available from http://lta.iwlearn.org/documents/the-convention-on-the-sustainable-management-of-lake-tanganyika-eng.pdf.
104 Text available from www.abn.ne/index.php?option=com_content&view=article&id=111:charte-de-leau-du-bassin-du-niger&catid=48:publications&Itemid=42&lang=en.

Public participation is an essential ingredient of the EIA procedure, as acknowledged in the UNECE convention on EIA. It entails the provision to the public concerned of information on a planned activity at a very early stage of the decision-making process. This will enable the public to express its opinion on the activity and to raise objections, as the case may be. A further UNECE convention emphasizes the citizens' right of access to information, public participation in decision making and access to justice in environmental matters.[105] The rules contained in these two conventions are increasingly being incorporated into international agreements relating to specific river basins in the UNECE region and elsewhere.

Many more examples could be cited in order to highlight present trends and place international water resources in the broader context of the management of natural resources. The interaction of these resources with the surrounding environment must be taken into consideration. Recent international agreements and the law-making activities of regional and international organizations such as the European Union and the United Nations reflect these trends.

11.7.3 The emerging rule

A number of principles may be derived from the recent developments of international law with regard to the management of natural resources and the protection of the environment:

(i) the first establishes the basis of a necessary community of interests;
(ii) the second relates to the obligation to prevent transboundary harm;
(iii) the third concerns the duty to consider water and related natural resource as ecosystem components;
(iv) the fourth emphasizes the obligation to cooperate, to inform and to consult;
(v) the fifth relates to the duty to provide advance notification, and conduct a transboundary EIA, when a planned activity may cause significant harm; and
(vi) the sixth relates to the qualification of states' responsiblity for wrongful acts.

These principles are still being elaborated, although it is true that nowadays, given the growing concern for the state of the environment and the occurence of accidents such as the disaster of Chernobyl, the Baia Mare and other industrial accidents affecting the Danube and other transboundary water resources, international customary law principles are developing more rapidly than they did before. As was recognized by the Permanent Court of Arbitration in its partial award on the Kishenganga Indus Waters dispute (2013), 'There is no doubt that States are required under contemporary customary international law to take environmental protection into consideration when planning and developing projects that may cause injury to a bordering State.'

---

105 Convention on Access to Information, Public Participation in Decision-Making and Access to Justice in Environmental Matters, Aarhus, 25 June, 1998. Text in: 38, ILM, 517 (1999).

11.8 The right to water in international law

11.8.1 Definitions

The right to water has been defined by the UN Committee on Economic, Social and Cultural Rights (CESCR)[106] in 2002, through General Comment No. 15, as the entitlement of everyone to 'sufficient, safe, acceptable and physically accessible and affordable water for personal and domestic uses' (Para. 2).

'Sufficient' means that the water available or made available must be enough in term of quantity to satisfy personal and household needs (drinking, sanitation, hygiene, washing, cooking, etc.); the term 'safe' refers to water quality; 'acceptable' means that water should be of an acceptable colour, odour and taste for each domestic and personal use; 'physically accessible' means that the water and sanitation facility or service must be in the proximity of the households, work place, health institution or educational facility; 'affordable' refers to the capacity of people to pay for having access to water, water facilities and water services.

11.8.2 Evolution

The human right to water was not explicitly recognized at the time of codification of human rights law in the aftermath of World War II. The UN Universal Declaration of Human Rights of 1948 reflected the aspirations of the global community after the atrocities of the war and provided a basis for the formulation of two international human right conventions. The first, adopted in 1966, was the International Covenant on Economic, Social and Cultural Rights.[107] The second, also adopted in 1966, was the International Covenant on Civil and Political Rights.[108] Neither of these two instruments recognizes the right to water as such. The first covenant places emphasis, *inter alia*, on the right to adequate standards of living, 'including adequate food, clothing and housing . . .' (Art. 11), on the right to health (Art. 12) and on the obligation of the parties to take steps, to the maximum of their available resources, with a view to the progressive realization of these (and other) rights 'by all appropriate means, including particularly the adoption of legislative measures' (Art. 2). The second covenant highlights the right to dignity and the right to life.

It was already clear when the two covenants were formulated that water plays a key role in the realization of these rights, but the debate on the right to water gained momentum only in 1977, on occasion of the UN Water Conference of Mar del Plata. The Action Plan produced at this conference declared that all people have a right 'to have access to drinking water in quantities and of a quality equal to their basic needs,' and called for international cooperation to ensure that water is available and equitably distributed among people. This principle was reasserted in Chapter 18, Agenda 21, of the UN Conference on Environment and Development, held at Rio de Janeiro in 1992. In the meantime, growing support for the right to water may be tracked in a number of international legal instruments, such as the

---

106 The CESCR was established to monitor the implementation by states of the International Covenant on Economic, Social and Cultural Rights (1966). The covenant entered into force on 3 January, 1976.
107 Text in: 993, UNTS, 3.
108 Text in: 999, UNTS, 171.

Convention on the Elimination of all Forms of Discrimination against Women (1979),[109] according to which rural women have the right to 'enjoy adequate living conditions, particularly in relation to... water supply,' and the Convention on the Rights of the Child (1989),[110] which recognizes the rights of children to clean drinking water.

Population growth, technological progress, industrialization and climate change have been a cause of increasing global concern in the past years, because of their impact on water availability, demand and quality. In 2002, the UN Committee on Economic, Social and Cultural Rights adopted General Comment No. 15, on the right to water,[111] for the stated purpose of interpreting and clarifying the provisions of Articles 11 and 12 of the International Covenant on Economic, Social and Cultural Rights. This document explains that this right, being one of the most fundamental conditions for survival, is essential for securing an adequate standard of living and is 'inextricably related to the right to the highest attainable standard of health' (Para. 3). General Comment No. 15 is not legally binding as such.

As to international water treaties, the UN Watercourses Convention fails to shed light on the question as to whether the right to water is a self-standing human right. The convention codifies rules of conduct in the field of transboundary water resources that, being based on consistent state practice and having been embodied in a number of bilateral and multilateral treaties, have become part of international customary law. The social and economic needs of the states concerned and the effects of the uses of a watercourse by a state on the other watercourse states feature among the factors to be taken into consideration to determine equitable and reasonable utilization (Art. 6). The weight to be given to each factor must be determined considering all the other factors, and a conclusion must be reached on the basis of the whole. No use enjoys inherent priority over other water uses in the absence of an agreement or custom asserting the contrary, but in the event of conflict, a solution is to be sought with reference to Articles 5 (equitable and reasonable utilization) to 7 (obligation not to cause significant harm), special attention having to be paid to vital human needs (Art. 10). Thus, it may happen that the consideration of the right to water of the population of a given state within a watercourse is overridden by the ability of that state to resort to alternative sources of supply, which is another factor to be considered. If this is not the case, the state cannot be obliged to deprive its own population of the water required to satisfy vital human needs.

Through Resolution 64/292 of 28 July, 2010,[112] the UN General Assembly has explicitly recognized the human right to water and sanitation and has acknowledged that clean drinking water and sanitation are essential to the realization of all human rights. The resolution calls upon states and international organizations to provide financial resources and promote capacity building and technology transfer to help countries, in particular developing ones, to provide safe, clean, accessible and affordable drinking water and sanitation for all. A further resolution, adopted by the UN Human Rights Council on 6 October, 2010,[113] reaffirms the obligation of states to ensure the full realization of all human rights, and adds that the delegation of safe drinking water supply and sanitation services to third parties does not exempt

---

109 Text available from www.un.org/womenwatch/daw/cedaw/text/econvention.htm.
110 Text available from www.ohchr.org/en/professionalinterest/pages/crc.aspx.
111 UN Document E/C.12/2002/11 of 20 January, 2003.
112 UN Document A/RES/64/292 of 3 August, 2010.
113 A/HRC/15/9.

states from this obligation. It calls upon states to develop appropriate tools and mechanisms, including legal and institutional mechanisms. Neither Resolution 64/292, nor the Human Rigthts Council resolution, have gained universal acceptance, since many countries have abstained from voting.

Although it is not uncommon to find the use of water to satisfy drinking and domestic needs ranking high in the order of priorities among water uses established through treaties, very few contemporary water agreements acknowledge the right to water as a self-standing human right. The Water Charter of the Senegal River (2002) and the Niger Basin Water Charter (2008) both recognize the right to water as guiding principle, and assign to it a clear priority as a factor for the determination of equitable utilization and participation. The fundamental right of each individual to access water also features in the preamble to the Memorandum of Understanding for the Establishment of a Consultation Mechanism for the Integrated Management of the Water Resources of the Iullemeden, Taoudeni/Tanezrouft Aquifer Systems (2014),[114] in addition to being a factor for the determination of equitable and reasonable utilization.

The UNECE Protocol on Water and Health (1999), the stated aim of which is to protect human health and well being through improved water management, was perhaps the first to place a number of positive obligations on the parties, which are of relevance to the realization of the right to water. These relate *inter alia* to the prevention, control and reduction of water-related diseases, the protection of the quality of water used for drinking purposes, the preservation of aquatic ecosystems, cooperation in water resources planning, resource monitoring and the setting of indicators and targets for measuring compliance. Without mentioning the right to water as a self-standing human right, the protocol places emphasis on the need to ensure equitable access to water, adequate in terms of both quantity and quality, and to sanitation services, for all, particularly for disadvantaged and marginalized groups.

11.8.3 The emerging rule

Given the paucity of international agreements recognizing the right to water as a self-standing human right, and the fact that the numerous resolutions and declaration on the subject are not binding, the question as to whether such right exists under international customary law is still being debated. Current international customary water law principles and obligations already provide a framework for its realization and protection, by requiring states to take into account the social needs of the populations, together with other factors, when making decisions as to equitable and reasonable utilization. The obligation not to cause significant harm and the obligation to prevent, control and reduce pollution provide additional safeguards. In any case, it cannot be denied that the satisfaction of drinking and domestic water needs ranks high in the order of priorities set in many agreements.

The right to water is now recognized in a number of national constitutions, whether explicitly or indirectly.[115] Moreover, an increasing number of countries have adopted legislation that acknowledges the human right to water as a matter of principle and protects it by prioritizing it in national planning and decision making and strengthening the framework for preventing, controlling and reducing pollution. Thus, if it has not attained the status of a rule

---

114 Not in force yet. See Chapter 12.
115 See Chapter 7.

of international customary law, the human right to water may be considered as a matter of general principles of law.

As to the obligations of states, the International Covenant on Economic, Social and Cultural Rights, as interpreted through General Comment No. 15, requires states to take steps towards the progressive realization of the right to water 'by all appropriate means, including particularly the adoption of legislative measures.' The covenant takes into consideration the fact that not all states might have sufficient resources to fulfill this obligation.

References

Boisson de Chazournes, L., Leb, C. & Tignino, M. eds. (2013) *International Law and Freshwater – The Multiple Challenges*. Cheltenham, U.K., Northampton, MA, Edward Elgar Publishing.

Bourne, C.B. (1997) *International Water Law, Selected Writings*. International and National Water Law and Policy Series. The Hague, London, New York, Kluwer Law International.

Cano, G.J. (1975) *A Legal and Institutional Framework for Natural Resources Management*. Legislative Study No. 9. Rome, FAO.

Caponera, D.A. (1957) *Diritto Internazionale della Navigazione: I Fiumi Internazionali*. Rome, Ed. Internazionali Sociali.

Caponera, D.A. (1976) *Outline for the Preparation of an Inventory of the Legal and Institutional Aspects of International Water Resources Basins*. Background Paper No. 11. Rome, FAO.

Caponera, D.A. (1980) *The Law of International Water Resources*. Legislative Study No. 23. Rome, FAO.

Caponera, D.A. (1981) International River Law. *Proceedings of the National Symposium on River Basin Development*. Dhaka, Bangladesh, 4–10 December 1981.

Caponera, D.A. (1985) Patterns of Cooperation in International Water Law, Principles and Institutions. *Natural Resources Journal*. Albuquerque, University of New Mexico School of Law.

Caponera, D.A. (2003) *National and International Water Law and Administration, Selected Writings*. International and National Water Law and Policy Series. The Hague, London, New York, Kluwer Law International.

Caponera, D.A. & Curti Gialdino, C. (1988) *Fiumi Internazionali*. Excerpt from Vol. XIV. Enciclopedia Giuridica. Rome.

Chauhan, B.R. (1981) *Settlement of International Water Disputes in International Drainage Basins*. Berlin, Erich Schmidt Verlag.

Dupuy, P.M. (1980) International Liability for Transfrontier Pollution. *Tendences actuelles de la politique et du droit de l'environnement*. Gland, IUCN.

Kiss, A. (1989) *Droit international de l'environnement*. Paris, Pedone.

Lammers, J.L. (1984) *Pollution of International Watercourses*, The Hague, Nijhoff.

Leb, C. (2012) The right to Water in a Transbounday Context: Emergence of Seminal Trends. *Water International* 37:6, 640–653.

Leb, C. (2013) *Cooperation in the Law of Transboundary Water Resources*. New York, Cambridge University Press.

McCaffrey, S.C. (2007) *The Law of International Watercourses*. 2nd Ed. Oxford University Press.

Rios y Lagos Internacionales (Utilización para Fines Agricolas e Industriales), 4a ed. rev., Pan American Union, OAS, Washington D.C., 1974.

Systematic Index of International Water Resources Treaties, Declarations, Acts and Cases by Basin, Legislative Study No. 15, FAO, Rome, 1978; and Legislative Study No. 34, FAO, Rome, 1984.

Salman, M.A.S. (2016) The Grand Ethiopian Renaissance Dam: The Road to the Declaration of Principles and the Khartoum Document. *Water International* 41:4.

Salman, M.A.S. & Boisson de Chazournes, L. eds. (1998) *International Watercourses, Enhancing Cooperation and Managing Conflict, Proceedings of a World Bank Seminar*. Technical Paper No. 414. World Bank, Washington, D.C.

Teclaff, L. (1967) *The River Basin in History and Law*. The Hague, Nijhoff.

United Nations (1976) *Management of International Water Resources, Legal and Institutional Aspects*. New York, United Nations.

United Nations Environment Programme (2010) *The Greening of Water Law: Managing Freshwater Resources for People and the Environment*. United Nations Environment Programme.

Van Eysinga, W.J.M. (1920) *Evolution du droit international fluvial du Congrès de Vienne au Traité de Versailles*. Leiden, A.W. Sijthoff.

Various authors (1967) *The Law of International Drainage Basins*. Institute for International Law, University School of Law. Dobbs Ferry, New York, Oceana Publications Inc.

Chapter 12

Developments in the law of transboundary aquifers

12.1 Introduction

Like surface water, underground water does not respect the political boundaries drawn by man. Many aquifers and aquifer systems are shared by two or more countries, sometimes in areas where surface water is scarce. Examples are the Nubian Sandstone Aquifer System, shared by Libya, Egypt, Chad and Sudan; the North-Western Sahara Aquifer System, underlying Algeria, Libya and Tunisia; the Arabian Peninsula Aquifer, which extends under Saudi Arabia, Bahrein, and perhaps Qatar and the United Arab Emirates; the Iullemeden Aquifer System, shared by Mali, Niger and Nigeria; and the Chad Aquifer, underlying Chad, Niger, Sudan, the Central African Republic, Nigeria and Cameroon. In South America, the Guaraní Aquifer System is shared by two large federal states, namely, Argentina and Brazil, and two unitary states, Paraguay and Uruguay. Many other examples of transboundary aquifers can be cited.

Groundwater in general, and international groundwater in particular, may be the cause of legal problems and potential conflicts which are more important and of more difficult solution than those raised by international surface waters, for various reasons. First, the 'invisibility' of this resource and the paucity of data constrain a full understanding of its characteristics and behaviour, and of the possible adverse effects of human activities. Second, aquifers are more vulnerable to pollution than surface water; groundwater moves slowly and self-purification is extremely limited. Thus, pollution may have long-lasting if not irreversible effects. Third, groundwater monitoring requires institutional capacity and financial resources which are not always available.

As in any other case where the need for adequate regulation arises, it is necessary to have a clear idea of what the subject of regulation is, in other words, what is meant by the expressions 'international underground waters' and 'transboundary aquifers.' International groundwaters, or groundwaters common to two or more states, or shared groundwaters, are not only those which underlie two or more states. An aquifer situated entirely in one state may be connected to another aquifer or to a stream belonging to the territory of another state. Is that aquifer merely national? By contrast, the expression 'transboundary aquifer' refers to a groundwater body which is itself intersected by a boundary.

From the legal viewpoint, the presence of an element of responsibility is necessary. For example, an aquifer may be considered of interest at the international level when activities or omissions taking place within the territory of one state cause, or are likely to cause, damage or injury to the rights or interests of another state sharing the same aquifer. If an aquifer situated in one state supplies water to a river which flows entirely within the territory of another state, underground water abstraction in excess will reduce the flow reaching the river. Likewise, the drawing of too much water from a stream on which an aquifer situated in another state is dependent may bring about the exhaustion of the aquifer.

It may be difficult, however, to identify this element of responsibility in the presence of a hidden, or little known, underground resource. Hence the need for legal and institutional mechanisms that, without a conflict being necessarily incipient or ongoing, allow the states concerned to improve their knowledge of the resource and, based on needs, to cooperate in the identification of the measures necessary to achieve its optimum utilization and protection.

The law of international water resources has undergone steady development and has enjoyed the attention of intergovernmental organizations, non-governmental organizations and scholars.[1] However, in spite of the wide acceptance of the concept of 'drainage basin' developed by the International Law Association in 1966,[2] which encompasses underground water, and of the extension of this concept to groundwater not related to surface water,[3] legal research has been mostly oriented towards surface waters. In its definition of the term 'watercourse,'[4] the UN Watercourses Convention refers to groundwater only in so far as it is connected to surface water. By the same token, the great majority of international treaties concerning water refer to surface water and, only incidentally, to groundwater. However, recent agreements, the adoption by the UN General Assembly of the Resolution on the Law of Transboundary Aquifers of 11 December, 2008,[5] and a number of studies, programmes and projects on transboundary groundwater resources sponsored and implemented by international organizations lead to the conclusion that more attention is now being paid to groundwater.

12.2 Sources and evolution of international groundwater law

Like international surface waters, international groundwaters are governed by international water resources law. Accordingly, pursuant to Article 38 of the Statute of the International Court of Justice, the sources of international law are:

(i) international conventions (general and particular);
(ii) international custom;
(iii) the general principles of law recognized by civilized nations;
(iv) judicial decisions and the teachings of the most highly qualified publicists (as subsidiary sources).

---

1 See Chapter 10.
2 The Helsinki Rules. Text in: The International Law Association, *Report of the Fifty-Second Conference*, Helsinki, 14–20 August, 1966, 484–532.
3 International Law Association, *Rules on International Groundwaters*, Seoul, 1986. See Chapter 10.
4 See Section 10.2.2 of Chapter 10.
5 This resolution was adopted by the UN General Assembly at its sixty-third session of 11 December, 2008 (A/RES/63/124, 15 January, 2009). The Draft Articles on the Law of Transboundary Aquifers, which are the result of the work of codification of groundwater-specific rules of the International Law Commission (ILC) of the United Nations, are appended to it.

Art. 38 does not mention the law-making activities of international regional bodies such as the European Union, the quasi-legislative activities of international organizations and the resolutions and recommendations of intergovernmental organizations, all of which have considered groundwater on a number of occasions.

A number of international water treaties refer to underground waters, among which the following may be mentioned:

(i) the Convention on the Law of Non-Navigational Uses of International Watercourses (UN Watercourses Convention), a global legal instrument adopted by the UN General Assembly at New York, on 25 May, 1997;[6]
(ii) regional legal instruments, including the UNECE Convention on the Protection and Use of Transboundary Rivers and International Lakes, Helsinki, 1992 (UNECE Water Convention), its Protocol on Water and Health, London, 1999, and the 2000 Revised SADC Protocol, which is modeled on the UN Watercourses Convention;
(iii) river basin-specific treaties, such as the Convention and Statute of the Lake Chad Basin, signed by Cameroon, Chad, Niger and Nigeria on 22 May, 1964; the 1994 Danube Protection Convention; the 1999 Convention on the Protection of the Rhine; the 2002 Sava River Basin Framework Agreement; the 2003 Protocol on Lake Victoria; the 2005 Great Lakes – St. Lawrence River Basin Sustainable Water Resources Agreement; and the 2012 Treaty on Cooperation in the Field of Protection and Sustainable Develoment of the Dniester River Basin;
(iv) particular conventions addressing specific situations between neighbouring states;[7]
(v) agreements dealing with boundary demarcation, which mention the use of wells and springs considered as groundwater,[8] or refer to groundwater use while dealing with mining activities.[9]

These international treaties, and many others, are tailored to surface water resources and deal with groundwater only incidentally; there are only a few groundwater-specific international legal instruments.

---

6 See Chapter 10.
7 Among these agreements, the following may be mentioned: Agreement between Egypt and Italy, concerning the Ramla well, 1925; Convention and Protocol between the USSR and Turkey, regarding the use of frontier waters, 1927; Treaty of Peace between the Allies and Italy, 1947; Agreement with Statutes of the Yugoslav-Hungarian Water Economy Commission, 1955; Agreement between Yugoslavia and Bulgaria concerning water economy questions, 1958; Agreement between Czechoslovakia and Poland concerning the use of water resources in frontier areas, 1958; Agreement between Poland and the USSR concerning the use of frontier waters, 1964; Agreement between Poland and the German Democratic Republic concerning groundwater, 1965; Convention between Italy and Switzerland concerning water pollution control, 1972. Reported in Caponera, D. & Alhéritière, D. (1978) Principles of International Groundwater Law. *Natural Resources Forum* 2, 279–290 and 359–371.
8 Among which an agreement of 1888 between France and Djibouti and another between Syria and Palestine of 1923.
9 This is the case, among others, of the agreement of 1934 between Tanganyika and Rwanda Urundi concerning water rights on the boundary and of the agreement of 1843 between Belgium and Luxembourg concerning mining.

The Convention on the Protection, Utilization, Recharge and Monitoring of the Franco-Swiss Geneva Aquifer, which was signed by the Community of the Annemasse region, the Community of the Geneva rural districts and the rural district of Viry on the one hand, and the Republic and Canton of Geneva on the other, on 18 December, 2007,[10] has been in force since January, 2008. It has replaced the Geneva Aquifer Agreement (*Arrangement*) between the Canton of Geneva (Switzerland) and the Préfet of Haute-Savoie (France) of 9 June, 1977,[11] on the same subject when it expired in 2008, and was until recently the only example of an international legal instrument focusing on groundwater resources management in a comprehensive manner. The convention provides for the establishment of a joint commission to facilitate cooperation in the monitoring of groundwater levels, groundwater abstraction and quality, to prepare annual water utilization programmes and to provide advice on groundwater protection measures. Furthermore, it contains provisions as to the rights and obligations of the parties with regard to the artificial recharge of the aquifer. The commission keeps an inventory of groundwater abstraction works in both states and, since these works are equipped with meters, it is in a position to know how much water is taken by each user.

The agreement on a permanent and definitive solution of the salinity problem of the Colorado River, which was concluded by the USA and Mexico in 1973,[12] offers another example of cross-border cooperation on groundwater issues, albeit on a smaller scale. Pending the conclusion of a permanent agreement on transboundary groundwater, which has not been arrived at yet, the two countries undertook, through this agreement, to introduce restrictions on groundwater pumping for a strip of land extending five miles on both sides of the Arizona-Sonora border, and to consult each other before engaging in any new resource development which might have adverse effects across the boundary. It is interesting to note that Minute No. 242 was agreed upon within an institutional mechanism having no specific jurisdiction on transboundary groundwater, i.e., the International Boundary and Water Commission.[13]

On 5 October, 2000, the Nubian Sandstone Aquifer System (NSAS) states, i.e., Egypt, Libya, Sudan and Chad, signed two 'technical' agreements on procedures for data collection and sharing and on access to their regional information system.[14] Also, in the early 1990's Egypt and Libya established a joint authority to facilitate the study and development of the aquifer system. Sudan and Chad became members later.[15] The authority, which is supported by the Center for Environment and Development of the Arab Region and Europe

---

10 Text available from www.internationalwaterlaw.org/documents/regionaldocs/2008Franko-Swiss-Aquifer-English.pdf
11 Reprinted in Teclaff, L.A. & Utton, A.E. (1981) *International Groundwater Law*. London, Rome, New York.
12 Exchange of Notes of 30 August, 1973, approving Minute No. 242 under the treaty between the United States of America and Mexico relating to the utilization of the waters of the Colorado and Tijuana Rivers, and of the Rio Grande from Fort Quitnam, Texas, to the Gulf of Mexico, signed at Washington on 3 February, 1944. Reprinted in Teclaff, L.A., & Utton, A.E, *op. cit.*
13 See Chapter 13, Section 13.2.2.2.
14 Programme for the Development of a Regional Strategy for the Utilization of the Nubian Sandstone Aquifer System (NSAS) – Terms of Reference for the Monitoring and Exchange of Groundwater Information of the Nubian Sandstone Aquifer System. Tripoli, 5 October, 2000. Text available from http://web.cedare.org/wrm-pages/transboundary-water-management/
15 Constitution of the Joint Authority for the Study and Development of the Nubian Sandstone Aquifer Waters, 1992. Text available from http://ewp.cedare.org/wp-content/uploads/2018/09/Constitution-of-the-Joint-Authority-for-the-Study-and-Development-of-the-Nubian-Sandstone-Aquifer-Waters-1992.pdf

(CEDARE), is in charge of data collection and administration, the conduct of studies, the formulation of water resources development programmes and plans and the implementation of common policies. It has a board of directors, a secretariat and a director. In 2013, the ministers in charge of water affairs of the four states and the chairperson of the joint authority adopted a regional strategic action plan for the NSAS.

Cooperation efforts are also ongoing with respect to the North-Western Sahara Aquifer System (NWSAS),[16] which extends across more than 1,000,000 km^2 and spans Algeria, Libya and Tunisia. In 2002, the three countries agreed on the establishment of a tripartite consultation mechanism[17] which, in 2008, became a permanent consultation mechanism hosted by the inter-governmental *Observatoire du Sahara et du Sahel* (OSS). The objective of the mechanism is to coordinate, promote and facilitate the rational management of the NWSAS water resources, with focus on designated 'hot spots', such as the Ghadames basin (Continental Intercalaire), the Artesian and Tunisian Outlet basins (also in the Continental Intercalaire) and the Chotts basin. Its main functions are to manage a hydrogeological database and a simulation model, to develop a reference observation network, to process, analyze and validate data and information about the aquifer and its use, and to develop the relevant indicators. The agreement on the NWSAS reflects an evolutionary approach to cooperation, but, like the Nubian Sandstone Aquifer agreements, is silent on general principles.

In 2009, Niger, Nigeria and Mali, signed a Memorandum of Understanding relating to the setting up of a consultation mechanism for the Iullemeden Aquifer System (IAS), which covers 500,000 km^2, with the objective to promote the integrated management and the sustainable development of the aquifer system, as well as cooperation and the joint identification and management of risks. *Inter alia*, the consultation mechanism was mandated to provide opinions on policies and projects, to coordinate programmes, to make recommendations to harmonize legislation and to settle disputes. The countries also undertook to take into consideration a number of principles relating to equitable and reasonable utilization, the duty not to cause significant transboundary harm, sustainability, public participation in decision making and implementation, precaution, and the polluter-pays and user-pays principles. The memorandum also sets out procedural obligations, including the obligation to exchange data and information and prior notification of planned measures with possible adverse effects. Finally, provision is made for dispute settlement.

The Memorandum of Understanding for the IAS was never ratified. Because of the recognition of the unity of the Iullemeden and Taoudeni/Tanezrout Aquifer Systems (covering 2,500,000 km2), it was replaced by the Memorandum of Understanding for the Establishment of a Consultation Mechanism for the Integrated Management of the Water Resources of the Iullemeden, Taoudeni/Tanezrouft Aquifer Systems (ITAS), adopted by Algeria, Benin, Burkina Faso, Mali, Mauritania, Niger and Nigeria on 28 March, 2014.[18] This memoran-

---

16 SASS is its acronym in French.
17 Minutes of the meeting of representatives of Algeria, Libya and Tunisia held at FAO Headquarters at Rome, Italy, on 19–20 December, 2002, subsequently endorsed by Algeria on 6 January, 2003, Tunisia on 15 February, 2003 and Libya on 23 February, 2003. The minutes and the subsequent letters of endorsement constitute the agreement to establish the consultation mechanism. They are reproduced in Burchi, S. & Mechlem, K. eds. (2005) *Groundwater in International Law. Compilation of Treaties and other Legal Instruments.* Legislative Study No. 86. Rome, FAO.
18 Text available from www.internationalwaterlaw.org/documents/regionaldocs/Iullemeden_MOU-2014.pdf.

dum of understanding, which has not yet been ratified by the states concerned, is similar to the previous one on the IAS, but highlights the ecological function of groundwater. After acknowledging in its preamble the contribution of the UN Watercourses Convention and of the International Law Commission (ILC) Draft Articles on the Law of Transboundary Aquifers to the codification and progressive development of international law, it reflects in its text the principles and procedural rules set out in these two instruments, like its predecessor. However, it enters into detail as to their implementation.

The Guaraní Aquifer System (GAS), which is shared by Argentina, Brazil, Paraguay and Uruguay, is one of the world's largest water reservoirs.[19] In 2004, recognizing its importance within the framework of the Southern Common Market (MERCOSUR), the governments of the four countries created a high-level *ad hoc* group of experts and charged it with the elaboration of an aquifer agreement. The group received inputs regarding the aquifer's characteristics from the project for the Environmental Protection and Sustainable Development of the Guaraní Aquifer System,[20] which was sponsored by the Global Environment Facility (GEF) and implemented in parallel with the agreement elaboration process. On 2 August, 2010, after about six years of intense negotiations,[21] Argentina, Brazil, Paraguay and Uruguay signed the Guaraní Aquifer Agreement, a framework instrument aiming to facilitate their cooperation even in the absence of major conflicts in the region.[22] The agreement, which refers explicitly to the ILC Draft Articles in its preamble, places emphasis on the sovereignty of the states involved over their respective portions of the aquifer. At the same time, it calls for equitable and reasonable use of the aquifer and sets out the obligation not to cause significant harm, the duty to cooperate, the duty to exchange information and monitor the aquifer and the duty to promote conservation and environment protection. The agreement also provides for the notification of planned activities with possible adverse effects (including the results of any environmental impact assessment) and for the identification of critical boundary areas requiring the implementation of special measures. A commission is to be established under the La Plata River Treaty to coordinate cooperation. The Guaraní Aquifer Agreement was ratified by Argentina and Uruguay in 2012, by Brazil in 2017 and by Paraguay in 2018.

---

19 The total surface of the Guarani Aquifer System is estimated at 1,190,000 km2. Organization of American States (2005) *Environmental Protection and Sustainable Development of the Guarani Aquifer System*. Water Project Series No. 7. Available from www.oas.org/dsd/Events/english/Documents/OSDE_7Guarani.pdf

20 The project (2002–2009) aimed at assisting the countries in the pursuance of the long-term objective of creating a common legal, institutional and technical framework for the system's management and preservation, keeping in mind present and future generations. It focused on technical issues and on localized problems, through pilot initiatives.

21 During the first phase of the negotiations, which started in 2004, the governments approved the principles underlying the agreement, but at the time of finalizing the text they failed to arrive at a common understanding of the arbitral procedure set out at Article 19. For this reason, and because the countries intended to wait for the outcome of the GEF-sponsored project, the negotiations were suspended. In 2010, the negotiations were resumed and a second paragraph was included in Article 19, stating that the arbitral procedure would be the subject of a protocol to the agreement, to be developed later. Thus, a compromise solution was opted for.

22 Thus, it is preventative in nature. See Sindico, F., Hirata, R. & Manganelli, A. (2018) The Guarani Aquifer System: From a Beacon of Hope to a Question Mark in the Governance of Transboundary Aquifers. *Journal of Hydrology: Regional Studies*.

Finally, on 30 April, 2015, Jordan and Saudi Arabia signed the Agreement for the Management and Utilization of the Groundwaters in the Al-Sag/Al-Disi Layer, a fossil aquifer which covers about 308,000 km². The need for bilateral cooperation in the management of the aquifer emerged as early as 1997, when Jordan submitted to the World Bank the proposed Disi-Amman Water Conveyor Project for its consideration, and the Bank indicated that Saudi Arabia should be notified of this plan.[23] Jordan notified Saudi Arabia, but did not receive an official reply. Thus, no further consultations took place at that time and the Bank did not process the project. The two countries continued to exploit the aquifer unilaterally for irrigation purposes, while Jordan pursued its plan to supply water to Amman. Data on extractions were not shared. Apparently, a memorandum of understanding was signed by the two countries in 2007, through which they agreed to protect the aquifer, limit extractions and set up a joint institutional mechanism. A formal agreement, however, was concluded only in 2015[24] and entered into force upon its signature. Recognizing the need for adequate management, utilization and sustainability of the waters of the aquifer, it provides for the creation of: a protected area within each country along the border, where all existing activities depending on groundwater extraction must be discontinued within five years and no new activities are to be authorized; and, a broader management area encompassing the prohibited area, where activities are permitted, but only for municipal water supply purposes. There are no numerical thresholds to extractions. In addition, the agreement contains an absolute prohibition to pollute groundwater in the management area. The agreement provides for the establishment of the Saudi/Jordanian Joint Technical Committee to supervise its implementation.

As regards the second and the third sources of international law, i.e., customary law and the general principles of law, the analysis relating to international water resources law in general applies. However, considering that the UN Watercourses Convention does not cover groundwater that is not connected to surface water resources, the ILC renewed its codification efforts and produced, in 2008, the Draft Articles on the Law of Transboundary Aquifers. These draft articles are illustrated at Section 12.4 here below.

As to the fourth source, i.e., the decisions rendered by tribunals and arbitral awards, apparently no state has submitted a dispute relating to international groundwaters to the International Court of Justice or to arbitration. The court has dealt with this subject only incidentally, in the judgment of 25 September, 1997, relating to the Gabcikovo-Nagymaros case (Hungary v. Slovakia).[25]

The teachings of qualified publicists are the last subsidiary means for determining legal rules governing the relationships between states. Many theories have been developed concerning the law of international water resources in general,[26] but transboundary groundwater issues have been dealt with by scholars only relatively recently.

A theory which has gained wide acceptance in international law is that of the equitable and reasonable utilization of the waters of an international drainage basin, both surface and underground. This theory, which was developed by the International Law Association (ILA)

---

23 Salman, M.A.S. (2011) The World Bank Policy and Practice for Projects Affecting Shared Aquifers. *Water International*, 36:5.
24 Jarvis, W. T. (2014) *Contesting Hidden Waters, Conflict Resolution for Groundwater and Aquifers*. UK, Routledge.
25 See Chapter 10, Section 10.3.8.1.
26 *Ibid.*, Section 10.3.

and is the main focus of the 1966 Helsinki Rules,[27] aims at weighing the benefit of one state in the use of water against the injury which might be caused to another state.[28] However, it applies to groundwater resources only in so far as these flow, together with surface water, into a common terminus. Aware of this shortcoming, in 1986 the ILA adopted the Rules on International Groundwaters (Seoul Rules), which expand the scope of the Helsinki Rules by making them applicable to all international groundwater resources, including those not connected to surface water bodies. The Seoul Rules also recognize the interaction between groundwater and other natural resources and the particular vulnerability of groundwater to pollution.

Further reference to groundwater is made *inter alia* by the Inter-American Bar Association, which in 1969 adopted a resolution on the legal aspects of the contamination of international rivers and lakes, including groundwaters related to them, by the Council of Europe's 1968 European Water Charter and by the Asian-African Legal Consultative Committee in 1975. In 1971 the Council of Europe adopted a recommendation concerning the fight against the pollution of the Rhine water table.[29]

The need for considering the legal and institutional aspects of groundwater management together with those of surface water management was stressed by the International Association for Water Law (AIDA) at its second conference held at Caracas in 1976.[30] In addition, various individual specialists in water law have dealt with the legal and institutional aspects of groundwater management at the international level[31] and have upheld the doctrine of reasonable and equitable utilization as applicable to both surface and underground water.

The Bellagio Draft Treaty concerning the use of transboundary groundwaters, which was prepared by a multidisciplinary group of experts over an eight-year period, represents an important scientific contribution in this field, because it adopts a practical approach to the solution of specific problems. This document, which is in the form of a model treaty, identifies concrete mechanisms for cooperation in the management of transboundary aquifers in critical areas. Although the model applies to the USA-Mexico border, it is of interest for aquifers elsewhere in the world, particularly in arid regions. The model treaty addresses aquifer contamination, planned depletion ('mining'), drought management and transboundary transfers, in addition to abstraction and recharge issues. It assigns a key role to a joint commission by vesting it with the power to declare transboundary groundwater conservation areas, prepare aquifer management plans, allocate, conserve and protect groundwater and prepare and approve aquifer depletion ('mining') plans and a drought management plan. Water quality protection remains a responsibility of the aquifer states, but based on the information provided by these states to the commission, the commission may proceed to the declaration of a transboundary groundwater conservation area. The provisions on the planned depletion of transboundary aquifers are particularly relevant in areas where groundwater

---

27 *Ibid.*, Section 10.3.9.2.
28 Caponera, D.A. & Alheritiere, D., *op. cit.*, 367.
29 Consultative Assembly of the Council of Europe, *Report on the Pollution of the Rhine Water Table* (doc. 2904).
30 *Annales Juris Aquarum*, Vol. II, Conclusions and Recommendations, *op. cit.*
31 Among others, Caponera, D.A. & Alheritiere, D., *op. cit.*; Barberis, J. (1986) *International Groundwater Resources Law*. Legislative Study No. 40. Rome, FAO; Hayton, R. (1978) Institutional Alternatives for Mexico-US Groundwater Management. *Natural Resources Journal*, Vol.18, 201–212; Hayton, R. (1982) The Law of International Aquifers. *Ibid.*, Vol.22, 71–93.

recharge is negligible, because it is there that groundwater mining should take place over a calculated period of time, sufficient for finding a 'way out' through resort to other water sources, changes in the economic production patterns and other means.[32]

Within the framework of international organizations, two United Nations conferences have focused on underground water issues. The first, held at Mar del Plata in 1977, recommended that 'Countries sharing water resources . . . should review existing available techniques for managing shared water resources and cooperate in the establishment of programmes, machinery and institutions necessary for the coordinated development of such resources.'[33] The principle of equitable utilization is referred to as follows: 'In relation to the use, management and development of shared water resources, national policies should take into consideration the right of each state sharing the resources to equitably utilize such resources as the means to promote bonds of solidarity and cooperation.'[34]

The United Nations Conference on Desertification, held in Nairobi in 1977, recommended the 'wise and efficient management of shared water resources for rational use.'[35] In addition, it suggested the execution of 'transnational projects for studying, screening, processing, interpreting and integrating available data, and for refining management guidelines for rational, economical and sustained exploitation of regional aquifers,' and for 'developing and strengthening regional activities concerning the assessment of surface and groundwater resources.'[36] Aquifer management is also called for in the Statement on Water and Sustainable Development, adopted at the International Conference on Water and the Environment, Dublin, 1992, and in Agenda 21, produced by the United Nations Conference on Environment and Development (UNCED) in Rio de Janeiro in 1992.

The UNECE has focused its attention on groundwater on a number of occasions. In 1989, it developed the Charter on Groundwater Management,[37] an instrument seeking to provide member states with policy support to cooperation in the field of groundwater protection. In 2000, with a view to assisting states in the implementation of the UNECE Water Convention, it produced a set of guidelines on the monitoring and assessment of transboundary groundwaters. In 2013, it issued model provisions on transboundary groundwaters.

Finally, through its International Hydrological Programme (IHP) and the Internationally Shared Aquifer Resources Management (ISARM) initiative, UNESCO has facilitated the codification work of the ILC, by providing a forum for the discussion and further elaboration of the relevant reports. In parallel with this, it has supported the mapping of the world's transboundary aquifers, thereby generating the scientific knowledge needed to raise awareness of these aquifers and to facilitate the discussion of the proposed draft articles. UNESCO continues to support studies, projects and programmes focusing on the enhancement of legal and institutional frameworks for the management of transboundary aquifers.[38]

---

32 Hayton, R. & Utton, A.E. (1989) Transboundary Groundwaters: The Bellagio Draft Treaty. *Natural Resources Journal*, Vol. 29.
33 United Nations (1977) *Report of the United Nations Water Conference*, 51. New York, United Nations.
34 *Ibid.*,53.
35 United Nations (1977) *Report of the United Conference on Desertification*, 20. New York, United Nations.
36 *Ibid.*, 20–21 and 56.
37 Adopted by the UNECE at its forty-fourth session. Document E/ECE/1997 ECE/ENWA/12.
38 See Burchi, S. (2018). Legal Frameworks for the Governance of International Transboundary Aquifers: Pre- and Post-ISARM Experience. *Journal of Hydrology: Regional Studies*.

12.3 The experience of federal countries

Although some specialists in international law do not recognize the validity of the experience of federal countries in an international context, this experience is important when considering that the choices made by the states of a federation in respect of groundwater management and the decisions of federal tribunals purport to be based on principles of international law.

To improve interstate cooperation and to find a better solution to interstate disputes, federal countries have developed three trends: the conclusion of interstate agreements regulating the relationship between the states, the exercise of a paramount federal power over the states concerned, and the decisions of federal tribunals.

While a number of interstate agreements or compacts in the USA regulate the management of water resources, very few of them refer to underground waters. Among these, there are the compact made by the Delaware River Basin Commission, composed of four states (New York, New Jersey, Pennsylvania and Delaware), which deals partly with pollution and empowers the commission to undertake investigations regarding both surface and underground waters; and the compacts concerning the Niobara River and the Ponka Creek, regulating water resources common to the states of Nebraska and South Dakota. The latter agreements recognize that underground water abstractions may be influenced by the utilization of surface waters.

Another remarkable example of institutional arrangement for groundwater management in a federal country is provided by the Great Artesian Basin (GAB) Coordinating Committee in Australia.[39]

Decisions by federal tribunals concerning underground waters are also available. These decisions apply the principle of reasonable and equitable utilization of the waters of a common underground basin, the corollary of which is the need for cooperation among the basin states. It is also worth mentioning the decision handed down by the German *Staatsgerichtshof*, a national court in a federal country, in 1927, in the *Donauversinkung* Case, dealing with interferences in the natural phenomenon by which the water of the Danube at a certain location flows underground and then from the aquifer flows into the Aach, a tributary of the Rhine. In this case the court, applying international law though the case involved three *Länder* (states) of Germany – Württemberg, Prussia and Baden – asserted the duty of states to abstain from injuring the interests of the other states.

12.4 The codification of the law of transboundary aquifers

The paucity of international agreements on groundwater and the few examples that can be quoted of international state practice do not warrant a statement in the sense that groundwater-specific rules of international law have become consolidated over the years. Nevertheless, efforts made by the ILC to codify a set of legal rules on transboundary aquifers have led to the adoption, by the UN General Assembly in 2008, of draft articles on the subject. Although not legally binding as such, these draft articles constitute the most authoritative statement on the law of shared groundwater resources, since they reflect widely accepted principles of international customary water law as applicable to transboundary aquifers.

---

39 See Chapter 6, Section 6.2.3.1.

The codification efforts derived from the recognition that the UN Watercourses Convention does not adequately cover groundwater.[40] On the one hand, its scope is limited to groundwater related to surface water resources and flowing with them into a common terminus. On the other hand, the convention does not consider activities other than water utilization as a possible source of significant harm, such as in the case of certain land uses bringing about groundwater pollution. In addition, the convention does not acknowledge the fact that the effects of certain human activities on groundwater may be irreversible. It is to be noted that, before submitting the draft articles that then became the convention, the special rapporteur had proposed to include 'confined' transboundary groundwater, as groundwater not related to surface water was initially termed,[41] into their scope. However, the ILC did not accept the proposal and instead submitted a resolution on confined transboundary groundwater which commended states 'to be guided by the principles contained in the draft articles on the law of the non-navigational uses of international watercourses, where appropriate, in regulating transboundary groundwater.'[42]

In 1999, at its fifty-fourth session, the UN General Assembly requested the ILC to select new topics for study and codification. In 2002, the ILC incorporated the topic 'shared natural resources,' including oil, gas and groundwater, into its programme of work and appointed a special rapporteur. The commission adopted a step-by-step approach, which began with the study of rules on groundwater as a logical follow-up to the codification of the law of international watercourses.[43] After about six years of work and consultations with various groups of experts, the commission produced the Draft Articles on the Law of Transboundary Aquifers and submitted them to the UN General Assembly. On 11 December, 2008, the General Assembly adopted a resolution[44] that took note of the draft articles and encouraged states 'to make appropriate bilateral or regional arrangements for the proper management of their transboundary aquifers, taking into account the provisions of these draft articles.' The draft articles figure as an annex to the resolution.

The draft articles apply to transboundary aquifers and aquifer systems, which are defined as aquifers (or aquifer systems), 'parts of which are situated in different states' (draft Article 2). This excludes domestic aquifers connected to international surface water bodies, which are covered by the UN Watercourses Convention. The scope of the draft articles is not limited to the utilization of transboundary aquifers,[45] but extends to other activities that may have an impact on transboundary aquifers, such as the use of chemi-

---

40 Mechlem, K. (2003) International Groundwater Law: Towards Closing the Gaps? 14 *Yearbook of International Environmental Law*, 47–80.
41 'Confined' is not synonymous with 'unrelated.' The term '*confined*' refers to the situation in which aquifers dip beneath much less permeable strata and their groundwater becomes pressurized and thus, to varying degrees, isolated from the *immediately overlying* land surface. Confined aquifers are usually replenished by lateral groundwater flows originating in the unconfined portions of the same aquifer system, which receive recharge from infiltrating rainfall and surface water bodies. Thus, confined groundwater resources are often connected to surface water. See Nanni, M. & Foster, S. (2005) Groundwater Resources – Shaping Legislation in Harmony with Real Issues and Sound Concepts. *Water Policy* 7:5.
42 International Law Commission (1994) *Yearbook of the International Law Commission*, Vol. 2, Part II, 135.
43 UN General Assembly, International Law Commission, *Shared Natural Resources: First Report on Outlines* (UN Doc. A/CN.4/533), 30 April 2003.
44 Resolution 63/124 of 11 December, 2008 (A//RES/63/124).
45 And aquifer systems.

cals in agriculture, and to measures for aquifer protection, preservation and management. Draft Article 3 makes explicit reference to the sovereignty of states over the portions of a transboundary aquifer located in their respective territories, but adds that this sovereignty must be exercized in accordance with international law and with the principles set forth in the draft articles.

The principle of equitable and reasonable utilization, which is the subject of draft Articles 4 and 5, is adapted to the specific characteristics of groundwater, in that it recognizes that a transboundary aquifer may be non-recharging and that therefore the states concerned must cooperate with a view to obtaining maximum benefits therefrom. According to the draft articles, to implement the principle aquifer states are required to establish comprehensive aquifer utilization plans, whether individually or jointly, and to refrain from utilizing a recharging transboundary aquifer in such a way as to prevent its effective functioning. Further, the draft articles recognize the fact that aquifers and the water contained therein play an important role in the preservation of related ecosystems, especially in arid regions. When balancing factors in the process of determining what is equitable and reasonable, this role is to be considered. In addition, in weighing these factors, special regard is to be given to vital human needs.

Draft Article 6 establishes the obligation not to cause significant harm, along the lines of Article 7 of the UN Watercourses Convention. It considers the utilization of transboundary aquifers *and* activities other than utilization, in so far as they may have an impact on other aquifer states *and* on non-aquifer states in which a discharge zone is located. The importance of aquifer recharge and discharge processes is also recognized by draft Article 11, which introduces a duty of aquifer states to take all appropriate measures to prevent and minimize detrimental impacts on such processes, as well as a duty of non-aquifer states in which a recharge or discharge zone is located to cooperate with the aquifer states.

The general obligation to cooperate is the subject of draft Article 7, which also recommends the establishment, by aquifer states, of joint mechanisms for cooperation. The choice as to the mechanisms to be resorted to is left to the states concerned. Under draft Articles 8 and 13, aquifer states have the duty to regularly exchange readily available data and information on the conditions of the aquifer, and to monitor the aquifer in conformity with agreed or harmonized standards and methodologies. Due to uncertainty as to the nature and extent of transboundary aquifers and their vulnerability, draft Article 12 places emphasis on the necessity to take a precautionary approach when dealing with the prevention, reduction and control of pollution. Draft Article 15 is about the notification of planned activities. In contrast to the UN Watercourses Convention, which devotes nine articles to the notification of planned measures, it takes a minimalistic approach, justified by the scarcity of state practice with regard to transboundary aquifers.

Finally, draft Article 14 establishes an obligation for states to develop and implement plans for the proper management of their transboundary aquifers, and to consult on aquifer management. In addition, it requires the establishment of joint management mechanisms, where appropriate. 'Management' is an over-arching term that encompasses all activities directed to the use, development, protection and conservation of a transboundary aquifer, including the management of the risks and uncertainties to which an aquifer may be exposed.

In a nutshell, these are the main elements of the draft articles. On a number of occasions after 2008, the UN General Assembly has encouraged states to make appropriate bilateral and regional aquifer arrangements taking the draft articles into account.[46] The final form of the draft articles, i.e., whether they should be taken as a basis of a convention on transboundary aquifers complementing the UN Watercourses Convention, is still being discussed. The item has been included in the provisional agenda of the seventy-fourth session of the UN General Assembly. In the meantime, the draft articles are a source of inspiration for governments negotiating legal and institutional arrangements for transboundary aquifers, as the examples of the ITAS and of the Guaraní aquifer system have demonstrated.

12.5 Institutional issues

The management of transboundary groundwater raises some questions which are not of easy solution. The first relates to international cooperation in groundwater management. If an international commission has to be established between two or more states, should it deal with international water resources in general, or only with groundwater?

A second problem concerns the powers that an international commission should have as regards the monitoring and control of underground water activities. Ideally, the commission should be able to exercise these powers sometimes well inside the territory of the states overlying a transboundary aquifer. States, however, may be reluctant to agree on this solution.

This gives rise to a third problem: that of the coordination between the regulatory provisions set forth in an international agreement establishing the commission and those of national legislation. Likewise, a problem of coordination between the international commission and national water institutions may arise.

There are neither precise agreements on the subject, nor precise legal norms, nor well-established and universally recognized principles to address these questions.[47] However, the question as to the choice of adequate institutional solutions is certain to become more pressing and frequent.

A number of international water commissions, committees and the like have been established in the course of the years. Their jurisdiction, however, normally covers groundwater only in so far as it forms part of a river basin. As a result, groundwater plays a secondary role. Until recently, the only known institutional mechanisms for cooperation in the management of international groundwater resources as such were the Franco-Swiss commission established under the Geneva Aquifer Convention, and the joint authority for the Nubian Sandstone Aquifer System.[48] Developments since 1999, however, show a growing understanding by states of the important role that aquifer-specific institutional arrangements may play in facilitating this cooperation, given that the existing water commissions do neither address aquifer vulnerability issues, nor do they contemplate dealing with the hydrogeological risk deriving from land use, mining activities and increasing groundwater extraction.

The consultation mechanism for the NWSAS was initially conceived as a light structure made up of a coordinating unit attached to the OSS and national focal points. Its objective

---

46 Resolutions adopted by the General Assembly on 9 December, 2011 (A//RES/66/104), 16 December, 2013 (A/RES/68/118) and 13 December, 2016 (A/RES/71/150), respectively.
47 The Bellagio Draft Treaty may provide some guidance in the development of international institutional arrangements for groundwater management.
48 See this chapter, Section 12.2.

was to facilitate cooperation in hydrogeological data collection and aquifer modelling for better domestic planning and decision making. The mechanism then evolved into a permanent arrangement and is now also vested with advisory powers with regard to the introduction of aquifer management measures. The present mechanism consists of a council of ministers, a steering committee, national committees which replace the national focal points, working groups and a coordinating unit, always hosted by the OSS.

The structure of the consultation mechanism for the IAS is more elaborated. It consists of a council of ministers, an *ad hoc* technical committee, national technical and scientific committees and an executive secretariat. The ITAS memorandum of understanding, which supersedes the IAS memorandum, adds to it a permanent scientific and technical committee and a coordinating unit to be hosted by the OSS. The functions of the consultation mechanism include the coordination of studies, projects and programmes that may have an impact on, or create risks for, the ITAS resources, the formulation of recommendations on the harmonization of the member states' legislation and institutions, the identification and mapping of risks and vulnerable areas, the development of the ITAS water resources management plan and of action plans, the definition of risk management measures, the definition of monitoring programmes and the coordination of the exchange of data and information, the mobilization of financial resources and the settlement of disputes. Plans, measures, programmes and projects are subject to the approval of the council of ministers. The consultation mechanism has the status of a legal person, which entails the capacity to enter into contracts, acquire and dispose of movable and immovable property, receive loans, grants and gifts, sue and be sued.

The Guaraní Aquifer Agreement provides for the establishment of a commission under the La Plata River Treaty, with the general mandate to coordinate cooperation among the parties in its implementation. Thus, Argentina, Brazil, Paraguay and Uruguay will have to work together within an existing river basin framework. Since the commission is still to be set up, it is now too early to make statements as to its legal status, its functions and the manner in which it will operate. The agreement only makes reference to regulations to be drawn up by the commission itself and to the commission's advisory powers in the case of unsettled disputes on its application or interpretation.[49]

Finally, the joint technical committee for the Al-Sag/Al-Disi Layer is responsible in particular for monitoring the quantity and quality of groundwater extracted from the aquifer and for the collection, assessment and exchange of data and information. The committee is to submit its findings to the competent authorities of Jordan and Saudi Arabia.

Institutional mechanisms for cooperation and supporting legal arrangements are now being the focus of programmes supported by UNESCO-IHP in Africa and the Americas, which have devoted their attention to aquifers underlying transboundary surface water resources and have tabled recommendations pointing to the integration of aquifer management considerations into the existing legal and institutional frameworks. Among others, this is the case of the Stampriet Aquifer System (Botswana, Namibia, South Africa), for which it is proposed to nest a management unit in the structure of the Orange-Senqu River Commission (ORASECOM).[50]

These cases illustrate the key role that institutional mechanisms play with regard to the enhancement of the knowledge of transboundary aquifers, the identification of present and

---

49 For a discussion of possible roles for the commission, see Sindico, F., Hirata, R. & Manganelli, A., *op. cit.*
50 See Burchi, S. (2018), *op. cit.*

potential threats, the identification of critical areas or 'hot spots', the management of risks and the prevention and control of resource depletion and degradation. Also, it is increasingly understood that, in addition to water sharing, emphasis is to be placed on the impact of certain land uses on an aquifer. Since these uses may take place well inside the national territories of the states concerned, there is the need to strengthen the capacity of the national institutions to implement groundwater management measures of transboundary relevance within national boundaries.

12.6 The emerging rules

Considering the progress made towards a definition of well established rules of international law applicable to transboundary aquifers, and the developments described in this chapter, the need should be stressed for incorporating basic legal principles and tools into future agreements on the subject, such as:

(i) the acceptance of the principle of equitable and reasonable utilization, taking into consideration the specific characteristics of aquifers and the behaviour of groundwater;
(ii) the obligation for transboundary aquifer states to monitor the aquifer on a regular basis in accordance with agreed standards and methodologies and to exchange hydrogeological data and information, as well as data relating to water quality;
(iii) the need for transboundary aquifer states to identify 'hot spots,' i.e., vulnerable areas, and to map these areas;
(iv) given the vulnerability of groundwater, the adoption of a precautionary approach to pollution control;
(v) the obligation for the states concerned to exchange information on, and provide advance notification of, projects which are likely to cause serious and permanent alterations of the quantity and/or quality of the waters of a transboundary aquifer, or significant drawdowns;
(vi) the need for aquifer states to proceed, as far as possible, to joint underground water resources and aquifer planning, and to cooperate in the implementation of groundwater management measures;
(vii) the recognition of the non-discrimination principle as regards decisions to be made with respect to groundwater pollution. In fact, if one state incurs financial costs in order to combat pollution while another state omits to do so, the position of the two states is not equal;
(viii) the establishment of international water institutions which should be made responsible for the integrated consideration of shared surface and underground waters whenever this is the case, or for exchanging data and information and cooperating in the development of measures to prevent that activities within a state overlying a transboundary aquifer cause damage to other states overlying the same aquifer; and, finally,
(ix) the possibility to submit disputes to a commonly agreed tribunal, to arbitration, or to other chosen settlement mechanisms.

12.7 Conclusion

Notwithstanding the progress made in its codification, the law of transboundary aquifers is still going through a process of evolution. A number of physical and technical elements remain unknown, and this determines a certain fluidity in its development.

The agreements dealing specifically with transboundary aquifers are still few, although, given the growing awareness of states of the importance of legal and institutional frameworks for cooperation, their number may well increase in a not too distant future. The ILC Draft Articles on the Law of Transboundary Aquifers have exerted a considerable influence on the formulation of the agreements on the Guaraní Aquifer System and on the ITAS, and are contributing considerably to the discussion of future cooperative arrangements for other transboundary aquifers. The number of initiatives, projects and programmes which are being implemented worldwide demonstrate that transboundary aquifers enjoy high priority on the international agenda.

References

Barberis, J. (1986) *International Groundwater Resources Law*. Legislative Study No. 40. Rome, FAO.
Burchi, S. (2018) Legal Frameworks for the Governance of International Transboundary Aquifers: Pre-and Post-ISARM Experience. *Journal of Hydrology: Regional Studies*.
Burchi, S. & Mechlem, K. eds. (2005) *Groundwater in International Law: Compilation of Treaties and other Legal Instruments*. Legislative Study No. 86. Rome, FAO.
Burke, J.J. & M.H. Moench (2000) *Groundwater and Society: Resources, Tensions and Opportunities*, Themes in Groundwater Management for the Twenty-First Century. New York, United Nations Publication ST/ESA/265.
Caponera, D.A. & Alhéritière, D. (1978) Principles of International Groundwater Law. *Natural Resources Forum*, 2, 279–290 and 359–371.
Eckstein, G. (2017) *The International Law of Transboundary Groundwater Resources*. Earthscan Water Text Series. Routledge.
Hayton, R. (1982) The Law of International Aquifers. *Natural Resources Journal*, Vol. 22, 71–93.
Hayton, R. & Utton, A.E. (1989) Transboundary Groundwaters: The Bellagio Draft Treaty. *Natural Resources Journal*, Vol. 29. Albuquerque, N.M., CIRT, University of New Mexico School of Law.
Jarvis, W. T. (2014) *Contesting Hidden Waters, Conflict Resolution for Groundwater and Aquifers*. UK, Routledge.
Mechlem, K. (2004) International Groundwater Law: Towards Closing the Gaps? 14 *Yearbook of International Environmental Law*, 47–80.
Movilla Pateiro, L. (2016) Ad Hoc Legal Mechanisms Governing Transboundary Aquifers: Current Status and Future Prospects. *Water International* 41:6, 851–865.
Nanni, M. & Foster, S. (2005) Groundwater Resources – Shaping Legislation in Harmony with Real Issues and Sound Concepts. *Water Policy* 7:5.
Nanni, M., Burchi, S., Mechlem, K. & Stephan, R. M. (2006) Legal and Institutional Considerations. *Non-Renewable Groundwater Resources*, A Guidebook on Socially-Sustainable Management for Water Policy-Makers (Foster, S. & Loucks, D.P. eds.). IHP-VI, Series on Groundwater No. 10. Paris, UNESCO.
Salman, M.A.S. ed. (1999) *Groundwater: Legal and Policy Perspectives*. Technical Paper 456. Washington, D.C., World Bank.
Salman M.A.S. (2011) The World Bank policy and practice for projects affecting shared aquifers. *Water International*, 36:5.
Sindico, F., Hirata, R. & Manganelli, A. (2018) The Guarani Aquifer System: From a Beacon of Hope to a Question Mark in the Governance of Transboundary Aquifers. *Journal of Hydrology: Regional Studies*.
Stephan, R.M. ed. (2009), *Transboundary Aquifers – Managing a Vital Resource. The UNILC Draft Articles on the Law of Transboundary Aquifers*. Paris, UNESCO.

Teclaff, L.A. (1982) Principles for Transboundary Groundwater Pollution Control. *Natural Resources Journal*, Vol. 22, 1065–1079.

Teclaff, L.A., & Teclaff, E. (1979) Transboundary Groundwater Pollution: Survey and Trends in Treaty Law. *Natural Resources Journal*, Vol. 19, 629–667.

Teclaff, L.A., & Utton, A.E. (1981) *International Groundwater Law*, London, Rome, New York.

Utton, A.E. (1983) La administración internacional de aguas subterraneas: el caso de la región fronteriza Mexico-Estados Unidos. *Boletín Mexicano de Derecho Comparado*, No. 47, 545–576.

Chapter 13

International water resources administration

13.1 Introduction

There are doubts as to the existence of an international legal rule of general application concerning the administration of international water resources, as the 1997 United Nations Convention on the Law of Non-Navigational Uses of International Watercourses (UN Watercourses Convention) does not mandate international watercourse states to establish international basin institutions. However, recent state practice in Europe and beyond warrants a statement, in the sense that states have become aware of the importance of institutional arrangements for the management of shared water resources, including groundwater.

In fact, a number of international water agreements concluded in recent years, not only in Europe, but on the four continents, contain provisions relating to institutions vested with responsibilities concerning aspects of international river basin management, ranging from data administration to the screening of programmes, projects and activities with cross-border impact.

Institutions responsible for the management of international water resources may take different names, such as, agency, commission, committee, authority or administration. In line with the definition given by the doctrine,[1] the term 'international water resources administration' refers to any form of institution or other arrangement which is established by agreement between two or more states sharing a common basin (surface or underground) for the purpose of dealing with its management.

While it is not possible to analyze the actual operation of the numerous individual basin institutions in existence today, reference can be made to the major relevant international agreements, and to literature on the subject. Institutional developments with regard to transboundary aquifers are reviewed in Chapter 12.

13.2 Institutional developments

13.2.1 Institutional developments in Europe

At the outset, international river commissions in Europe were set up to regulate the use of waterways for navigation purposes; this was the case of the Rhine, Danube, Netze, Kudov,

---

1 Caponera, D.A. (1976) Administration of International Water Resources. *International Law Association Report of the 57th Conference*. Madrid, Art. 1.

Oder, Warthe, and Lower Niers Rivers. One of the earliest commissions was established in 1755 between Germany and the Netherlands to regulate navigation on the Rhine River.

The Final Act of Vienna of 1815 introduced the principle of freedom of navigation on European international rivers. Initially, this freedom was recognized to all states, riparian and non-riparian; later, it was restricted to riparian countries only. Other questions arising in connection with navigation refer to the obligation of the riparian states to maintain and/or improve waterways and to the imposition and collection of tolls. These questions were the major concern for the European river commissions until the First World War. No other water uses were considered, in view of climatic conditions and of the fact that water was plentiful.

Generally speaking, the establishment of international navigation commissions followed the conclusion of major wars or other international political crises. This is the case of the Congress of Vienna of 1815, after the fall of Napoleon, which led to the creation of the Central Commission for Navigation on the Rhine in 1831; the Treaty of Paris of 30 March, 1856, following the Crimean War and the defeat of the Ottoman Empire, which gave birth to the Danube Commission; the Treaties of Versailles of 1919 and of Paris of 1947, following the First and Second World Wars, which reorganized existing international river commissions and set up new ones.

Subsequently, states realized that matters relating to common water uses could be better dealt with through international institutional mechanisms. Thus, in the course of the last century, international commissions were established for hydropower generation and distribution purposes, and for dealing with the harmful effects of water. After World War II, several commissions were created to deal with water quality and pollution control on international rivers.[2]

More recently, a number of international commissions have been created for several international river basins in order to facilitate the protection of water quality and the control of water pollution, in line with the international water law principles prevailing in the UNECE region[3] and with European Union legislation.[4] Some of these commissions are the result of a reorganization of pre-existing ones, as in the case of the Commission for the Protection of the Rhine against Pollution. *Inter alia*, these institutions are responsible for facilitating river basin planning and flood management, in addition to pollution control. Some of them have been vested with the specific task of implementing the European Union (EU) Water Framework Directive of 2000,[5] as the directive mandates EU member states to coordinate their efforts within international river basin districts in order to arrive at single basin plans for these districts.[6] In some cases, these institutions coexist with navigation commissions predating their establishment.

---

2 Convention of 20 April, 1972, between Italy and Switzerland for the Protection of Waters from Pollution; Berne Convention of 1963, between the Federal Republic of Germany, France Luxembourg, the Netherlands and Switzerland, establishing the commission for the protection of the Rhine against pollution; Additional Protocol of 1976, through which the European Economic Community adhered to the 1963 convention.
3 See Chapter 11.
4 See Chapter 6, Section 6.4.5.
5 *Ibid.*
6 When some of the basin states are members of the EU and others are not, the production of a single basin plan is not obligatory; however, EU member states must formulate a plan for the portion of the basins situated within their territory.

13.2.1.1 The Rhine Commissions

The Central Commission for Navigation on the Rhine was created in 1831 after a lengthy process dating back to 1785 and 1816. It has essentially consultative and technical functions in matters primarily affecting navigation; it can undertake research studies and recommend to member states the provisions to be adopted in their respective municipal systems. It includes representatives from France, the Federal Republic of Germany, the Netherlands, Switzerland, Belgium and Great Britain. Ordinary decisions are made by majority vote but are not binding upon dissenting countries. Major policy decisions therefore require unanimity.

In the 1950's, with industrialization growing at a remarkable pace, water pollution became an issue. Thus, in order to promote cooperation in the control of pollution, a separate commission was created in 1949–50, by an exchange of notes. In 1963, Germany, France, Luxembourg, the Netherlands and Switzerland signed the Berne Convention on the Commission for the Protection of the Rhine. The European Economic Community joined the commission in 1976. In the same year, two conventions were entered into by the Rhine basin states, one concerning chemical pollution and the other pollution by chlorides.[7]

Further to the Sandoz chemical spill in 1986, which caused the massive death of fish between Basel and Koblenz, the governments of the Rhine basin states acted to reinforce their cooperation through a broader legal framework. Thus, in 1987 they adopted the Rhine Action Programme and in 1999 they signed a new convention to replace the 1963 Convention.[8] This convention entered into force in 2003.

The new convention has a broader scope, and since its implementation is largely left to the International Commission for the Protection of the Rhine, it provides for an expansion of the mandate of this commission to the preparation of monitoring programmes, the analysis of the Rhine ecosystem, the elaboration and evaluation of programmes of measures and the coordination of warning and alarm plans. After 2000, the commission has been vested with the task of implementing the EU Water Framework Directive.

As to its structure, the commission follows the pattern of the International Commission for the Protection of the Danube River, that is, it comprises the delegations of the member states, a secretariat with headquarters at Koblenz and working groups dealing with various issues.[9] By decision of the Rhine ministers in 2001, a committee has been created to deal specifically with the coordination of the implementation of the Water Framework Directive. The commission takes unanimous decisions, which have the legal force of recommendations.

A Rhine Hydrology Commission was created in 1970, with functions that are limited to the collection and administration of hydrological data.

---

7 See Chapter 11.
8 Convention on the Protection of the Rhine, signed at Berne on 12 April 1999. Parties: European Community, France, Germay, Luxembourg, Switzerland and The Netherlands. Text in: EU, *Official Journal* L 289, 16/11/2000, 31.
9 Working groups have been established to deal with water quality and emissions, ecology, groundwater, economy and GIS.

13.2.1.2 The Danube Commissions

The Danube Commission was established by the Belgrade Convention of 1948 and is an operating, administrative body which enacts a uniform set of navigation rules, facilitates flood control operations, sponsors integrated energy planning and encourages irrigation projects. It includes representatives of Austria, Bulgaria, Croatia, Germany, Hungary, Moldova, Slovakia, Romania, Russia, Ukraine and Serbia. The navigation régime of the Danube is still regulated by this convention and by a series of special agreements. Recommendations on ordinary matters are taken by majority vote, but on important questions the states directly involved have veto power. Another organization has been created for the administration of the lower Danube and the Iron Gates. It deals primarily with hydropower and navigation projects.

Following concern for the state of the environment and, in particular, for the increasing deterioration of water quality, in 1994 the Danube basin states and the European Community signed a convention[10] to establish, *inter alia*, the International Commission for the Protection of the Danube River. The convention entered into force in 1998, but the member states decided to implement its provisions soon after signature and established an interim secretariat.

This commission is vested with advisory functions and powers with regard to the management of the entire basin. In particular, it is responsible for elaborating proposals and recommendations with regard to activities that are likely to produce transboundary impact, such as, wastewater discharges, pollution from non-point sources, certain water uses, the construction and operation of hydraulic works and structures, the handling of hazardous substances and the prevention of accidents. In addition, it has been vested with coordination functions with regard to the preparation of the Danube Basin Management Plan based on the guidelines provided by the EU Water Framework Directive.

The structure of the commission is the same as that of the International Commission for the Protection of the Rhine; that is, it comprises the delegations of the parties, a secretariat and working groups, permanent and *ad hoc*. The commission makes decisions by consensus, which become binding on the parties voting for them on the first day of the eleventh month following them.

The International Commission for the Protection of the Danube River is not meant to replace the Danube Commission.

13.2.1.3 Other commissions

Other international joint technical commissions have been created from time to time to deal with specific issues affecting international rivers, such as flood control, water pollution or specific projects. Usually of a bilateral nature, they have rarely dealt with water apportionment questions.

---

10 Convention on Cooperation for the Protection and Sustainable Use of the River Danube, signed at Sofia, Bulgaria, on 29 June, 1994, by Austria, Bulgaria, Croatia, Germany, Moldova, Czech Republic, Slovak Republic, Romania, Slovenia, Ukraine, Hungary and the European Community. The text is to be found in EU, *Official Journal*, L 342, 12/12/1997, 19.

Recently, commissions with a broader scope have been established for many European international rivers. Generally speaking, they follow the same pattern as the International Commission for the Protection of the Danube River, in that their focus is on the prevention, mitigation and control of transboundary impacts, including the impacts deriving from pollution and natural calamities such as floods, and the preservation of aquatic ecosystems. As of the year 2000, they are also called upon to facilitate river basin planning and the implementation of the Water Framework Directive. Thus, they do not deal with water resources allocation *per se*.

As to their structure, these commissions normally comprise the member states' delegations, a secretariat and working groups dealing with various issues, which may be permanent or *ad hoc*. Among them, the following may be mentioned:

- the Sava River Basin Commission, established by the Framework Agreement on the Sava River Basin (2002) between Bosnia and Herzegovina, Croatia, Slovenia, Federal Republic of Yugoslavia;
- the Scheldt International Commission, established by the International Agreement on the Scheldt (2002) between Belgium, France and the Netherlands;
- the Meuse International Commission, established by the International Agreement on the Meuse (2002) between Belgium, France, Germany, Luxembourg and the Netherlands;
- the International Commission for the Protection of the Oder River against Pollution, which was established by the Agreement on the International Commission for the Protection of the Oder River against Pollution (1996) between Germany, Poland, the Czech Republic and the European Community;
- the Transboundary Water Commission set up by the Agreement on Cooperation in the Field of Protection and Sustainable Use of Transboundary Waters (1997), between Estonia and the Russian Federation;
- the Luso-Spanish commission created by the Convention on Cooperation for the Protection and Sustainable Use of the Waters of the Spanish-Portuguese Hydrographic Basins (1998) between Portugal and Spain;
- the Commission on Sustainable Use and Protection of the Dniester River Basin, established by the Treaty on Cooperation in the Field of Protection and Sustainable Development of the Dniester River Basin (2012), between Moldova and Ukraine.

13.2.2 Institutional developments in the Americas

13.2.2.1 The International Joint Commission between the USA and Canada

The International Joint Commission (IJC) exercises jurisdiction over all boundary waters of the USA and Canada and often is taken as an example for the use of waters for general purposes. It was created by the border treaty between Great Britain and the USA of 1909[11] after the principal questions relating to navigation between the two states had already been solved. It was realized, however, that the use of the shared waters could give rise to frictions

---

11 Treaty between Great Britain and the United States Relating to Boundary Waters, and Questions Arising between the United States and Canada, signed at Washington on 11 January, 1909. Text in: ST/LEG/SER.B/12, 260.

in other sectors, and for this reason the commission was conceived as a permanent instrument to prevent any type of controversy whatsoever connected with the use of common waters.

The commission is organized in conformity with the principles embodied in the treaty of 1909. It is composed of six members, three for each side, meeting twice a year alternately in Ottawa or Washington. In practice the IJC acts as an organ common to two states. Decisions may be taken by majority vote, but until now they have always been taken unanimously. In the case of parity of vote, the question is to be submitted to the governments.

According to the treaty, the commission must give its consent when a new project of utilization can alter the natural level or the flow of the water (Articles 3 and 4). Furthermore, the commission must undertake studies and enquiries and present recommendations when requested by either state (Article 9). The IJC has been involved in such questions as the level of the Great Lakes and the pollution of Lake Ontario, Lake Erie and other waters of the system, in spite of the fact that water quality aspects were not included in its original mandate.[12] Its powers of enquiry are often undertaken by special auxiliary commissions which are set up according to need.[13]

Canada and the United States have often had recourse to these auxiliary commissions for the supervision of specific projects and for verifying compliance with pre-established standards. Special organisms have been charged to examine the consequences of the exploitation of the Columbia River, to suggest criteria for the apportionment of the energy produced and related costs, and to furnish any other elements useful for the negotiations between the two states. For its efficiency and practicality in the accomplishment of its duties, the IJC has developed an important role in the elaboration of agreements for the exploitation of the St. Lawrence and Columbia Rivers.

Should differences between the parties arise, the IJC has been entrusted with real arbitration functions that had not been vested in it under the treaty of 1909. The commission should be considered as one of the most efficient instruments for international cooperation in the field of shared water resources.

It is to be asked whether the success of the IJC is to be attributed to the high technical level and the impartiality of the commission, or to the procedures and the mechanism utilized. In any case, the political climate existing between the two states is also a factor.

13.2.2.2 The International Boundary and Water Commission between USA and Mexico

The Boundary Commission of 1889 between the United States and Mexico, after years of negotiation and following disputes over the use of surface water by the USA, became the International Boundary and Water Commission in 1944, by virtue of the Treaty Concerning

---

12 Bourne, C.B (1997) *International Water Law, Selected Writings*. International and National Water Law and Policy Series. The Hague, London, New York, Kluwer Law International, 354 ff.
13 The 1978 Great Lakes Water Quality Agreement, as amended on 16 October, 1983, on 18 November, 1987 and on 7 September, 2012, has expanded the scope of work of the IJC to include water pollution issues affecting the Great Lakes, as well as the protection of the lakes' ecosystems.

the Utilization of the Waters of the Colorado and Tijuana Rivers and of the Rio Grande.[14] These disputes gave birth to the 'Harmon Doctrine.'[15]

Contrary to the IJC, this commission undertakes and executes works. It is composed of two national sections, each headed by a commissioner who must be a licenced engineer and is assisted by two principal engineers, a legal advisor and a secretary. The two sections are headquartered near the border, i.e., in El Paso (USA) and Ciudad Juarez (Mexico), but are integrated into the national government's foreign offices. Each member state pays for its own section; the common expenses are apportioned in equity. The decisions of the commission, consisting of agreements between the two commissioners, are recorded in minutes which must be approved by the two governments before coming into effect. However, if not refused within the term of 30 days from the date of signature, they are assumed to have entered into force.

In the exercise of its functions under the treaty of 1944, the commission must take into consideration the order of priorities among the different uses established by the treaty: domestic water supply, agriculture, watering of animals, production of electric energy, other industrial uses, navigation, fisheries and hunting and other uses determined by the commission (Article 3).

The problem of shared groundwaters between the USA and Mexico is still outstanding; studies are being conducted to seek an institutional solution to it. On 30 August, 1973, the Governments of the United States and Mexico approved Minute No. 242 of the commission concerning the control of salinity of the waters of the Colorado River and requiring the two countries to limit groundwater pumping in their respective territories along the Arizona-Sonora border[16] pending the conclusion of an agreement on transboundary groundwater. Minute No. 242 calls for consultations and the exchange of information on groundwater. The question remains as to whether the Boundary and Water Commission, which is responsible for the management of surface waters only (both quantity-and quality-wise), be also made responsible for the management of shared groundwater resources. Since groundwater management implies the monitoring of activities well inside the territories of the states, the issue arises as to how this power can be exercised and, in particular, as to how it can be carried into effect within the USA, where each state is responsible for its own water management.

In addition, environmental issues are gaining in importance in the border area between the two countries. Thus, the question arises as to whether the commission should also be vested with responsibilities with regard to environmental matters. To deal with these matters, the commission has established a binational working group including representatives of a binational coalition of non-governmental organizations. Minute No. 323, adopted on 27 September, 2017,[17] and superseding Minute No. 319 of 20 November, 2012,[18] requires the two countries to consult and to exchange information on the impact of flow deliveries on the Colorado River basin and its environment, among other things. Moreover, it indicates the actions to be taken by the two countries to generate water for the environment and to imple-

---

14 Text in: ST/LEG/SER.B/12, 236.
15 See Chapter 11, Section 11.3.2.
16 See Chapter 12, Section 12.2.
17 Available from www.ibwc.gov/Files/Minutes/Min323.pdf.
18 Available from www.ibwc.gov/Files/Minutes/Min319.pdf.

ment environment restoration and water conservation projects, having particular regard for the river delta.

It follows from this brief review that the present institutional situation may raise the question as to whether there should be one or more than one water commission dealing with different aspects of water resources management on the same river basin.

13.2.2.3 The Plata River Basin

The Statute of the Intergovernmental Committee for Coordination of 1968 is part of a series of agreements between Argentina, Bolivia, Brazil, Paraguay and Uruguay, signed the same year and modified in 1969 and 1971.[19] Among its institutional objectives, the committee has that of promoting the harmonious and integrated development of the resources of the basin. The ministers of foreign affairs of the countries involved form the highest body of this organ. They meet once a year ordinarily or in extraordinary session upon request of three member states. The committee is composed of one delegate per part, usually the ambassador of the country accredited to the government of Argentina. It decides on the guidelines for the exploitation of the Plata basin, evaluates the results and provides the committee with directives for action. It may, with special procedures, amend the statutes of the committee.

In 1974 a fund was established for the joint administration of the common rivers and the common projects within the framework of the Plata basin.

Various technical commissions have been created under the committee, including:

- the Administrative Commission for the Uruguay River (26 February, 1975), between Argentina and Uruguay;
- the Joint Technical Commission for the Salto Grande project (30 December, 1946), also between Argentina and Uruguay;
- the Binational Commission for the Development of the Upper Basins of the Bermejo River and the Rio Grande de Tarija (9 June, 1995), between Argentina and Bolivia;
- the Itaipú Binational Entity (26 April, 1973), between Brazil and Paraguay, for the Paraná River;
- the mixed technical commission for the Paraná River (16 June, 1971), between Argentina and Paraguay;
- the Yaciretá Binational Entity (13 December, 1973), between Argentina and Paraguay, for the Paraná River.

13.2.2.4 Other commissions

Other commissions dealing with specific basins are described here below:

(i) the Commission of the Mirim Lagoon between Brazil and Uruguay, for the joint development of the Mirim Lagoon and for the joint management of hydraulic works

---

19 For an overview of the history and of the development of the legal and institutional framework of the Río de la Plata basin, see Del Castillo, L. (2005) *El regimen jurídico del Río de la Plata y su frente maritimo*, Buenos Aires.

(26 April, 1963, 5 August, 1965, 7 July, 1977). This commission is composed of representatives of both governments and is responsible for administering technical and financial assistance from national and international sources. It has autonomy and administrative authority for this purpose. Recommendations concerning the execution of projects are to be approved by the central authorities of the riparian countries. The director of the commission is appointed by an international organization and the deputy director by the governments. This commission constitutes an authentic development agency whose officers enjoy diplomatic immunities in the exercise of their functions;
(ii) the institutional mechanism set up under the Treaty of Amazonian Cooperation, signed by Bolivia, Brazil, Colombia, Ecuador, Guyana, Peru, Surinam and Venezuela on 3 July, 1978, consisting of a Council of Ministers of Foreign Affairs, a Council of Amazonian Cooperation and a Secretariat with rotation among the states. Decisions are unanimous;
(iii) numerous other commissions and authorities have been entrusted with the study and development of various rivers and projects, among which the following may be mentioned:

- between Bolivia and Peru, for Lake Titicaca (1996);
- between Ecuador and Peru for the Rivers Puyango-Tumbas and Catamayo-Chira (27 November, 1971, and 14 June, 1972);
- between Colombia and Venezuela for the river basins of the Zulia, Catatumbo, Meta, Aranca and Orinoco (1976);
- between Guatemala and El Salvador (1938);
- between Guatemala and Honduras (1956);
- between the Dominican Republic and Haiti for the Lilion and Artibonite Rivers (21 January, 1929).

13.2.3 Institutional developments in Africa

After the Treaty of Berlin of 1885, which followed an international political crisis of the European powers over their colonial influence in Africa, two important African rivers were declared international: the Niger and the Congo. Under this internationalized régime, the principle of freedom of navigation was introduced for all countries, riparian and non-riparian.

13.2.3.1 The Nile Commission

The Permanent Joint Technical Committee of the Nile, created between Egypt and Sudan by the Agreement for the Full Utilization of Nile Waters (1959),[20] settled the apportionment of water between the two countries and provided for the construction of the Aswan Dam (Saad el Aali Dam). The function of this committee includes the planning, supervision of studies and execution of projects, in addition to the supervision of water distribution. As regards its composition, the committee is formed by four engineers per part. The agreement provides for the full utilization of the Nile waters. It further provides that a common front be formed between the two countries, should another country claim a share of such waters.

---

20 Text in: 453, UNTS, 51.

It is to be noted that another organization on the same basin was created independently from the Nile committee, that is, the Kagera Basin Authority. Since it was never effectively operational, it was dissolved in 2004. A further organization, also independent from the committee, was created for Lake Victoria in 2003 as a specialized institution of the East African Community.

A project called Hydromet, concerning all the riparian states of the White Nile,[21] was implemented to collect data concerning the river. Hydromet ceased to function in 1992, when the TECCONILE was established. In 1998, the ministers of water resources of the basin states agreed to establish, within the framework of the Nile Basin Initiative (NBI), which was launched in 1999 and superseded the TECCONILE, an interim institutional mechanism for cooperation among all the Nile countries, consisting of a council of ministers of water resources, a technical advisory committee and a secretariat headquartered in Entebbe, Uganda.[22] Among other duties, the NBI was mandated to prepare and facilitate the conclusion of a cooperative framework agreement reflecting the interests of all basin states.

The Agreement on the Nile River Basin Cooperative Framework was adopted in 2010, after over ten years efforts and negotiations. It is not in force, since it requires ratification or accession by at least six basin states.[23] It provides for the establishment of a Nile River Basin Commission composed of a Conference of Heads of State and Government, a Council of Ministers of water resources, a Technical Advisory Committee, sectorial advisory committees and a secretariat. The purpose of the commission is to promote and facilitate the implementation of the principles, rights and obligations set forth in the agreement, that is, *inter alia*, the principle of cooperation, the principle of equitable and reasonable utilization, the obligation to prevent the causing of significant harm, the duty to protect and conserve the basin and its ecosystems, the duty to exchange data and information on a regular basis, the duty of basin states to inform each other of planned measures and to carry out an environmental impact assessment of their possible adverse effects, and the peaceful settlement of disputes.

Once established, the commission will have juridical personality and will inherit all rights, obligations and assets of the NBI.

13.2.3.2 Post-1960's basin institutions

From a legal and institutional point of view the basin institutions established in Africa after 1964 are probably the most advanced in the world. In this connection, mention can be made of the following institutions:

- the Senegal River Basin Development Organization (OMVS), established in 1972 between Senegal, Mauritania and Mali;[24]

---

21 That is, to the exclusion of Ethiopia.
22 The Nile Basin Initiative Act 2002 confers legal status in Uganda to the NBI.
23 The agreement has been signed by Burundi, Ethiopia, Kenya, Tanzania, Rwanda and Uganda, and has been ratified only by Ethiopia, Rwanda and Tanzania.
24 The text of the relevant convention is reproduced in FAO (1997) *Treaties Concerning the Non-Navigational Uses of International Watercourses – Africa*. Legislative Study No. 61. Rome, FAO, 24. Guinea acceded to the convention in 2006.

- the Gambia River Basin Development Organization (OMVG), 1978, comprising the Gambia, Senegal, Guinea and Guinea Bissau;[25]
- the Kagera Basin Authority (KBA), established in 1977 between Rwanda, Burundi, Tanzania and Uganda and dissolved in 2004;[26]
- the Lake Chad Basin Commission (LCBC), which was set up in 1964 between Cameroon, Chad, Niger and Nigeria;[27]
- the Niger Basin Authority (NBA), 1980, which comprises the nine co-basin states: Benin, Burkina Faso, Cameroon, Chad, Côte d'Ivoire, Guinea, Mali, Niger, Nigeria;[28]
- the International Commission of the Congo-Ubangi-Sangha Basin (CICOS), established in 1999 between Cameroon, the Central African Republic, the Democratic Republic of Congo and the Republic of Congo[29] to deal with river navigation issues. Its mandate was expanded in 2007[30] so as to cover the integrated management of the water resources of the basin;
- the Volta Basin Authority (VBA), created in 2007 between Benin, Burkina Faso, Côte d'Ivoire, Ghana, Mali and Togo.[31]

The strength of these institutions lies in their constitutional and decision-making powers. Each one of them is composed of:

(i) a *Conference of the Heads of States and Governments*, with full powers to take major strategic and policy decisions, ordinarily meeting every two years and extraordinarily as necessary. Unanimity is the rule, and the decisions are binding on the member states;

(ii) a *Council of Ministers* of the basin states, the decisions of which must be unanimous and are binding on the member states. The Council of Ministers holds ordinary meetings annually. It refers to the Conference of the Heads of States any matter on which a decision has not been agreed upon;

(iii) an *Executive Secretariat* which proposes and services the meetings and prepares the documents to be discussed by intergovernmental technical committees of experts before submission to the Council of Ministers.

These African basin institutions have been established for the purpose of:

- facilitating cooperative arrangements among the basin countries;
- avoiding interstate water conflicts;

---

25 *Ibid.*, 47.
26 *Ibid.*, 35.
27 The text of the relevant convention may be found in: *Journal Officiel de la République Fédérale du Cameroun*, 15 September 1964, 1003.
28 The text of the relevant convention, revised in 1987, is reproduced in FAO, *Treaties Concerning the Non-Navigational Uses of International Watercourses – Africa; op. cit.*, 62.
29 Agreement to Establish a Uniform River Regime and Creating the International Commission of the Congo-Ubangi-Sangha Basin, signed on 6 November, 1999. On file with Author.
30 Addendum to the Agreement, 22 February, 2007. Gabon and Angola became parties to the agreement in 2011 and 2015, respectively. On file with Author.
31 Convention on the Status of the Volta River and the Establishment of the Volta Basin Authority, signed at Ouagadogou on 19 January, 2007.

- facilitating international financing;
- creating larger markets; and
- promoting regional economic integration.

However, political differences between co-basin states, different degrees of development and, often, inadequate availability of financial and manpower resources have not facilitated the smooth achievement of the original aims, although the OMVS, due to its capacity to steer development and adjust to changes, is regarded as a successful organization. Through a convention of 1978 concerning the legal status of common works[32] and a subsequent convention concerning the financing of these works,[33] the member states of this organization succeeded in securing financing for the Diama and Manantali Dams and for other infrastructure on the Senegal River. The first convention provides for the joint and indivisible ownership of these works, and for the sharing of costs, which shall be proportional to the benefits accruing to each state from their exploitation. Each owner guarantees the reimbursement of the loans received by the OMVS from the donor community for the construction and operation of the joint works. In 1997, the OMVS has established two autonomous interstate agencies to manage and operate the two dams. A further autonomous agency was created in 2011 to manage and improve navigation on the river.

In 2002, the member states of the OMVS adopted the Senegal Water Charter, which sets out principles and modalities for sharing the basin waters among different water using sectors, as well as rules for environment protection and preservation and a framework for the participation of water users in decision making.

13.2.3.3 The Liptako-Gourma Authority

A Protocol between Burkina Faso, Mali and Niger, signed on 3 December, 1970, has created and approved the statutes of the Liptako-Gourma Authority for the integrated development of a geographic zone including the Niger River and two of its tributaries, namely the Tilemsi and Sirba Rivers. The main purposes of the authority are to promote the development of mining, energy, water, crop, livestock and fisheries resources and to seek project financing.

13.2.3.4 Institutional framework for cooperation in Southern Africa

In the 1990's, the fourteen states of the Southern African Development Community (SADC) decided to establish a framework for cooperation in the management of international river basins, and in 1995 they signed the Protocol on Shared Watercourse Systems (SADC Protocol). The protocol was revised in 2000[34] in order to bring it in line with the UN Watercourses Convention. A number of international basin commissions were established in line with the protocol, even before its adoption, including,

---

32 Convention between Mali, Mauritania and Senegal relating to the Legal Status of Common Works, signed at Bamako on 21 December, 1978.
33 Convention between Mali, Mauritania and Senegal relating to the Modalities for Financing Common Works, signed at Bamako on 12 May, 1982.
34 The text of the revised protocol is to be found in: 40, ILM, 31 (2001).

- the Permanent Water Commission between Namibia and South Africa (1992);
- the Permanent Okavango River Basin Water Commission (OKACOM), comprising Angola, Botswana and Namibia (1994);
- the Orange-Senqu River Commission (ORASECOM), between Botswana, Lesotho, Namibia and the Republic of South Africa (2000);
- the Limpopo River Commission (LIMCOM), between Botswana, the Republic of South Africa, Mozambique and Zimbabwe (2003);
- the Incomati Maputo River Commission, between the Republic of South Africa, Mozambique and Swaziland (2003);
- the Zambesi Watercourse Commission (ZAMCOM), between Angola, Botswana, Malawi, Mozambique, Namibia, Tanzania, Zambia and Zimbabwe (2004);
- the Cuvelai Watercourse Commission (CUVECOM), between Angola and Namibia (2014).

Other international water management institutions such as the Tripartite Permanent Technical Committee, comprising the People's Republic of Mozambique, the Republic of South Africa and Swaziland (1983), pre-date the establishment of the cooperation framework just mentioned.

13.2.3.5 The contribution of regional economic integration organizations

The role played by regional economic integration organizations in recent years in respect of the development of legal and policy frameworks for the management of transboundary water resources should not be overlooked. Many of these organizations are active in Africa, including the East African Community (EAC), the Economic Community of West African States (ECOWAS), the SADC and the Intergovernmental Authority on Development (IGAD).

The EAC was established in 1999 between Kenya, Tanzania and Uganda to facilitate cooperation in various economic areas and the creation of a common market. Burundi and Rwanda became members in 2007, while South Sudan joined in 2016. The EAC has been instrumental in the development and negotiation of the Protocol on Sustainable Development of Lake Victoria (2003). The Lake Victoria Basin Commission, which coordinates the sustainable development and management of Lake Victoria, is a specialized institution of the EAC.

ECOWAS was founded in 1975, but its functioning is now governed by the Revised Treaty of 1993. Its members are Benin, Burkina Faso, Capo Verde, Côte d'Ivoire, The Gambia, Ghana, Guinea, Guinea Bissau, Liberia, Mali, Mauritania, Niger, Nigeria, Senegal, Sierra Leone and Togo. Its objective is to promote cooperation and integration, leading to the establishment of an economic union. The harmonization and coordination of national natural resources and environmental policies, as well as the establishment of enabling legal frameworks, fall within its mandate. In 2008, the ECOWAS Heads of State and Government adopted the West African Water Resources Policy, the aim of which is, *inter alia*, to support the harmonization of national policy and legal frameworks for water resources management, as well as the development of international legal instruments. More recently, ECOWAS has developed directives on the management of shared water resources and the development of hydraulic infrastructure in West Africa.

The creation of SADC dates back to 1992. The members of the organization, of which the main objectives are to achieve development, peace, security and economic growth, are

Angola, Botswana, Comoros, the Democratic Republic of Congo, Lesotho, Madagascar, Malawi, Mauritius, Mozambique, Namibia, Seychelles, South Africa, Swaziland, Tanzania, Zambia and Zimbabwe. In 1995 SADC member states adopted the Protocol on Shared Watercourse Systems, which was largely based on the 1966 Helsinki Rules of the ILA and was revised in 2000, as was mentioned earlier. In 2005, five years after the Revised Protocol, SADC adopted a regional water strategy.

IGAD was established in 1986 to address drought-related problems affecting the Greater Horn of Africa. Ten years later, its mandate was expanded and its current mission is 'to assist and complement the efforts of the member states to achieve, through increased cooperation, food security and environmental protection; promotion and maintenance of peace and security and humanitarian affairs, and, economic cooperation and integration' (agreement establishing IGAD). The member states of IGAD are Djibouti, Eritrea, Ethiopia, Kenya, Somalia, South Sudan, Sudan and Uganda. In 2012, aware of the connection between water scarcity and conflict, IGAD launched a programme to support the development of a regional policy and legal framework for the management of transboundary water resources. A regional water resources policy was endorsed in 2015 by the ministers responsible for water of the member states. IGAD is now facilitating the elaboration and negotiation of a regional water resources protocol and supporting member states in the improvement of their national legal frameworks. In 2015 IGAD integrated a Water Unit into its structure.

13.2.4 Institutional developments in Asia

13.2.4.1 The Mekong River Commission

The Mekong River Commission is the successor of the Committee for Coordination of Investigations of the Lower Mekong Basin or, in short, the Mekong Committee. This committee was established in 1957 under the aegis of the United Nations to promote joint cooperation between Cambodia, Laos, Thailand and Viet Nam. It has continued to operate in spite of political differences and even during the Viet Nam war, when investigations and data collection were carried on, sometimes even in enemy-occupied territories. In 1975 a joint declaration of principles was approved, setting forth the basic rules concerning the common use of water resources and the apportionment of the benefits and costs deriving from the common projects. The joint declaration was based on the principles of the 1966 Helsinki Rules of the International Law Association.[35] During the period in which Cambodia (then called Kampuchea) could not participate in the work of the committee, the other basin states operated under an 'Interim Mekong Committee,' composed of three of the four lower riparian states. These operations were made possible because of a strong desire to cooperate – the so-called 'Mekong spirit' – and the fact that the United Nations continued to provide an umbrella to the organization in the form of institutional and financial support. The executive agent of the committee was an official of the United Nations Development Programme (UNDP), while the permanent secretariat staff was composed almost exclusively of riparians. It is worth mentioning that the Mekong Secretariat promoted studies and training in national and

---

35 See Chapter 10, Section 10.3.9.2.

international water law and administration for the benefit of its member countries, financed by the EEC and the Asian Development Bank.

With the request by Cambodia to resume participation in the Mekong Committee, it became clear that the time had come to create an enhanced legal and institutional framework for Mekong cooperation. Thus, through an agreement concluded on 5 April 1995,[36] the four riparian states of the lower Mekong Basin established a new Mekong River Commission to replace the committee. The commission consists of three permanent bodies, i.e., the council at the interministerial level, which makes policy decisions, the joint committee, which is responsible for the implementation of these decisions, and the secretariat, which provides technical and administrative support to the commission. The secretariat is headed by a chief executive officer. The agreement provides for cooperation among the member states in the sustainable development, utilization, management and conservation of the basin's water resources, in the prevention of harmful effects, in the protection of the environment and in matters relating to inter-basin diversions. Freedom of navigation on the course of the Mekong is safeguarded.

Studies on equitable water utilization are presently being carried out at the secretariat, with a view to the elaboration of rules for water utilization and inter-basin diversions, as required under Article 26 of the agreement, consistent with the Helsinki Rules and with the UN Watercourses Convention. A first set of these rules, concerning the exchange of information, was agreed upon by the Council of the Mekong River Commission in 2001. In 2003, the council adopted procedures for notification, prior consultation and agreement, and procedures for the monitoring of wate uses. In 2006, the council endorsed procedures for the maintenance of minimum mainstream flows and, in 2011, procedures for the protection of water quality.

13.2.4.2 The Indus Commission

A Joint Commission for the Indus basin was set up on the basis of the Indus Water Treaty, signed by India and Pakistan on 19 September, 1960,[37] after years of negotiations under the auspices of the World Bank, following a long-standing dispute over the use of the basin waters for irrigation and hydro-power generation purposes.[38] The commission is composed of two commissioners, one for each country, with particular competence in the field of hydraulics. Its objective is to maintain cooperative arrangements for the implementation of the treaty. Its functions and powers include inspection, research and the resolution of controversies pertaining to the interpretation and the application of the treaty. In the case of disputes not settled by the commission, it is foreseen that a court of arbitration handle the question in last resort.[39]

---

36 Agreement on Cooperation for the Sustainable Development of the Mekong River Basin, signed by Cambodia, Laos, Thailand and Viet Nam on 5 April, 1995; text in: 34, ILM, 864 (2005).
37 The text of the treaty may be found in 419, UNTS, 126.
38 The treaty provided for the allocation of the waters of the basin's Eastern Rivers (Sutlej, Beas and Ravi) to India and of those of the Western Rivers (Indus, Jhelum and Chenab) to Pakistan.
39 For a description of the Indus River dispute and of the present legal and institutional framework, see Salman, M.A.S. & Uprety, K. (2002) *Conflict and Cooperation on South Asia's International Rivers – A Legal Perspective*, Law, Justice and Development Series. Washington, D.C., The World Bank, 37.

13.2.4.3 Joint commissions between Nepal and India

Similar commissions exists between Nepal and India for the exploitation of the Rivers Kosi,[40] Gandak[41] and Mahakali.[42] In broad terms, they are in charge of overseeing the process of water resources development and benefit-sharing between the two countries, but this process has been somewhat delayed due to the fact that the relations between the two states have not always been smooth. These three rivers form part of the Ganges basin.[43]

13.2.4.4 The India-Bangladesh Joint Commission

After the creation of the state of Bangladesh following its separation from Pakistan in 1971, the Indo-Bangladesh Joint Rivers Commission was set up for the River Ganges,[44] with focus on water utilization and flood control. The pressing issue, however, was the operation of the Farakka Barrage, in India, and water apportionment, which was left to a series of partial agreements of an *ad hoc* nature. The commission had advisory powers and functioned only intermittently, its last meeting being held in 1990. It was, however, revived in the form of a joint committee through a treaty concluded in 1996 to establish a 30-year water discharge régime. The joint committee is responsible for treaty implementation and, in particular, for observing and recording daily flows in given areas. The data are to be submitted to the two states.[45]

13.2.4.5 Institutional arrangements for the Aral Sea basin

Soon after the disbandment of the Soviet Union, the Aral Sea basin states, that is, Kazakhstan, Kyrgyzstan, Tadjikistan, Turkmenistan and Uzbekistan, took action to replace the Moscow authorities with new institutional arrangements for the management of the basin's water resources. On 18 February, 1992, they signed the Agreement on Cooperation in the Management, Utilization and Protection of Water Resources in Interstate Sources. This agreement provides for the establishment of the Interstate Water Management Coordination Commission (ICWC), and vests this commission with regional water management policy-making, the approval of annual water withdrawal limits for each republic and the approval of schedules

---

40 Agreement between the Governement of India and the Governement of Nepal on the Kosi Project, done at Kathmandu on 25 April, 1954. Text in: ST/LEG/SER.B/12, 290; The agreement was revised on 21 December, 1966.
41 Agreement on the Gandak Irrigation and Power Project, done at Kathmandu on 4 December, 1959. Text in: ST/LEG/SER.B/12, 295.
42 Treaty between His Majesty's Government of Nepal and the Government of India Concerning the Integrated Development of the Mahakali River Including Sarada Barrage, Tanakpur Barrage and Pancheshwar Project, done on 12 February, 1996. Text in: 36, ILM, 531 (1997).
43 For an in-depth analysis of legal and institutional arrangements for water resources development and management on these three rivers, see Salman, M.A.S. & Uprety, K. *Op. cit.*, 65 ff.
44 Statute of the Indo-Bangladesh Joint Rivers Commission, signed by India and Bangladesh at Dhaka on 24 November, 1972. Reproduced in: FAO (1995) *Treaties concerning the Non-Navigational Uses of International Watercourses – Asia.* Legislative Study No. 55. Rome, FAO. It is to be noted that Nepal, Bhutan and China are also riparian to the Ganges, but do not participate in this arrangement.
45 An analysis of the international legal and institutional framework of the Ganges River and of its evolution is to be found in Salman, M.A.S. & Uprety, K., *op. cit.*, 125 ff.

for the operation of reservoirs. The ICWC meets on a quarterly basis and whenever the need arises, and makes unanimous decisions which are immediately binding on all water users. Two basin water management organizations, BVO Amu Darya and BVO Syr Darya, which have been in existence since 1986, are vested with executive functions with regard to the operation of hydraulic works, structures and installations on the Amu Darya and Syr Darya Rivers, tributaries of the Aral Sea. Both the ICWC and the BVOs must ensure compliance with the water withdrawal limits, and that a guaranteed annual volume of water be supplied to the Sea. The ICWC has a secretariat and a scientific information center.

In the years following the establishment of the ICWC, other institutions have been created: the Interstate Council on the Aral Sea Basin Problems (ICAS), the International Fund of the Aral Sea (IFAS) and the Commission on Sustainable Development. ICAS was vested mainly with programme management functions, although its functions under the governing statutes partly overlapped with those of the ICWC. IFAS was conceived as a fund for the financing of the programmes managed by ICAS.[46] The Commission on Sustainable Development was vested with the administration of ecological programmes. ICAS and IFAS were merged into one institution, IFAS, in 1997.

13.2.4.6 The Helmand River Commission

The Helmand River Commission was set up for the Helmand delta by virtue of the agreement of 7 September, 1950, between Afghanistan and Iran. It is composed of three engineers coming from states without any vested interest in the area, with powers of recommendation only; their duties include those of documentation, information and planning.

13.2.4.7 Other commissions

A number of joint commissions and committees have been established over the years throughout Asia, among which the following may be mentioned:

- between China and USSR for the Amur River (1956);
- between the USSR and Turkey for boundary rivers (1927);
- between the USSR and Afghanistan (1946);
- between Kazakhstan and the Russian Federation for transboundary water bodies (1992);
- between Mongolia and China for transboundary waters (1995);
- between Kazakhstan and China for transboundary rivers (2001);
- the Chu-Talas Joint Rivers Commission, between Kazakhstan and Kyrgyzstan (2006).

---

46 For a review of the Aral Sea basin institutions, see Nanni, M. (1996) The Aral Sea Basin: Legal and Institutional Issues. *Review of Community and International Environmental Law* (RECIEL), Issue 2. London, Blackwell Publishers, 130; for an in-depth analysis of environmental issues and institutional responses in Central Asia, see Weinthal, E. (2002) *State Making and Environmental Cooperation – Linking Domestic and International Politics in Central Asia*. Cambridge, Massachusetts, MIT Press.

13.3 Evaluation of existing arrangements

Unfortunately, the lack of institutionalized cooperative arrangements is a cause of concern among co-basin states in several parts of the world. This concern relates either to a question of boundary demarcation on the course of an international river,[47] or to the quantitative aspects of the use of shared water resources, either surface, such as in the case of the Nile, Ganges/Brahmaputra and Tigris-Euphrates Rivers, or underground, such as in the case of the USA and Mexico, Israel and the occupied territory of Palestine. During the past decades, questions relating to the quality of water resources, either surface or underground, have arisen. This is one of the reasons why river basin commissions have been established, among others, for the Rhine, the Danube and a number of other European river basins.

A comparative evaluation of the existing administrative structures of international river commissions is not an easy task because of the varying geographical conditions, purposes, size and responsibilities to be considered. However, on the basis of the institutional arrangements described above, an attempt of an evaluation follows.

Commissions for the Danube and the Rhine Rivers were initially set up for the purpose of regulating navigation only. Subsequently, water pollution aspects were taken into consideration, and a commission was established for the Rhine for the stated purpose of controlling pollution. More recently, in concomitance with a process of reform initiated in the late 1990's by European member states, new commissions have been established for these two river basins to cover broader aspects, including water pollution control and flood management. With the adoption of the EU Water Framework Directive, the development or coordination of basin plans has become a major task for these commissions and for the commissions that have been established for other European river basins.

The Indus commission and the Nile Permanent Joint Technical Committee were established for the purpose of settling major questions on water sharing. In particular, the Indus commission could only function after the intervention of the World Bank, leading to the conclusion of the 1960 treaty. The Nile committee does not really represent a basin institution, as only two of the co-basin states are party to it. However, the Nile basin states have acknowledged the need for a basin-wide institutional arrangement for water resources management, and have negotiated a framework agreement for the basin.

Among all the commissions, the two North American commissions seem to be the most effective with regard to the settlement of disputes. They have operated with success both for problems which arose in connection with the Columbia River and in the Great Lakes for pollution aspects. In addition, they have been able to respond to needs even when they did not have the authority to do so under their governing treaties.

The Mirim Lagoon Commission between Brazil and Uruguay has also been successful, probably because it only involves two countries and is relatively small.

The African commissions could be more efficacious if specialized manpower and financial means were made available to them. Nevertheless, the OMVS has proved to be functional, due to its flexibility and its capacity to steer development. Due to its ability to

---

47 This was one of the original causes of the war between Iraq and Iran (1980–1988) as regards the border on the Shatt el Arab. It is still a cause of concern among riparians of the Lake Malawi in Africa, the Ussuri in Asia, and of many other rivers.

establish relations with the donor community and to its approach to financial issues, the OMVS has attracted external financing for the execution of major water resources development works, such as the Diama and Manantali Dams.

The Mekong Committee was prevented by war from exercising its activities fully, but it demonstrated that, because of its flexibility, it could perform a large part of its functions. The committee's successor, the Mekong River Commission, has embarked in the study of rules on the utilization of the basin's waters, on the exchange of information, on prior notification of planned measures and on the protection of water quality. Until recently, it was headed by a UN official, without vested interests. Since 2016, the chief executive is a national from one of the riparian states.

In general, the above-mentioned commissions have performed the important function of bringing their member states to cooperative water resources management and planning.

Ideally, optimum cooperation in international river basin or aquifer management is better achieved when all the states sharing the waters effectively participate in the corresponding administrative institution. However, this is not always possible for a number of political, technical, economic or geo-physical reasons.

From a purely international legal standpoint, states sharing water resources are under no obligation to participate in an international water management institution. For instance, the Nile Permanent Joint Technical Committee comprises only two of the ten riparian states, i.e., Egypt and Sudan, while, on the same basin, three states, Rwanda, Burundi and Tanzania, have set up the Kagera Basin Authority on the sources of the White Nile before it enters Lake Victoria. The authority was disbanded in 2004, but the basin states are considering the possibility to revitalize it. The Mekong River Commission includes only four out of the six basin states; the People's Republic of China and Myanmar are not members, although the 1995 agreement would allow them to join. When it was established in 1964, the Lake Chad Basin Commission included only four of the basin states: Cameroon, Chad, Niger and Nigeria. The Central African Republic and Libya joined it in 1996 and 2008, respectively.

It is worth noting that, in some instances, a basin commission has served as an economic pole of attraction for a non-basin state; this is the case of Guinea, which has joined the OMVG.

In other instances the necessity to cooperate in shared water resources questions has prompted basin countries with very different political and socio-economic systems to seek institutional arrangements among themselves. This is the case, *inter alia*, of the Joint Permanent Commission which was established between the Republic of South Africa, the People's Republic of Mozambique and the Kingdom of Swaziland in 1983.

Most international river basin commissions have the following common characteristics:

(i) generally, they issue recommendations and are advisory to governments, but do not have executive powers;
(ii) by their nature, they are of indefinite or long duration, thus providing them with time to adapt to changes and new requirements;
(iii) most commissions have authority to undertake studies, conduct or coordinate investigations and planning and make recommendations for development, either on their own initiative or upon instruction. In so doing, the commissions have an important influence during early stages of development when the need for a river basin management plan is crucial;
(iv) the technical nature of the secretariat staff precludes the domination of political influences on decisions. Political influence has tended to be contained at the commissioner's

level. This is an important element to be reckoned with, as the influence of one commissioner may be offset by that of another. This is why it is necessary to have the participation of most, if not all, of the states of a basin;
(v) the commissions and their secretariats are provided with permanent headquarters and adequate financing on a proportional basis, together with diplomatic immunities for the performance of their functions;
(vi) the possession of judiciary powers to settle disputes, decide on apportionment of waters, on costs and benefits, and on other questions which may be submitted to it is a definite advantage for any commission.

It should be kept in mind that not all of the main requirements for successful basin administration may be applicable in every case, as each basin has its history, tradition and characteristics.

13.4 Objectives and purposes[48]

As regards the objectives and purposes (*ratione materiae*) of an international water resources administration, these may cover technical, economic, financial, legal and institutional aspects. In order to facilitate the achievement of the stated objectives, adequate powers should be conferred to the international administration through an agreement.

As regards the functions and activities which may be attributed to an international water resources administration, a tentative list of these is indicated here below.

13.4.1 Technical responsibilities

Technical responsibilities include:

(i) collection and exchange of hydrological, technical and other data, which may be undertaken by member states separately or jointly, and their standardization;
(ii) basin plan formulation, or the coordination and harmonization of plans prepared separately by member states;
(iii) fixing of priorities between different water uses, projects and areas;
(iv) construction, control and protection of waterworks and hydraulic structures, which may be undertaken by member states separately or jointly, or which may be entrusted to a non-member state or to some other institution;
(v) operation and maintenance of waterworks and hydraulic structures, which may be entrusted to each member state, separately, or to a joint administration;
(vi) coordinated management of hydraulic structures, which may be entrusted to the basin institution or to another joint organization;
(vii) control of one or more beneficial uses of water, which may include,

- domestic and community uses;
- agricultural uses, including the watering of animals and agro-allied industrial uses;
- industrial uses, including cooling;

---

48 For the methodology followed in Sections 13.4 to 13.8, see the guidelines annexed to Caponera, D.A. (1976) Administration of International Water Resources. *International Law Association Report of the 57th Conference*. Madrid.

- hydropower generation and transmission;
- navigation;
- timber floating;
- fishing;
- other beneficial uses of common interest;

(viii) control of water uses, diversions and abstractions, which ought to be entrusted to a regulatory basin institution;
 (ix) control of one or more harmful effects of water, which may include,

- flood control measures such as flood warning, flow regulation and river drainage;
- embankment construction and maintenance;
- drought warning, prevention and mitigation;
- soil erosion control;
- land reclamation, including salinity control and drainage;
- dredging, maintenance and improvement of the navigable sections of an international watercourse;
- siltation control;
- other harmful effects of common interest;

 (x) water quality control, including such coastal sea areas of the member states as might be adversely affected, and which may include,

- prevention and abatement of water pollution resulting from one or more beneficial uses or harmful effects and the measures to be taken separately or jointly by member states;
- health preservation, including human beings and genetic resources (animals and plants), and the measures to be taken separately or jointly by member states;
- monitoring of water quality;
- environment protection, with reference to the waters of the basin, including the setting of minimum standards and objectives, and measures to be taken separately or jointly by member states;
- protection of coastal areas.

13.4.2 Economic and financial responsibilities

Economic and financial responsibilities include:

 (i) internal financing of the administration, including cost sharing and sharing criteria;
(ii) financing of particular projects and works, including,

- constitution and management of joint international funds;
- cost sharing and criteria for sharing;
- procedures and criteria for compensation;
- sharing of benefits;
- payment of interest and repayment of debts;
- assessment and collection of revenues (at the basin, national and local levels) and criteria for sharing;

(iii) external financing, with particular reference to the powers of the institution necessary to enter into agreement for this purpose; institution and member state financial liability.

13.4.3 Legal and administrative responsibilities

Legal and administrative responsibilities include:

(i) administration of the right to use water at the basin level and coordination with the national administrations and agencies of the member states;
(ii) prevention and settlement of water disputes;
(iii) drafting and implementation of the required legislation (international agreements, ministerial resolutions, parallel or harmonized legislation);
(iv) monitoring water uses through national administrations;
(v) other legal and legislative advice and responsibilities.

13.4.4 Possible options

When establishing an international basin institution, thought should be given to which functions and responsibilities must be attributed to:

(i) an existing basin administration;
(ii) a new basin authority, commission or committee;
(iii) one or more specialized management or development institutions (for instance, navigation, electricity, coordinating institution, etc.);
(iv) national administrations;
(v) other regional or basin administrations or institutions.

13.5 Duration, constitution and decision-making procedures

The constitution, duration, procedures and legal status of an international water resources administration will depend on many factors.

13.5.1 Duration

An international water resources administration may be set up on a permanent or *ad hoc* basis. A permanent institution is one which works on a continuing basis with a permanent secretariat, while an *ad hoc* one generally meets only occasionally, either on a pre-established schedule or whenever called by one of the parties. Examples of permanent administrations are the Danube Commissions in Europe, the OMVS in Africa, the Mekong River Commission in Asia, the Salto Grande organization and the USA-Mexico Boundary and Water Commission in the Americas. *Ad hoc* commissions constitute the great majority of international water administrations. These are in existence on practically every international or boundary river, under the name of 'joint technical committee,' 'mixed river commission,' 'international water commission,' etc.

13.5.2 Constitution

The constitution of an international water resources administration may take various forms, such as:

(i) a single basin institution vested with supra-national decision-making powers. This is the case of some of the African basin institutions such as the OMVS, the OMVG, the NBA and the LCBC;
(ii) separate national commissions or agencies, such as in the case of the USA-Mexico Commission;
(iii) a joint commission or institution composed of national representatives;
(iv) a mixed commission or agency, such as in the case of most international water resources administrations;
(v) a joint commission or institution composed of national representatives, where such national representatives are an agency; this was the case of the Mekong Committee of 1957, where the representative of Thailand was the National Energy Administration (NEA).

13.5.3 Procedures for decision making

Generally speaking, the procedures for decision making include:

(i) a quorum, for the validity of the meetings, which might depend on the importance of the decisions to be taken; and
(ii) the principle of unanimity or a simple or qualified majority or any other combined form of decision making. Most basin commissions have adopted the principle of unanimity in the decision-making process.

13.5.4 Legal status

The legal status of an international water resources administration vis-a-vis both its member states and other states which are not members of the administration, as well as vis-a-vis international and other organizations, should be defined. This legal status may cover:

(i) the managing body;
(ii) the staff;
(iii) the assets, equipment and other properties; and
(iv) the whole administration as such, including the power to sue and to be sued.

International water resources administrations have addressed these legal issues in a variety of ways; generally speaking, permanent basin or river institutions have recognized diplomatic protection and immunities to the institution itself and to its staff, in varying degrees. While diplomatic immunities are indispensable in order to allow the staff members of the international institution to operate in a way which protects their independence and personal security, it should be understood that immunities are granted not for the personal interest of the officials or for their economic benefit, but in the interest of the organization for which they work.

13.6 Territorial competence

The territorial competence (*ratione loci*) of an international water resources administration needs to be defined in the law establishing the administration.

By territorial competence is meant the area over which the international administration has the power to exercise its jurisdiction or its activities. In this case also, the decision will vary and may include:

(i) the *whole drainage basin*, including surface water, underground waters or both. In some cases a state may be comprised totally within the area of an international basin; this is the case of Cambodia within the Mekong basin, of most of Hungary within the Danube basin, and of Uruguay within the Rio de la Plata basin. It is obvious that if an international basin administration is granted supra-national decision-making powers, the comprised co-basin state could result as a 'province' of the administration, something which any state would be rightfully unwilling to accept;

(ii) *multi-basin* (more than one drainage basin). This is the case where a number of basins are included in the area of jurisdiction of the international administration, such as in the case of the commission for the Chu and Talas basins (Kazakhstan and Kyrgyzstan);

(iii) *sub-basin* (part of a drainage basin). There are many examples under this category, such as, the Indus Commission, which includes India and Pakistan but not China, the Nile Permanent Joint Technical Committee, which formally includes only Egypt and Sudan with the exclusion of the other riparian states, and the Lake Victoria Basin Commission, of which only Burundi, Kenya, Rwanda, Tanzania and Uganda are members. The same applied to the competence of the Mekong Committee, which extended only to the lower Mekong basin, with the exclusion of the upstream riparians, Myanmar and China;

(iv) a *conventional area*, i.e., an area otherwise defined and clearly delimited. This is the case of the Lake Chad Basin Commission, which was initially responsible for the 'conventional' basin around the Lake Chad (an internal basin), with the exclusion of other countries contributing water to the basin, i.e., Algeria, Libya, the Central African Republic and Sudan;

(v) *boundary waters*, whenever the institution is responsible for all or part of boundary waters. In many countries, in order to settle questions arising in connection with shared water resources, boundary river commissions have been set up. There are a number of these boundary river commissions in Europe (Finland/Sweden; Sweden/Norway; the former USSR and neighbouring states) and in the Americas (USA/Canada; USA/Mexico; Guatemala/Mexico). These commissions are usually bilateral and are responsible for all water questions along the boundaries of the two countries.

Other factors which are likely to be considered in this connection include the contribution of water by each basin state to the hydrology of the basin, the economic and social requirements of the basin states, local interests and the existing national legal and institutional framework.

A question of increasing importance is that concerning the quality control of international waters, i.e., the monitoring and control of pollution, which, in turn, brings with it issues concerning the protection of the environment. Often, it is not clear whether the territorial competence of existing international water resources administrations includes underground waters. In addition, the question arises as to whether international administrations which differ from those responsible for surface water have to be set up in order to control water quality of both surface and groundwaters and environment protection in general, or whether the existing water administrations should also be made responsible for the quality control of water and/or the environment. Each solution has advantages and disadvantages.

13.7 Functions and powers

The functions and powers of an international water resources administration should be defined. These may vary from case to case, depending upon factors which include the kind of cooperation envisaged, the desired degree of involvement in international administration and the specific fields for which it is proposed to establish the administration.

Some functions and powers may include, without being limited to, one or more of the following:

(i) *advisory, consultative* or *coordinating* functions. In this case the agreement should specify the procedural rules for deciding on conflicting rights and interests, including notification, objections and timing. Most of the existing commissions have the above-mentioned functions; the Plata River basin coordinating committee may be cited as an example. In this case the institution does not have autonomous policy decision-making powers but only advisory ones. Its decisions need subsequent ratification by the governments to become binding on the states;

(ii) *executive*, or *operational* functions, which may include the carrying out of studies, explorations, investigations and surveys; of feasibility studies, plan formulation, project inspection and control; construction, operation, maintenance or financing. Most international water administrations carry out one or more of these functions. Examples are the OMVS, the LCBC and the NBA;

(iii) *regulatory* functions, including the implemementation of decisions of the administration, as well as law-making. Decisions in these matters may take effect directly or after acceptance by member states. Not many river commissions have the necessary powers to exercize regulatory functions. An example is the Danube Commission as regards its powers to regulate the navigation of the river;

(iv) *judicial* functions, which may include arbitration or final dispute settlement. Only a few international water administrations have these judiciary powers. A noteworthy example is the judiciary function of the Rhine Navigation Commission established in 1831.

13.8 Form

By 'form' of an international water resources administration is meant the type of institution as regards the degree of government participation or involvement. This participation may be only of a financial nature or it may be of a different type.

As regards the form, two main solutions are possible:

(i) *separate national water institutions*, fully governmental; this is the case of many river basin commissions;
(ii) a *multinational river basin institution*, which may be:

- *intergovernmental*, i.e., a multinational agency or corporation, composed only of representatives of the governments of the basin states. Most basin institutions are of this type, particularly those responsible for overall policy formulation and implementation.
- *mixed*: a multinational corporation or agency which would include third parties and/or the private sector. This type of institution is generally created for the purpose

of operating particular economic activities of an international nature, such as a multinational hydropower company. The Cahora Bassa on the Zambesi basin, operated by agreement between Mozambique and Portugal with the participation of a private company, is one example. Another may be an international river navigation company.

13.9 Major institutional requirements for rational international water resources administration

There are two sorts of institutions concerned with the development of water resources: one is *regulatory* and the other *development*. The needs and objectives of a regulatory institution at the national level have already been discussed.[49] In the international context these may be briefly indicated as follows:

(i) establishment of technical cooperation for the formulation of common policies and for the joint or separate, but coordinated, planning for the rational management of the shared basin waters;
(ii) the coordination of those studies and schemes prepared by member states which may sensibly affect the hydrologic régime or quality of the waters of a basin;
(iii) the centralization and exchange of all hydrologic and related data;
(iv) the determination of the respective rights and duties of member states with regard to the development, utilization, protection and conservation of the basin water resources;
(v) the prior submission for examination and approval of proposed activities, schemes or plans which would modify the quantity or quality of the waters, or would unfavourably affect existing projects, the hydrologic régime of the water or related resources of another basin state, or the basin ecosystem;
(vi) the prevention and settlement of, or the decision on, conflicts concerning the use and equitable sharing, or the pollution, of waters; and finally,
(vii) the monitoring and control of the activities of basin states and reporting for discussion of such activities.

It is to be noted that the existence of regulatory institutions does not preclude the existence of other development institutions carrying responsibilities with regard to the actual development of certain areas or the utilization of the basin water and other resources.

It is indispensable that the activities and schemes of a development institution which might modify the hydrologic régime or the quality of the basin waters, or impair the basin ecosystem, be submitted to the regulatory basin institution beforehand for study, suggested modifications and approval. What is more, the water schemes of development institutions which might affect the rights or the legitimate interests of other basin states must be closely coordinated by the regulatory institution.

Conflicts arising between institutions may be avoided by clearly defining their respective statutory functions and powers.

---

49 Chapter 9, Section 9.3.4.

On the other hand, it is not advisable that a regulatory basin institution go beyond its original regulatory functions to become a development or executing institution; nor should it be given the operational functions and powers entrusted to development institutions. In such a situation the regulatory institution would undoubtedly encounter increasing resistence, which would prevent it from functioning. It is difficult to conceive of a regulatory institution being at the same time both the judge of the activities and an interested party thereto.

Likewise, a development institution to be set up within an international basin, whether it be created at the national level or between two or more countries, should not be given powers and functions of a regulatory nature which would conflict with those of the regulatory basin institution.

A development institution should have the necessary powers to plan, construct and operate schemes, but these should first be submitted to the regulatory basin institution for study and approval.

13.10 Economic and financial requirements

As to the economic and financial requirements of an international water resources administration or basin institution, the statutes should provide for the internal financing of the administration, including cost sharing and the criteria for sharing. Treaties or other regulatory provisions should govern the capital formation for the whole of the international administration and differentiate between internal and external resources, i.e., the contribution of each basin state, and grants and loans to the enterprise coming from outside sources. Provisions governing the apportionment of financial liabilities among basin states, criteria for cost/benefit in general and the constitution and management of joint international funds, if relevant, should be clearly specified.

The same is required with regard to third-party financial participation in development financing and related operations which include the constitution and management of joint international funds; the sharing of costs and benefits and criteria for sharing; procedures and criteria for compensation, payment of interests and repayment of debts, assessment and collection of revenues.

With regard to external financing, a basin institution, and, in any case, a development institution, should also have the necessary powers to negotiate or receive grants and loans, have a legal personality and financial autonomy, the capability to sue and to be sued, and be otherwise in a position to provide the donors or creditors with all required financial guarantees. Possibly, financial guarantees should be backed by one or more governments of the basin states.

In order to provide the external financing institution with greater guarantees when financing one project in one country, a basin institution could adopt the principle of 'joint and several guarantees,' whereby if one basin country defects, the other co-basin states are jointly responsible for the payment. This innovative practice has been adopted by the OMVS.

13.11 Prevention and settlement of disputes

One of the major functions which an international water resources administration should perform is that relating to the prevention and settlement of disputes between the co-basin states.

Disputes may refer to questions concerning the sharing of the water resource of a transboundary basin, to the determination of the principles of equity and reasonability of water uses, to the respective liabilities in the case of damages due to polluting activities or to the interpretation and implementation of a treaty or other legal provisions concerning the international water resources administration.

For these reasons, the agreements establishing the administration should contain the necessary provisions for the prevention and settlement of disputes.

13.12 Conclusion

It may be said that, as to the 'ideal' type of institution required for international basin management, in view of the varying physical, economic, social and political conditions prevalent in each individual transboundary river basin or aquifer, it is not possible in the abstract to suggest a particular type of administrative institution. Each international hydrological or hydrogeological unit has its own characteristics and its own problems, and must take into account the needs of the participating states. Often, compromise solutions are opted for, rather than optimum arrangements. Many possibilities and alternatives may be envisaged, and any decision necessarily depends upon, *inter alia*, the functions and activities to be performed and the political will to cooperate.

References

Böge, V. (2006) *Water Governance in Southern Africa – Cooperation and Conflict Prevention in Transboundary River Basins*. Brief 33. Bonn, International Center for Conversion.

Bourne, C.B. (1997) *International Water Law, Selected Writings*. International and National Water Law and Policy Series. The Hague, London, New York, Kluwer Law International.

Caponera, D.A. (1976) Administration of International Water Resources. *International Law Association Report of the 57th Conference*. Madrid.

Caponera, D.A. (2003) *National and International Water Law and Administration, Selected Writings*. International and National Water Law and Policy Series. The Hague, London, New York, Kluwer Law International.

Del Castillo, L. (2005) *El regimen jurídico del Río de la Plata y su frente marítimo*. Buenos Aires.

FAO (1997) *Treaties Concerning the Non-Navigational Uses of International Watercourses – Africa*. Legislative Study No. 61. Rome, FAO.

Salman, M.A.S. & Uprety, K. (2002) *Conflict and Cooperation on South Asia's International Rivers – A Legal Perspective*. Law, Justice and Development Series. Washington, D.C., The World Bank.

United Nations (1975) *Management of International Water Resources, Institutional and Legal Aspects*. New York, United Nations.

United Nations (1981) *Experiences in the Development and Management of International River and Lake Basins*. Natural Resources Water Series No. 10. New York, United Nations.

Index

Absolute Monarchy (ancient Rome) 33, 40, 48
absolute ownership (doctrine) 162
absolute territorial integrity 279
absolute territorial sovereignty 278
access to information 190, 305
Adad 20
adat law 104–6, 109
aedileship (ancient Rome) 35, 39, 47
Afghanistan 13–14, 29, 77, 119, 126, 127, 345
Aga Khan 77
Agenda 21 306, 319
agreements (*see* treaties, conventions and agreements)
Alberta (Canada) 95
Algeria 80, 84–5, 118, 189, 311, 315, 352
Al-Sag/Al-Disi Layer 317, 324
Andhra Pradesh 122, 258
Andorra 231
Anglo-Saxon laws 53
Angola 86, 115, 189, 341, 342
appropriation 3, 74, 119, 124–5, 155, 157–8, 160–4, 167, 175, 180–1, 184
aqueducts 14–5, 36–7, 39–40, 42, 44, 46, 50–1
aquifer(s) 3, 5–8, 42, 80, 87, 120, 121, 126, 132–3, 135–6, 141, 163, 172, 177–8, 180, 184, 191–3, 196–8, 200–1, 214, 218, 226, 231, 233–4, 237–8, 241, 245, 249, 251, 311–26, 329, 347, 356
  artificial recharge of 196, 224, 314
  recharge areas 201 (*see also* groundwater recharge areas)
  vulnerability 322–3
Arabian Peninsula Aquifer 311
Arabs 77, 86, 118
arbitral awards 65, 68–9, 257–8, 265, 317
  Faber Case (1903) 258
  Gut Dam Case (1968) 258
  Helmand River Delta Case (1872 and 1905) 69, 257, 280
  Kishenganga Indus Waters Arbitration 258, 305
  Kushk River Case (1893) 257
  Lake Lanoux Case (1957) 69, 258, 280, 282, 295
  Milestone 62-Mount Fitz Roy Case (1994) 258
  San Juan River Case (1888) 257
  Tacna-Arica Case (1925) 258
  Trail–Smelter Case (1938 and 1941) 69, 258, 282, 295
  Zarumilla River Case (1945) 258
Arizona 164, 314, 335
armed conflict 70, 265, 297–9
Argentina 67–9, 132, 134–6, 189–90, 230, 248, 256, 259, 295, 311, 316, 324, 336
Armenia 103
Asian-African Legal Consultative Committee 254, 262, 318
Asian Development Bank 4, 123, 343
Assyro-Babylonese civilization 13 (*see also* Mesopotamian civilization)
Aswan Dam 337
Australia 67, 91, 92, 95–6, 120–1, 123–4, 128, 226, 237, 320
Austria 57, 86, 140, 142, 267, 272
Austria-Hungary 276
authorization(s) 55, 58, 82, 87–8, 89, 94, 114, 133, 134, 135, 138, 140, 175, 177, 179, 198 (*see also* licences, permits)
awig-awig 105, 109
Aztec 28, 76

Babylon 21, 25, 27
Bahrain 126
Baia Mare 305
Bali 76, 103–9
Bangladesh 77, 85, 119, 121–2, 127, 279, 344
Barbados 132
basin agencies 84, 89, 97, 116, 124, 136, 147–8, 225–6, 236, 238
basin committee(s) 128, 136, 148
basin institution(s) 84, 90, 136, 141, 224, 226
basin organization(s) 103, 117–8, 134–5, 237–8

basin plan(s) 138, 155, 214, 220–1, 293, 330, 348 (*see also* planning of water resources, river basin management plans)
Bavaria 53
Bayern 276
Belgium 68–9, 85–6, 139–40, 147, 153, 189, 255, 274, 277, 279, 331, 333
Belize 96, 132, 227
Bellagio Draft Treaty 318
beneficial use 74, 126, 157, 159, 161–3, 181
beneficial uses 62, 161, 185, 202, 261–2, 348–9
   agricultural uses 1–2, 10, 91, 94, 116, 130, 185–7, 213, 225, 284, 348
   domestic and municipal uses 185, 225
   industrial uses 1–2, 62, 116, 130, 161, 185–6, 213, 215, 225, 278, 282–4, 335, 348
   hydropower 1, 8–9, 62, 64, 66, 91, 111, 116, 130, 143, 157, 171–2, 177, 179, 184–5, 187, 202, 204, 212, 215, 228, 233, 242, 249, 283, 330, 332, 343, 349, 354
   mining 6, 9, 62, 64, 89, 91, 130, 155, 177, 179, 207, 313, 323, 340
Bengala 287
Benin 113, 315, 339, 341
Bible 25
Bolivia 68, 69, 96, 129–30, 132, 133, 135, 189, 248, 257, 336, 337
Bosnia and Herzegovina 333
Botswana 118, 231, 324, 341, 342
boundary demarcation 242, 265–6, 313, 346
Brahma 22
Brahmanism 21
Braila 257, 276
Brazil, 68, 86, 132–6, 231, 248, 311, 316, 324, 336, 337, 346
British Columbia (Canada) 95
Brunei 123
Bruntland Report 303
Buddhist law 73
Bulgaria 57, 141, 145, 149, 154, 193, 292, 332
Burgundians 80–1
Burkina Faso 113–4, 189, 315, 339, 340, 341
Burma 76 (*see also* Myanmar)
Burundi 87, 114, 339, 341, 347, 352
BVO Amu Darya 345
BVO Syr Darya 345
Byzantines 52

cabotage 272–3, 275–7
Cahora Bassa 283, 354
California 163–4
Cambodia 29, 76, 86, 119, 231, 342–3, 352
Cameroon 113, 311, 313, 339, 347
Canada 67, 86, 91–2, 137, 226, 293, 295, 333–4, 352
Capo Verde 86, 115, 341

censorship (ancient Rome) 35, 38, 46
Center for Environment and Development of the Arab Region and Europe (CEDARE) 314–5
Central African Republic 113, 189, 311, 339, 347, 352
Ceylon 29, 123 (*see also* Sri Lanka)
Chad 62, 311, 313, 314, 339, 347
Chad Aquifer 311
Chamizal (territorial dispute) 268
Charlemagne 52–3, 55
Chernobyl 305
Chile 69, 132–5, 257
Ch'in Dynasty (China) 24
China 14, 23, 30, 76, 86, 119, 124, 127, 287, 294, 345, 347, 352
   Chinese water law 23–5
   People's Republic of 124, 127, 347
Ch'ing Dynasty (China) 25
city states 29
civil law 61, 63, 73, 132
civil law countries 74, 85–86, 88–90, 96, 137–8, 173
civil law system 86, 87, 111, 113, 119, 132, 197
classification of water bodies 193, 293
climate change 8–9, 130, 152, 167, 170–1, 173, 184, 211–2, 214–6, 302, 307
Cochabamba 96n30, 129
code or laws of Manu 50–1
Colombia 132–5, 196, 337
Colorado 164, 258
Columbus 28
combined approach 141, 150
COMECON 291
Comitia (ancient Rome) 35, 38
Commerce Clause (USA) 155–6
commerce, freedom of 270–1, 273
common law 70, 73, 85, 86, 91–5, 119, 132, 155, 159, 167
common law countries 74, 90, 91, 95–6, 174, 197
common law system 91–3, 95–6, 111, 115–16, 120, 174
Commonwealth of Independent States 98
community of interests 76, 85, 244, 250, 256, 270–2, 279, 280, 286, 305
Compact Clause (USA) 155–6
concession(s) 37, 39–42, 47, 50–1, 56, 58, 68, 87–8, 114–5, 124, 129, 133–4, 139–40, 146, 167, 175–81, 183–8, 190–2, 194–5, 198, 201, 203, 205–7, 213, 224, 228, 230–2, 236
Confucianist theories 24
Congo 113, 339
Congo, Democratic Republic of 87, 113–4, 189, 339, 342

Congress of Rastadt (1798) 270
Congress of Vienna (1815) 243, 270–2, 330
consorzi di bonifica 76
consorzi di irrigazione 76
Constantine, Roman emperor 48–9, 51
consulship (ancient Rome) 35, 39
contiguous rivers (international) 243, 266–8
contingency plans 289–90
conjunctive use (of surface water and
    groundwater) 164
conventions (*see* treaties, conventions and
    agreements)
Corpus Iuris Civilis 49–51
correlative rights 74, 162–3
Costa Rica 69, 132–5, 189, 196, 256–7, 304
Côte d'Ivoire 113, 339, 341
Council of Europe 67, 254, 262–3, 291, 301, 318
Court of Justice of the European Communities
    68–9 (*see also* European Court of Justice)
Croatia 154, 292, 332, 333
Curaçao 132
customary law 57, 65, 67–8, 73–6, 83–5, 92,
    111–3, 118, 137, 174, 206–7, 228
Czechoslovakia 142, 144, 256–7

Danube Basin Management Plan 332
decentralization 62, 121, 141, 144, 146–8, 218,
    229–31, 234
decisions of national tribunals 65, 255, 258–9
declaration (of water uses) 88, 176
deconcentration 218, 229, 230–1
Delaware 320
Delaware River Basin Commission 320
delegation of powers 230
Denmark 68, 139–41, 144
desa 104–9
desa-kala-patra 107
development functions 103, 225, 238
development institution(s) 127, 167, 225, 232–4,
    350, 354–5
Dharma 22
Diama and Manantali Dams 340, 347
dictatorship (ancient Rome) 35
Diet of Roncaglia 52, 56
Digest 34, 49, 56
Diocletian, Roman emperor 33, 47–9, 51
directives, EU 67, 102, 137, 140, 149, 151–2,
    252–3, 293
  Drinking Water Directive (1998) 150, 252
  directive on groundwater (2006) 151–2,
    252
  Directive on Integrated Pollution Prevention
    and Control (IPPC) (1996) 150
  Directive on the Assessment and Management
    of Flood Risks (2007) 151–2, 252, 289
  Nitrates Directive (1991) 149–50, 252–3

  Urban Wastewater Treatment Directive (1991)
    149, 252–3
  Water Framework Directive (2000) 69, 90, 96,
    102, 138, 141, 145, 147–54, 171, 194, 197,
    200, 220, 252–3, 289, 293, 330–3, 346
disputes (water), prevention and settlement of
    261, 350, 355–6
  Kansas v. Colorado 258
  Krishna River dispute 258, 286
  Narmada dispute 258
Dominican Republic 337
Donauversinkung case 259, 320
Draft Articles on the Law of Transboundary
    Aquifers 4, 245, 251, 316–7, 321–3, 326
drainage basin 4–5, 62, 70, 101, 112, 241, 243–4,
    261–2, 282, 285, 291, 296, 312, 317, 352
driller 121, 199
drillers' licences 199
due diligence 281
duty to cooperate 250, 285, 304, 316 (*see also*
    obligation to cooperate)
duty to exchange data and information 284, 288,
    338
duty to notify 304 (*see also* notification)

East African Community (EAC) 338, 341
East Punjab 279
ecological balance 6, 87, 117, 178, 184, 195
ecological flow(s) 133–4, 172, 182, 195, 212
Economic Community of West African States
    (ECOWAS) 341
ecosystem(s) 1, 9, 117, 121, 131, 133, 148,
    153, 184, 187, 193, 195–6, 200, 215, 216,
    301–2, 305, 308, 322, 331, 333, 338, 354
Ecuador 132–6, 189, 337
effluent standards 139, 147–8, 150, 194 (*see also*
    emission limit values)
Egypt 16, 17, 84–5, 91, 116–8, 234, 269, 279,
    281, 283, 287, 311, 314, 337, 347, 352
Egyptian civilization 13
EIA (*see* environmental impact assessment)
Elam and Suziana civilizations 13, 29
El Salvador 132, 135, 337
emergency situations 119, 289–90, 303 (*see also*
    harmful effects)
emission limit values 141, 150, 194 (*see also*
    effluent standards)
Empire (Roman) 33, 43, 48, 50–2, 54, 57
enforcement 24, 100, 121, 143, 169–70, 195,
    200, 204, 231
England 53, 70, 73, 85–6, 91–5, 155, 159, 267
England and Wales 94–5, 97, 142, 152–3
environmental impact assessment (EIA) 102,
    196, 257, 295, 301, 304–5, 316, 338
environmental objectives 102, 141, 150–2, 200,
    220, 293

environment protection 3, 62, 89, 92, 96, 98, 144–5, 169, 172, 180, 185, 195, 211, 216, 225, 233, 235, 301–5, 316, 340, 342, 349, 352
Equatorial Guinea 86, 115
equitable and reasonable utilization 68, 251, 280–3, 286, 307–8, 315, 317, 322, 325, 338 (*see also* equitable utilization; reasonable and equitable utilization, share)
equitable apportionment 253, 258–9, 281
equitable participation 281
equitable utilization 257, 259, 261, 281, 296, 301, 308, 318–20 (*see also* equitable and reasonable utilization)
Eritrea 342
erosion 1, 5–6, 99, 111, 129, 180, 190, 196, 202, 213, 225, 241–2, 249, 287, 288–9, 349
Estonia 103, 137, 139, 154
Ethiopia 116, 119, 281, 283, 338, 342
European Commission 69, 149–50, 151, 153, 252
European Court of Justice 69, 153–4 (*see also* Court of Justice of the European Community)
European Economic Community 246, 252, 331
European Union (EU) 4, 65, 67, 69, 90, 96, 102–3, 137–8, 140–2, 144–5, 147–54, 171, 192, 194, 197, 200, 238, 251–3, 289, 291, 305, 313, 330 (*see also* directives, EU)
candidate countries 90, 102, 137, 151, 253
European Water Charter 67, 262, 318
externalities 11

fa 24
Farakka Barrage 344
Federal Republic of Germany 67, 68, 137, 142–3, 292, 331 (*see also* Germany)
federal states (or countries) 141, 142, 230, 258, 259, 281, 311, 320
federation 69, 225, 320 (*see also* federal states)
feudal period 54–5, 269
Finland 68, 139, 141, 145–6, 193, 207, 293, 352
First World War 114, 274, 276, 299, 330 (*see also* World War I)
fishing 1, 9, 16, 20, 24–5, 41–2, 44, 50, 54, 56, 58, 79, 84, 91–4, 96, 98–100, 107, 113, 133, 146, 212, 225, 233, 242, 249, 278, 349
floating (of timber) 9, 25, 177, 187, 207, 242, 249, 266, 278, 282, 349
flood risk management plan(s) 152, 289
flood control 25, 66, 70, 148, 155, 287–8, 332, 344, 349
Florence 287
Fluvial Danube 276
Food and Agriculture Organization (FAO) 4, 70, 123, 291

France 52–3, 56–7, 69, 84–90, 114, 119, 130, 138–41, 148, 152, 189, 225–6, 228, 267–70, 276–7, 280, 292, 314, 331, 333
Franks 52–3
free use (of water) 115, 176, 186
French Equatorial Africa 113–14
French Guyana 132
French Indochina (*see* Indochina)
French Napoleonic Code 74, 85–6, 137
French Revolution 54, 85–6, 270–1
French West Africa 113
Fundamentals of Water Legislation of the USSR 97–9

Gabon 113
Galatz 257, 276
Gambia 231, 339
General Comment No. 15 189, 306–9
general principles of law 175, 246, 253, 259, 267–8, 271, 288, 298–9, 300, 309, 312, 317
Geneva Aquifer 314, 323
German Democratic Republic 137
German Empire 267, 270
Germany 53, 56–7, 69, 86, 139–40, 141, 152, 153, 230, 259, 266, 269, 274, 282, 292, 320, 330, 331, 332, 333
Gezira Board 234
Ghana 113, 115, 339, 341
Global Environment Facility (GEF) 316
Glossators 56–7
good faith, principle of 287
good neighbourliness, principle of 253, 288, 290, 295
Goths 52
Grand Ethiopian Renaissance Dam (GERD) 283
Great Britain 84, 85, 274, 277, 331, 333
Greece 29, 57, 277
Great Artesian Basin (GAB) 121, 226, 237, 320
Coordinating Committee 121, 226, 237, 320
Great Manmade River 114
'greening' of water law 195
Grotius 269
groundwater(s) 1–3, 5, 6–8, 11, 42, 62–4, 70, 74, 78–9, 81, 84, 102, 106, 116, 121, 122, 124, 125, 126, 127, 129, 132–3, 139, 140–1, 147, 149–52, 154, 157, 159, 162–4, 174–6, 183, 190, 192–4, 196, 197–201, 211, 213–4, 215, 217, 220, 223, 224, 225, 226, 229, 237, 241, 244–5, 252, 258, 260, 280, 292, 311–25, 329, 335, 352
vulnerability of 318, 325 (*see also* aquifer vulnerability)
groundwater conservation districts 163
groundwater management districts 163–4, 226
groundwater ownership 7, 63, 79, 197–8 (*see also* water ownership)

groundwater recharge area 141 (*see also* aquifer recharge area)
Guadalupe 132
Guaraní Aquifer System 132–3, 311, 316
Guatemala 132–3, 189, 267, 337, 352
Guinea 113, 119, 189, 338
Guinea Bissau 86, 114, 115
Gujarat 258
Gulf War 299
Guyana 132, 337

Hadiths 77
Haiti 132, 337
hak guna air 106
hak milik 105
Hammurabi Code 18–9, 21
Hammurabi, King 18–9
Han Dynasty (China) 24
Hanbalites 77, 80
Hanifites 77, 80
harim 78–9, 80, 82
harmful effects of water 14–5, 20, 33, 38, 45, 51, 61–2, 99, 111, 146, 190–1, 194, 202, 213, 225, 242, 249, 287–90, 294, 330, 349
Harmon, M. 278
Harmon Doctrine 335
hazardous substances 141, 291, 332 (*see also* priority substances)
Helmand civilization 13, 29
Helsinki Rules (ILA, 1966) 70, 244, 260–2, 263, 280, 282, 285, 286–7, 290–1, 296, 299, 318, 342–3
Hindu civilization 13, 21
Hindu doctrine 21–2, 104–5
Hittite civilization 18
Hittite laws 19
Holy Koran 77
Honduras 132–6, 267, 337
Hong Kong 124
Horemheb (codification) 16
human environment 300
Hungary 69, 86, 140–1, 256–7, 276, 292, 317, 332, 352
hydraulic civilizations 13–16, 19, 21, 28–30, 287
hydrologic cycle 4–5, 6–7, 8, 102, 125, 131, 133, 159, 170, 197, 216, 241
Hydromet (Nile) 338

Ibadites 77
Iceland 141
Imamites 77
implementation of water legislation 204, 228
Inca (civilization) 28, 76
India 67, 76, 91, 122–3, 128, 182, 189–90, 228, 230–1, 258, 279, 286, 287, 343–4, 352

Interstate Water Disputes Act (1956) 122
Supreme Court 189, 221, 286
Indochina 86, 119
Indonesia 77, 80, 85, 87, 104–6, 119, 126, 128, 230–1
innocent passage (right of) 269
innocent use of rivers 258
Institute of International Law 70, 259–60, 285, 295, 298
Institutes (*Institutiones*) 49, 56
Inter-American Bar Association 259, 261–2
Inter-American Economic and Social Council 262
inter-basin diversions 214, 232, 343 (*see also* inter-basin water transfers)
inter-basin water transfers 237 (*see also* inter-basin diversions)
Interdicta (ancient Rome) 43–5
Intergovernmental Authority on Development (IGAD) 341–2
International Association for Water Law (AIDA/IAWL) 70, 259, 263, 318
international basin commissions, committees, authorities
  Administrative Commission for the Uruguay River 336
  Binational Commission for the Development of the Upper Basins of the Bermejo River and the Rio Grande de Tarija 336
  Central Commission for Navigation on the Rhine 272, 330–1 (*see also* Rhine Navigation Commission)
  Chu-Talas Joint Rivers Commission 345, 352
  Commission for the Protection of the Rhine against Pollution (1963) 330
  Commission on Sustainable Development (Aral Sea) 345
  Commission of the Riparians 272–3
  Cuvelai Watercourse Commission (CUVECOM) 341
  Danube Commission 330, 332, 346, 353
  European Commission (Danube) 272–3, 276–7
  Gambia River Basin Development Organization (OMVG) 67, 339, 347, 351
  Helmand River Commission 345
  India-Bangladesh Joint Commission 344
  Indus Commission 343, 346, 352
  Incomati Maputo River Commission 341
  Intergovernmental Committee for Coordination of the Plata Basin 336, 353
  International Boundary and Water Commission (USA/Mexico) 294, 314, 334–5, 350
  International Commission of the Congo-Ubangi-Sangha Basin (CICOS) 339
  International Commission for the Protection of the Danube River 332–3

International Commission for the Protection of the Oder River against Pollution 292, 333
International Commission for the Protection of the Rhine (1999) 331
International Fund for the Aral Sea (IFAS) 345
International Joint Commission (USA/Canada) (IJC) 293, 333–5
Interstate Council on the Aral Sea Basin Problems (ICAS) 345
Interstate Water Management Coordination Commission (ICWC) 344–5
Joint Technical Commission for the Salto Grande project 336
Kagera Basin Authority (KBA) 338, 339, 347
Lake Chad Basin Commission (LCBC) 339, 347, 351–3
Lake Victoria Basin Commission 341, 352
Limpopo River Commission (LIMCOM) 341
Liptako-Gourma Authority 340
Luso-Spanish Commission 333
Mekong Committee 342–3, 347, 351, 352
Mekong River Commission 342–3, 347, 350
Meuse International Commission 333
Mirim Lagoon Commission 336, 346
mixed technical commission for the Paraná river 336
Permanent Joint Technical Committee of the Nile 337, 346–7, 352
Permanent Water Commission (Namibia/South Africa) 341
Niger Basin Authority (NBA) 339, 351, 353
Nile River Basin Commission 338
Okavango River Basin Water Commission (OKACOM) 341
Orange-Senqu River Commission (ORASECOM) 324, 341
Rhine Hydrology Commission 331
Rhine Navigation Commission 353 (*see also* Central Commission for Navigation on the Rhine)
Sava River Basin Commission 333
Scheldt International Commission 333
Senegal River Basin Development Organization (OMVS) 67, 246, 338, 340, 346–7, 350–1, 353, 355
Tripartite Permanent Technical Committee (Mozambique, South Africa, Swaziland) 341
Volta Basin Authority 339
Zambesi Watercourse Commission (ZAMCOM) 341
International Conference on Water and the Environment (Dublin, 1992) 9, 254, 319
international conventions 71, 246–7, 253, 286, 297, 300–1, 312
International Court of Justice (ICJ) 68–9, 246–7, 250–1, 254–7, 263, 295 , 304, 317
advisory opinion 257
statute 246–7, 250–1, 254–5, 300, 312
international custom 246, 250, 300, 312 (*see also* international customary law; international customary water law)
international customary law 68, 188, 259, 288, 297, 305, 307–9 (*see also* international custom; international customary water law)
international customary water law 250, 284, 308, 320 (*see also* international custom; international customary law)
International Law Association (ILA) 70, 244, 259–63, 280, 282, 285–6, 296, 312, 317, 342
Resolution of Dubrovnik (1956)260, 285, 296
Helsinki Rules (1966) (*see* Helsinki Rules)
Articles on Flood Control (1972) 260
Resolution on International Water Resources Administration (1976) 260
Articles on Regulation of the Flow (1980) 260
Articles on Water Pollution (1982) 260
Rules on International Groundwaters (1986) 260, 318
Complementary Rules Applicable to International Water Resources (1986) 260, 285
Berlin Rules on Water Resources (2004) 70, 260–1
International Law Commission (ILC) 4, 244–5, 251, 255, 261, 280–1, 285–6, 316–7, 319–21, 326
International Maritime Organization (IMO) 291
international public law 242
international watercourses 66, 244–5, 280, 283–4, 288, 294, 299, 302, 321, 349
Internationally Shared Aquifer Resources Management (ISARM) 319
inventory of water resources 47, 62, 172, 220, 224, 230, 232–3
inventory of water uses (or users) 176, 214
Iran 13, 29, 77, 84–5, 127, 277, 345
Iraq 84–5, 127, 266–7,
Iron Gates 276, 332
irrigation 1, 3, 8–11, 13, 15, 17–20, 25–9, 39, 41–3, 50–1, 54, 56, 59–60, 62, 66, 68, 78–9, 81, 83, 86, 89, 91, 94, 96, 99, 104–9, 114, 116–8, 120, 123, 126, 130, 132–4, 136, 144–5, 156, 161, 164, 168, 172, 177, 184, 186–7, 192–3, 203, 225–9, 233, 241–2, 249, 278, 287, 317, 332, 343
irrigation districts 226, 229
Ismailites 77
Islamic law 23, 57, 73, 82, 86, 126 (*see also* Moslem law)

Islamic schools 77–80
Israel 26, 346
Itaipú Binational Entity 336
Italy 37, 39, 47, 52, 56, 58, 68–9, 85–6, 90,
    138–9, 146, 148–9, 153–4, 259, 274,
    276–7, 292
Iullemeden Aquifer System 311, 315
Iullemeden, Taoudeni/Tanezrouft Aquifer
    Systems 308, 315

Jamaica 132
Japan 119, 125, 128, 137, 225, 228
Jewish law 25
Jews 25, 27
Jordan 24, 84–5, 127, 317, 324
judicial decisions (PCIJ and ICJ)
    Corfu Channel Case 295
    diversion of water from the Meuse 69, 256,
        279–80
    Gabcikovo-Nagymaros case 69, 256–7, 287,
        317
    territorial jurisdiction of the International
        Commission of the River Oder 69, 250,
        255–6, 257, 271, 279
    Oscar Chinn Case 255–6
    Pulp Mills Case 69, 256, 295, 304
Justinian, Roman emperor 33, 35, 48–52
    compilation of 49, 52, 56 (*see also Corpus
        Iuris Civilis*)

Kabupaten 109
Kampuchea 342 (*see also* Cambodia)
Kansas 163, 258
Karma 22
Karnak 16
Karnataka 122
Kazakhstan 103, 151, 294, 344–5, 352
kelian subak 105–6, 108–9
Kenya 113, 115, 189, 206, 234, 341–2, 352
Kerala 122
Kharijites 77
Khmer civilization 13, 29
Kitakami River Basin Authority 225, 228
Korea, Republic of 119, 128–9, 137
kudurru 19
Kurdistan 19
Kuwait 77, 85, 126
Kyrgyzstan 103, 344–5, 352

Lake Nasser Development Authority 234
Lake Victoria Development Authority 234
Länder (Germany) 68, 137, 142–3, 153, 320
Laos 76, 86, 119, 128, 342
La Pampa (Argentina) 259
Latvia 154
Law of the Twelve Tables 35, 38

Laws of the Indies 59–60, 130
League of Nations 257, 274
Lebanon 84–5, 127
Legalists 24
Lex Visigothorum 52, 54
li 24
Liberia 116, 341
Libya 84–5, 87, 113–4, 311, 314–5, 347, 352
licence(s) 92–5, 97, 115, 121, 124, 126–7,
    139–40, 163, 175, 177, 179, 199 (*see also*
    authorizations, permits)
Li-Chi 24–5
limited territorial sovereignty 258, 279, 281
Lithuania 103, 139
Longobards 52
Louisiana 86

Macedonia 154
Madagascar 113–4, 342
Madhya Pradesh 122, 258
magistracies (ancient Rome) 35, 39–40, 47
Maharashtra 122, 258
Mahaweli Authority of Sri Lanka 123, 225,
    227–8
Maimonides 26–7
Malaysia 77, 124, 128
Mali 114, 311, 315, 338–40, 341
Malikites 77, 80
Manchu Dynasty (China) 25
Manu 22–3
Mari 20
Maritime Danube 276–7
Martinique 132
Marx, Karl 97
Master of Canals 16
Master of Lakes 16
Master of Water Castles 16
Mauritania 84, 189, 315, 338, 341
Mauritius 116, 117, 231, 342
Maya (civilization) 28, 76
median line 267–8
Mejelle, Ottoman Civil Code 80–2, 84–5, 118
Mendoza (Argentina) 259
Menes (pharaoh) 16
Meso-America 28, 59
Meso-American (Pre-Columbian) civilizations
    13
Mesopotamia 18, 287
Mesopotamian civilization 13–4, 17–19
metering 199
Mexico 86, 132–5, 137, 189, 226, 268, 278, 289,
    294, 314, 318, 334–5, 346, 350–1
mining (groundwater) 163, 214, 317–9
Mohan-Jo-Daro civilization 21
Moldova 292, 293, 332, 333
Monaco 231

Mongolia 294, 345
monitoring 121, 140, 142, 144, 146, 147, 149, 150, 152, 171, 186, 200, 220, 225, 233, 236, 247, 252, 292, 293, 308, 311, 314, 319, 323, 324, 331, 335, 343, 349, 350, 352, 354
Montesquieu 73
Morocco 80, 84–5, 118, 189
Moslem law 59, 127 (*see also* Islamic law)
Mozambique 86, 115, 341, 342, 347, 354
Myanmar (Burma) 76, 123–4, 347, 352

Namibia 118, 189, 206, 324, 341, 342
natural flow (theory or doctrine) 91, 93, 159–60, 279
natural resources 1–2, 6, 9, 62–5, 70, 100, 102, 106, 112, 115, 117, 126–7, 131, 167, 171, 185, 195–6, 207, 211, 213, 215–7, 223, 236, 242, 244–5, 254, 257, 280–1, 286, 299–302, 305, 318, 341
navigability and floatability (criterion) 87, 119, 174
navigation 9, 15, 24–6, 29, 38–9, 44–5, 54–6, 62, 64, 66, 92–3, 129, 133, 146–7, 171, 177, 184, 187, 212, 233, 242–3, 245, 247, 249, 262, 265–7, 269–77, 278–80, 282, 329–33, 335, 340, 346, 349, 353
  Navigation Act (1857) 273
  freedom of 44, 54, 66, 243, 255, 256, 270–7, 330, 337, 343
Nebraska (USA) 320
Nebuchednezzar I 19
Neghev 14, 29
Nepal 103, 127, 344
Netherlands 68, 85–6, 139–40, 147, 266, 277, 279, 291–2, 330–1, 333
New Jersey 320
New Mexico 164
New South Wales (Australia) 95, 120–1, 226
New York (City) 226, 231
New York (state) 320
New Zealand 91, 123–4, 128, 137
Nicaragua 69, 132, 189, 256
Niger 113–4, 311, 313, 315, 339, 340–1, 347
Nigeria 113, 115, 116, 225, 311, 313, 315, 339, 341, 347
Nile Basin Initiative (NBI) 338
Nilometers 17
Ninive 19
non-domanial waters 87–8 (*see also* private waters)
non-point pollution (*see* non-point sources)
non-point sources (of pollution) 134, 150, 193, 195, 294, 332
North Atlantic Treaty Organization (NATO) 291
Northern Territory (Australia) 95, 120–1, 226, 237

North-Western Sahara Aquifer (NWSAS) 311, 315, 323
Norway 140, 151, 352
notification 256, 285, 289, 304–5, 315–6, 322, 325, 343, 347, 353 (*see also* duty to notify)
Nubian Sandstone Aquifer System 311, 314, 323

Observatoire du Sahara et du Sahel (OSS) 315, 324
obligation to cooperate 281, 283–4, 305, 322 (*see also* duty to cooperate)
Octavian (1st Roman emperor) 33, 39
OFWAT 96
Oman 77, 84, 85, 126–7
order of priorities 27, 78, 83, 99, 107, 133–4, 308, 335
Organization for Economic Cooperation and Development (OECD) 67, 137, 254, 291
Orsova 276
Osiris, god 16
Ostrogoths 52
Ottoman Empire 84–5, 111, 117, 118, 272, 330
Ottoman Civil Code 80 (*see also* Mejelle)
ownership of water 24, 33, 36, 41, 49, 57, 66, 76, 91, 97–8, 113, 120, 122, 138, 167, 174 (*see also* water ownership)

Pakistan 77, 85, 119, 127, 182, 279, 343–4, 352
Palerm Stone 16
Palestine 27, 84, 346
Panama 132, 135, 136
Pan American Union 262
  Declaration of Montevideo (1933) 286, 295
panchayats 123
Papua New Guinea 128
Paraguay 68, 132, 134–5, 189, 248, 311, 316, 324, 336
participation (public, stakeholder, community) 9, 66, 102, 113, 118, 122, 126, 133, 135–6, 151–3, 168, 171, 190, 219, 221, 228, 315, 340
  in EIA 305
  private sector 122, 129
payment for ecosystem services (PES) 196
penalties 40, 54, 73, 100, 102, 105, 138, 179, 204, 206–7
Pennsylvania 320
Permanent Court of Arbitration 305
Permanent Court of International Justice (PCIJ) 69, 250, 255, 257, 271, 274, 279
permit(s) 46, 58–60, 87–8, 90, 98–9, 114–5, 117, 119, 123, 125–7, 133, 136, 138–40, 144–7, 149, 158, 163, 175–88, 190–2, 194–8, 201–3, 205–7, 213, 224, 229–33, 236, 295 (*see also* authorizations, licences)

exploration or prospecting permits 198, 199
permit system 88–90, 95, 120, 124, 139, 160, 162, 176–8, 181, 185–7, 190, 201
Persia 269
Peru 28, 132–6, 189, 196, 206, 337
pharaoh(s) 16–7
Philippines 86, 119, 125–6, 128
planning of water resources 146, 172, 212, 216, 219 (*see also* water resources planning)
   aquifer depletion (mining) plan(s) 318
   aquifer management plan(s) 318
   drought management plan 318
   contingency plans 289–90
   river basin management plan(s) 90, 102, 142, 145, 148, 150–3, 194, 220–1, 236, 293, 347 (*see also* basin plans)
planned depletion 318
planned measures 304, 315, 322, 338, 347 (*see also* notification)
Poland 141, 149, 193, 292, 333
policy (water, water resources, management) 2, 4, 62, 65, 73, 75–6, 113, 123, 127, 136, 141, 150, 170–1, 199, 203, 212, 215, 218, 223, 230, 232, 235–6, 342
'polluter pays' principle 9, 89, 140, 203, 292, 297, 315
Ponka Creek 320
Portugal 69, 85–6, 154, 303, 333, 354
praetorship (ancient Rome) 35
precautionary approach (*see* precautionary principle)
precautionary principle 292, 297, 322, 325
Pre-Columbian civilizations 13, 86
Principate (ancient Rome) 33, 36, 38, 39–45, 51–2
prior appropriation (*see* appropriation)
priority substances 141, 151, 153, 293 (*see also* hazardous substances)
private waters 3, 37, 42–4, 49, 51, 57, 59, 74, 86–8, 90, 125–6, 138, 174, 176, 181, 183
privatization (of water services) 96, 129–30, 133, 146, 189
Property Clause (USA) 155–7
Prophet Mohammed 77–8
protected areas 26, 78, 102, 133, 140–1, 183, 188, 201–3, 220, 300
public domain 26, 58, 65, 77, 80–1, 86–9, 113–4, 118–9, 125, 132, 173, 206
public waters 23, 36, 39, 41, 49, 51, 58, 74, 86–9, 106, 114–5, 167, 173–7, 181, 187, 203
Pufendorf 269
Punjab 258

Qatar 77, 85, 126, 311
quaestorship (ancient Rome) 35
Queensland (Australia) 120–1, 226

reallocation (of water or water rights) 2–3, 47, 106, 170, 172, 180, 182, 189
reasonable and equitable utilization (or share) 257, 261–2, 282, 318, 320 (*see also* equitable and reasonable utilization)
reasonable use theory or doctrine 91, 159–60, 162–3
recycling (of water) 3, 9, 100, 138, 177, 185–6, 191–2, 199
Regional Seas Programme 303
regulatory functions 103, 148, 353, 355
regulatory institution(s) 167, 218, 232–4, 238, 354–5
Republican period (ancient Rome) 33, 35–9, 42, 47
Republic of Venice 57, 58
Reservation Doctrine (USA) 156–7
reuse of wastewater 138, 140, 177, 199 (*see also* wastewater reuse)
Revolutionary War (1775) 154
Rhine Action Programme 331
right of thirst (*chafa*; *hakki chefe*) 78–9, 81
right of irrigation (*chirb*; *hakki chirb*) 78–9, 81
right of way 23, 26, 42, 79, 88, 107, 183, 205 (*see also* servitude)
right to water (human) 130, 133–4, 188–90, 280, 306–9
riparian doctrine 3, 85, 87, 91–3, 137, 155, 158–60, 162–3, 181
river basin district(s) 102, 142, 147, 149–50, 152, 154, 171, 194, 220, 289, 293, 330
   international 220, 289, 293, 330
River Murray Waters Agreement 120
rivers declared international 243
rivers, lakes, seas and their basins
   Aach 320
   Amazon 231, 283
   Amu Darya 345
   Amur 345
   Aral Sea 283, 344–5
   Aranca 337
   Arno 287
   Artibonite 337
   Atuel 259
   Bansagar 122
   Bermejo 336
   Betwa 122
   Bodrog 149
   Bratis 29
   Calderon 28
   Catamayo-Chira 337
   Catatumbo 337
   Chambal 67
   Chu 352
   Colorado (USA-Mexico) 289, 294, 335

Colorado (Argentina) 67
Columbia 334, 346
Congo 243, 273, 283, 294, 337, 339
Cuvelai 341
Danube 53, 62, 66–7, 149, 226, 256–7, 272–4, 276–7, 283, 286, 292–3, 305, 313, 320, 329–30, 332–3, 346, 350, 352, 353
Dez 13, 29
Dniester 293, 304, 313, 333
Ems 277
Euphrates 13, 17, 21, 278, 287, 346
Gambia 67, 286, 339
Gandak 344
Ganges 66, 231, 278, 279, 283, 344, 346
Great Lakes 67, 293, 334, 346
Helmand 13, 14, 29, 69, 257, 280, 345
Hornag 149
Hron 149
Huang-Ho 13
Indus 13, 21, 62, 66, 258, 278–9, 283, 287, 305, 343, 346, 352
Kagera 286
Kosi 344
Krishna 258, 286
Kudov 329
Lake Chad 294, 313, 339, 347, 352
Lake Erie 334
Lake Ontario 334
Lake Peipsi 293
Lake Prespa 293, 304
Lake Tanganyka 304
Lake Titicaca 337
Lake Victoria 234, 294, 313, 338, 341, 347, 352
Lilion 337
Limpopo 294
Lower Niers 330
Mahakali 344
Mahaweli 29, 123
Mekong 13, 29, 62, 66, 226, 231, 283, 286, 342–3, 347, 350–2
Meta 337
Meuse 66, 69, 147, 256, 270–1, 279, 293
Mirim Lagoon 336, 346
Mississippi 226, 231, 267, 287
Mohan-Jo-Daro 21
Montagua 267
Murray 67, 120
Murray-Darling 121
Narmada 258
Netze 255, 329
Niger 226, 243, 273, 283, 294, 304, 308, 337, 339–40,
Nile 13, 16–7, 21, 62, 226, 278, 281, 283, 287, 294, 303–4, 337–8, 346–7, 352
Niobara 320

Oder 66, 69, 250, 255–6, 266, 271, 274, 279, 292–3, 330, 333
Omo-Turkana 283
Orinoco 337
Paraná 336
Plata 62, 66, 226, 283, 336, 352, 353
Po 68, 146
Puyango-Tumbas 337
Rhine 53, 66, 69, 147, 226, 263, 266–8, 270–4, 277, 283, 286, 291–3, 301, 313, 318, 320, 329–32, 346, 353
Rio Grande 278, 289, 294, 335
Rio Grande de Tarija 336
Rio Tinto 267
Saar 277
Sava 67, 286, 293, 303, 313, 333
Scheldt 66, 147, 269–71, 277, 293
Senegal 62, 66–7, 226, 246, 294, 304, 308, 338, 340
Shatt-el-Arab 267–8
Silala/Siloli 69, 257
Sirba 340
St. Lawrence 313, 334
Syr Darya 283, 345
Talas 352
Tiber 39, 46, 47, 51, 278
Tigris 13, 17, 19, 21, 278, 287, 346
Tilemsi 340
Uruguay 69, 226, 256, 294–5, 336
Ussuri 265n1
Vistula 271
Volta 339
Warthe 255, 330
Weser 271
White Nile 338, 347
Zambesi 226, 283, 294, 354
Zulia 337
Romania 57, 140, 149, 154, 276, 292, 332
Roman law 33–5, 38, 49–54, 56–7, 59, 64, 70, 73–4, 86, 162, 197, 268–9
Rome, city of 33–5, 39, 43, 46–8, 51, 53, 68, 207, 231
rule of capture 162, 164 (*see also* absolute ownership)
Russia 57, 271, 332
Russian Federation 103, 139, 149, 293, 294, 333, 345
Rwanda 87, 114, 281, 339, 341, 347, 352

safe yield principle 7, 199
Salto Grande 283, 336, 350
São Tomé and Principe 69, 86, 115
Saskatchewan (Canada) 95
Sassanids 27
Saudi Arabia 24, 77, 84–5, 126, 207, 311, 317, 324

Second World War 246, 299, 230 (*see also* World War II)
Senate (ancient Rome) 35, 39–40, 46, 48
Senegal Water Charter (*see* water charter, Senegal)
Serbia 57, 137, 276, 332
servitude(s) 26, 37–8, 43–6, 59, 62, 79, 88, 183, 186, 269 (*see also* right of way)
Seychelles 342
Shafi'ites 77, 80
shared natural resources 280, 281, 301, 321
Shari'a 77, 80, 85, 115, 118, 127
Shi'ites 77
Sicily 68, 86
Sierra Leone 115–6, 341
Siete Partidas 59
significant harm 68, 281–4, 289–90, 294, 296, 302, 305, 307–8, 316, 321–2, 338
Sind 258
Singapore 123, 124
Slovakia 69, 144, 154, 256, 292, 317, 332
Slovenia 189, 292, 333
S. Marino 231
Somalia 84, 87, 114, 324
South Africa 9, 68, 87, 117, 187, 189–90, 206, 324, 341–2, 347
South Australia 95, 120–1, 226,
South Dakota 320
Southern Africa Development Community (SADC) 248, 287, 340, 341–2
South Sudan 341, 342
Soviet law 73
Soviet system 97
Soviet Union 97, 101–3, 137, 139, 142, 144, 193, 248, 344
Spain 52, 59, 68–9, 85–6, 90, 115, 125, 139, 147–8, 225–6, 267, 280, 303, 333
Sri Lanka 76, 87, 119, 123, 225, 228 (*see also* Ceylon)
Stampriet Aquifer System 324
St. Lucia 132
subak system 103–4, 108–9
substantial damage 261, 295–6, 299
successive rivers 243, 265
Sudan 84, 113, 116, 234, 281, 283, 311, 314, 337, 341–2, 347, 352
Sultan of Johore 124
Sumerians 18
Sunna 77
Sunnites 77
Surinam 132, 337
sustainable development 303
Sustainable Development Goal No. 6 189
Sweden 68, 145, 352
Switzerland 67, 86, 142–3, 225, 230, 259, 267, 277, 282, 292, 314, 331
Syria 84–5, 127

Tadjikistan 103, 344
Taiwan 119, 124
Talmud 25
Talmudic law 26–7
Tamerlane 14
Tamil Nadu 122
T'ang Dynasty (China) 25
Tanzania 113, 115, 116, 189, 225, 281, 339, 341–2, 347, 352
TECCONILE 338
Tecomaltepec 28
Tennessee Valley Authority 225, 228
Teotihuacan 29
Tepanec Empire 29
Texas 162, 164
Thailand 76, 128, 342, 351
thalweg 267–8
Togo 113, 231, 339, 341
Toltec 29
Torah 25, 27
transboundary aquifers 245, 251, 311, 318–26
Trans-Jordan 84
treaties, conventions and agreements
    Act of Asunción on the Use of International Rivers (1971) 248
    Act of Berlin (1885) 243, 273, 274
    African Convention on the Conservation of Nature and Natural Resources (1968 and 2003) 248, 301
    Agreement for the Full Utilization of Nile Waters (1959) 281, 337
    Agreement for the Management and Utilization of the Groundwaters in the Al-Sag/Al-Disi Layer (2015) 317
    Agreement on the International Commission for the Protection of the Oder River from Pollution (1996) 292, 333
    Agreement on the Nile River Basin Cooperative Framework (2010) 281, 303, 338
    Agreement on the Protection and Sustainable Development of the Prespa Park Area (2010) 304
    Agreement of Sinaia (1938) 277
    ASEAN Agreement on the Conservation of Nature and Natural Resources (1985) 248
    Barcelona Convention and Statute (1921) 66, 243, 247, 274–6
    Belgrade Convention (1948) 277, 332
    Berne Convention on the Commission for the Protection of the Rhine (1963) 292, 331
    Boundary Waters Treaty (1909) 293, 333–4
    Campoformio Peace Treaty (1797) 270
    Convention and Statute of the Lake Chad Basin (1964) 313
    Convention of Dresden (1922) 274

Index

Convention of Mayence (1831) 272
Convention on Biological Diversity (1992) 282, 302
Convention on Cooperation for the Protection and Sustainable Use of the Danube River (1994) 292–3, 313
Convention on Cooperation for the Protection and Sustainable Use of the Waters of the Spanish-Portuguese Hydrographic Basins (1998) 303, 333
Convention on the Elimination of all Forms of Discrimination against Women (1979) 307
Convention on the Law of Treaties (1969) 246–7
Convention on the Protection and Use of Transboundary Watercourses and International Lakes (1992) (*see* UNECE Water Convention)
Convention on the Rights of the Child (1989) 307
Convention on Sustainable Use and Protection of the Dniester River Basin (2012) 304, 313, 333
Convention on the Law of the Non-Navigational Uses of International Watercourses 66, 245, 248, 251, 280, 313, 329 (*see also* UN Watercourses Convention)
Convention on the Protection of Lakes Lugano and Maggiore against Pollution (1972) 292
Convention on the Protection of the Rhine against Pollution from Chlorides (1976) 292
Convention on the Protection of the Rhine (1999) 292, 313, 331
Convention on the Protection, Utilization, Recharge and Monitoring of the Franco-Swiss Geneva Aquifer (2007) 314, 323
Convention on Transit Trade of Land-Locked States (1965) 248
Convention on Wetlands of International Importance (1971) 301
Convention relating to the Development of Hydraulic Power Affecting more than one State (1923) 66, 248, 284, 286
Convention relating to Fishing in the Waters of the Danube (1958) 292
Convention relating to the Legal Status of Common Works (1978) 340n32
Convention relating to the Modalities for Financing Common Works (1982) 340n33
Convention to Combat Desertification (1994) 248, 302
European Agreement on the Restriction of the Use of Certain Detergents (1968) 248

Final Act of Vienna (1815) 243, 271, 330 (*see also* Treaty of Vienna)
Fourth Geneva Convention (1949) 298
Fourth Hague Convention (1907) 298
Framework Agreement for the Sava River Basin (2002) 286, 303, 313, 333
Framework Convention on Climate Change (1992) 8, 302
Franco-Swiss Treaty concerning the Rhone (1825) 267
Geneva Aquifer Agreement (1977) 314
Geneva Protocol (1925) 298
Global Environment Facility (GEF) 316
Great Lakes – St. Lawrence River Basin Sustainable Water Resources Agreement (2005) 313
Great Lakes Water Quality Agreement (1978) 293, 334
Guaraní Aquifer Agreement (2010) 316, 324
Indus Water Treaty (1960) 279, 343, 346
International Agreement on the Meuse (2002) 333
International Agreement on the Scheldt (2002) 333
International Covenant on Civil and Political Rights (1966) 306
International Covenant on Economic, Social and Cultural Rights (1966) 189, 306, 309
London Convention (1933) 300
Mannheim Convention (1868) 273–4, 277
Memorandum of Understanding for the Establishment of a Consultation Mechanism for the Integrated Management of the Water Resources of the Iullemeden, Taoudeni/Tanezrouft Aquifer Systems (2014) 308, 315
Minute No. 242, USA-Mexico (1973) 314, 335
Minute No. 319, USA-Mexico (2012) 335
Minute No. 323, USA-Mexico (2017) 335
Protocol on Sustainable Development of Lake Victoria (2003) 313, 341
SADC Protocol, Revised (2000) 248, 287, 294, 313, 340, 342
Statute of the Intergovernmental Committee for Coordination (1968) 336
Treaty of Amazonian Cooperation (1978) 337
Treaty between France and the Batavian Republic (1795) 270
Treaty of Luneville (1801) 267
Treaty of Münster (1648) (*see* Treaty of Westphalia)
Treaty of Osnabruck (1648) 266
Treaty of Paris (1814) 270
Treaty of Paris (1763) 267
Treaty of Paris (1856) 272–3, 330

Treaty of Rome (1957) 252
Treaty of St. Germain-en-Laye (1919) 274
Treaty of Versailles (1919) 268, 274–5, 277, 330
Treaty of Vienna (1815) 247 (*see also* Final Act of Vienna)
Treaty of Westphalia (1648) 267, 269 (*see also* Treaty of Münster)
Treaty on Cooperation in the Field of Protection and Sustainable Development of the Dniester River Basin (2012) 304, 313, 333
USA-Mexico Treaty (1944) 294, 334–5
Washington Convention (1940) 300
Treaty Clause (USA) 156
tribuneship (ancient Rome) 35, 39
Trinidad and Tobago 132
Tunisia 80, 84–5, 113, 118, 311, 315
Turkey 84, 137, 272, 345
Turkmenistan 344
Turnu Severin 276

UAE 77, 127 (*see also* United Arab Emirates)
Ukraine 103, 292, 293, 332, 333
Ulm 276
underground water(s) 5, 7–8, 87, 92, 97–8, 101, 119, 121, 125, 138–40, 145, 153, 163–4, 175, 177, 193, 196–8, 241, 244, 311–3, 318–20, 323, 325, 352
UNECE (*see* Economic Commission for Europe)
  Charter on Groundwater Management 319
  Convention on Access to Information, Public Participation in Decision-Making and Access to Justice in Environmental Matters (1998) 305
  Convention on Environmental Impact Assessment in a Transboundary Context (1991) 304
  Water Convention (1992) 248, 282, 291–3, 295, 302–4, 313, 319
  Protocol on Water and Health (1999) 291, 303, 308, 313
unitary states 141, 144
United Arab Emirates 85, 126, 311 (*see also* UAE)
United Kingdom 92–3, 120, 137, 139–41, 148, 162, 255, 276–7
United Nations (UN) 4, 70, 249, 254, 291, 301, 305, 319, 342
  Charter 251
  Committee on Economic, Social and Cultural Rights 189, 306–7
  Conference on Desertification (Nairobi 1977) 254, 301–2, 319
  Conference on Environment and Development (Rio de Janeiro 1992) 254, 301–2, 306, 319
  Conference on the Human Environment (Stockholm 1972) 254, 299, 301

Conference on Human Settlements (Vancouver 1976) 301
Development Programme (UNDP) 342
Economic and Social Council (ECOSOC) 254
Economic Commission for Asia and the Far East (ECAFE) 4
Economic Commission for Asia and the Pacific (ESCAP) 4
Economic Commission for Europe (UNECE) 4, 151, 248, 254, 291, 293, 305, 319, 330
Economic Commission for Latin America and the Caribbean (ECLAC) 4
Educational, Scientific and Cultural Organization (UNESCO) 4, 68, 291, 319, 324
Environment Programme (UNEP) 291, 301
General Assembly 189, 245–6, 251, 254–5, 261, 263, 280, 298, 301, 303, 307, 312–3, 320–3
Human Rights Council 307
Millennium Declaration 254
Millennium Development Goals 188
Security Council 246
Universal Declaration of Human Rights 306
Water Conference (Mar del Plata 1977) 62, 246, 254, 301, 306, 319
Watercourses Convention (1997) 66, 248, 251, 257, 261, 263, 280–2, 284–8, 290–1, 294–6, 299, 302, 304, 307, 312–3, 316–7, 321–3, 330, 340, 343
United States (of America) 8, 74, 154–7, 159–60, 278, 295, 334–5 (*see also* USA)
  Bureau of Reclamation 155–6
  Corps of Engineers 156
  Environmental Protection Agency 156
  Supreme Court 258
'*urf* 77, 84
Uruguay 69, 132–6, 189, 248, 256, 295, 311, 316, 324, 336, 346, 352
USA 67, 73, 86, 91–2, 116, 119, 124–6, 137, 156, 159, 167, 225–6, 228, 230, 237, 268, 277–8, 289, 293–5, 314, 318, 320, 333–5, 346, 350–2 (*see also* United States)
USSR 97–8, 100–1, 141, 277, 292, 345, 352 (*see also* Soviet Union)
Uttar Pradesh 122, 123
Uzbekistan 283, 344

Valencia, water tribunal 68, 207
Veda 22
Venezuela 68, 132–6, 189, 337
Victoria (Australia) 95–6, 120, 123
Viet Nam 76, 86, 119, 127, 342
Visigoths 52–3

wadi 83
waqf 23, 78

War of Crimea 272
wastewater discharge(s) 89, 134, 139, 142, 146, 149, 194, 213, 231, 292, 332
  permit(s) 140, 196, 213, 230, 232–3
wastewater reuse 140, 191 (*see also* reuse of wastewater)
water administration 2, 4–5, 16–7, 20–1, 23, 25, 28–30, 38–9, 46–7, 52, 62, 73, 75, 89, 96, 101, 114, 115–6, 119, 121, 127–8, 135, 143, 146–7, 161, 169, 175–88, 191–2, 195–5, 198–9, 201, 203–7, 213, 228–30, 232, 235–7
  customary 74–5, 82, 237
  Islamic 82, 84
water authorities 95–6, 122, 127, 148
water boards 90, 96, 116, 127, 143, 147
water charges (*see* water rates)
Water Charter, Lake Chad 294
Water Charter, Niger 294, 304, 308
Water Charter, Senegal 294, 304, 308, 340
water courts 68, 118, 204, 206, 208 (*see also* water tribunals)
water equalization principle (ancient China) 24–5
Water Framework Directive (*see* directives, EU)
wateringues 68, 76, 147
water markets 133, 182
water master(s) 82–3, 105, 109
water ownership 15, 24–5, 37, 63, 78, 86, 90–1, 97, 173–5, 204–5, 207 (*see also* ownership of water)
water quality
  control (of) 64, 89, 141, 144, 147, 154, 156–7, 112, 191, 225, 238, 249, 349
  objectives and criteria 138, 292, 294
  standards 64, 69, 119, 138, 140–3, 145, 147, 150–1, 157, 193–4, 212
water rates 62, 171, 173, 179–80, 199, 203–4, 208
water resources planning 76, 114–5, 120, 123–4, 138, 144, 148, 168, 172, 184, 198, 211, 213, 216–8, 220, 224, 229, 232, 308 (*see also* planning of water resources)

water rights 3, 9, 20, 40, 42–5, 47, 50–1, 54, 60, 68, 74–5, 77–83, 85, 97, 106–8, 113–5, 117–8, 120–1, 124, 126, 131, 133–4, 138–9, 145–6, 155–8, 160–2, 164, 167, 170, 175–7, 178, 180–2, 186, 188, 190, 195–6, 202–8, 212, 224, 228, 230–3, 237
water rights administration 97, 117, 120, 145, 186, 188, 228, 230–3
water royalty 56, 58
waterschappen 68, 76, 147
water tribunals 29, 146, 204, 206, 225 (*see also* water courts)
water users' associations 68, 113, 117, 145, 156, 192, 201, 208, 226, 227, 228–9, 234, 235, 237
waterways of international concern 66, 243, 275
waterworks and structures 33, 38, 45–6, 51, 78, 108, 201, 206, 232, 348
Wattel 269
Western Roman Empire 52
World Bank 4, 317, 343, 346
World Commission on Environment and Development 303
World Health Organization (WHO) 200, 291
World Meteorological Organization (WMO) 291
World War I 95, 243, 268, 274 (*see also* First World War)
World War II 277, 300, 306, 330 (*see also* Second World War)
Württemberg 259, 276, 320

Xerp 17

Yaciretá Binational Entity 336
Yemen 77, 85, 126–7
Yugoslavia 142–4, 292, 333

Zaidites 77
Zaire 87
Zambia 113, 115, 118, 341, 342
Ziegler 269
Zimbabwe 115, 116, 189, 341, 342
zoning 62, 184, 201, 202 (*see also* protected areas)